Genetic Improvement and Conservation Practices of Medicinal Plants

This book explores the most recent advances in medicinal plant conservation and improvement through genetic transformation. It presents a compendium of topics related to conservation for sustainable utilization of important medicinal and aromatic plants, plant tissue culture interventions, genetic engineering tools for plant transformation, and transgenics for improved traits in the medicinally active plant species. Advancements in the areas of medicinal plants' nuclear and chloroplast engineering, stress tolerance, metabolite production, DNA barcoding, etc. have been carefully intertwined to offer novelty for the readers. The book caters to the interests of plant biologists, biotechnologists, ecologists, chemists, and pharmacologists and will be helpful to researchers, academicians, and students in the areas of medicinal plants' conservation, propagation and genetic improvement.

Salient Features:

- Provides a detailed and up-to-date account of the role of *in vitro* methods in the sustainable conservation of medicinally important plants.
- Serves as a comprehensive guide for different methods of genetic engineering of medicinal plants.
- Demonstrates application of genetic transformation strategies for improved characters, bioactive production, and stress tolerance in medicinal plant species.
- Elaborates the conservation and genetic engineering methods for priority medicinal plants.
- Discusses the application of advanced techniques like bioinformatics, genomics, and DNA banks for medicinal plant conservation and improvement.

Series Name: Exploring Medicinal Plants
Exploring Medicinal Plants
Series Editor: Azamal Husen, Wolaita Sodo University, Ethiopia

Medicinal plants render a rich source of bioactive compounds used in drug formulation and development; they play a key role in traditional or indigenous health systems. As the demand for herbal medicines increases worldwide, supply is declining as most of the harvest is derived from naturally growing vegetation. Considering global interests and covering several important aspects associated with medicinal plants, the Exploring Medicinal Plants series comprises volumes valuable to academia, practitioners, and researchers interested in medicinal plants. Topics provide information on a range of subjects including diversity, conservation, propagation, cultivation, physiology, molecular biology, growth response under extreme environments, handling, storage, bioactive compounds, secondary metabolites, extraction, therapeutics, mode of action, and healthcare practices.

Led by Azamal Husen, PhD, this series is directed to a broad range of researchers and professionals consisting of topical books exploring information related to medicinal plants. It includes edited volumes, references, and textbooks available for individual print and electronic purchases.

Medicinal Plants for the Management of Neurodegenerative Diseases
Edited by Jamal Akhtar, Maryam Sarwat and Fouzia Bashir

Nutraceuticals Inspiring the Contemporary Therapy for Lifestyle Diseases
Edited by Mala Trivedi, Sachidanand Singh, Parul Johri, and Pedro López-Sánchez

Plants as Medicine and Aromatics: Uses of Botanicals
Edited by Mohd Kafeel Ahmad Ansari, Mushtaq Ahmad, and Gary Owens

Exploring Traditional Wild Edible Plants
Edited by Vibhor Agarwal, Sachidanand Singh and Rhaul Datta

Propagation to Pharmacopeia: Modern Approaches in Medicinal Plants
Edited by Abha Manohar K, Gopal Shukla, Saikat Gantait, A P Das, and Sumit Chakravarty

Exploring Medicinal Orchids
Edited by Azamal Husen and Manu Pant

Symbiotic Association of Microorganisms with Medicinal and Herbal Plants
Edited by Allah Ditta, Sajid Mehmood, and Azamal Husen

Tissue Culture Techniques and Medicinal Plants: Enhancing Propagation and Production
Edited by Azamal Husen and Manu Pant

Comprehensive Guide to Hallucinogenic Plants
Edited by Noureddine Chaachouay, Abdelhamid Azeroual, and Lahcen Zidane

Genetic Improvement and Conservation Practices of Medicinal Plants
Edited by Manu Pant, Sayed Sartaj Sohrab and Azamal Husen

For more information about this series, please visit: www.routledge.com/Exploring-Medicinal-Plants/book-series/CRCEMP

Contents

Propagation. He has conducted research sponsored by the World Bank, the National Agricultural Technology Project, the Indian Council of Agriculture Research, the Indian Council of Forest Research Education, and the Japan Bank for International Cooperation. Husen has published extensively (over 300, which includes research papers, books, book chapters etc.) and served on the Editorial Board and as reviewer of reputed journals published by Elsevier, Frontiers Media, Taylor & Francis, Springer Nature, RSC, Oxford University Press, Sciendo, the Royal Society, CSIRO, PLOS, MDPI, John Wiley & Sons, and UPM Journals. He is on the advisory board of Cambridge Scholars Publishing, UK. He is a fellow of the Plantae group of the American Society of Plant Biologists, and a member of the International Society of Root Research, Asian Council of Science Editors, and International Natural Product Sciences. He has given numerous talks and presentations in many conferences and workshops. Husen has also given various orientation courses for academic staff, prepared different modules, developed BSc, MSc and PhD curriculum. Moreover, he has received a recognition award from the University of Gondar, Ethiopia for a decade of service in teaching, research and community services. He is Editor-in-Chief of the *American Journal of Plant Physiology*, and a Series Editor of *Exploring Medicinal Plants* (Taylor & Francis Group, USA); *Plant Biology, Sustainability, and Climate Change* (Elsevier, USA); and *Smart Nanomaterials Technology* (Springer Nature, Singapore). He is also working as a Handbook Series Editor, namely *Nanotechnology and Its Applications* (Springer Nature, Singapore). He has achieved the distinguished honour of being recognized as one of the "World's Top 2% Scientists" for the years 2022, 2023, and 2024 by Stanford University, USA. This recognition has also been prominently featured in the Elsevier Data Repository. Moreover, he has also worked as a special issue guest editor on various topics such as Advances in Plant Metabolomics: Insights, Applications, and Integration (*Plant Physiology and Biochemistry),* Exploring Nanotechnology in Medicinal Plants (*Journal of Herbal Medicine),* Integrating Nanotechnology with Plant Compounds: Enhancing Abiotic Stress Resilience and Crop Improvement via Signaling Molecules (*Frontiers in Plant Science),* etc.

Genetic Improvement and Conservation Practices of Medicinal Plants

Edited by
Manu Pant,
Sayed Sartaj Sohrab
and Azamal Husen

CRC Press
Taylor & Francis Group
Boca Raton London New York

CRC Press is an imprint of the
Taylor & Francis Group, an informa business

Designed cover image: Shutterstock IDs: 1856580655, 2344012545

First edition published 2025
by CRC Press
2385 NW Executive Center Drive, Suite 320, Boca Raton FL 33431

and by CRC Press
4 Park Square, Milton Park, Abingdon, Oxon, OX14 4RN

CRC Press is an imprint of Taylor & Francis Group, LLC

ISBN: 9781032773605 (hbk)
ISBN: 9781032773629 (pbk)
ISBN: 9781003482659 (ebk)

DOI: 10.1201/9781003482659

Typeset in Times New Roman
by Newgen Publishing UK

Dedication

Dedicated to Rajeshwari Badoni and D.P. Badoni (Parents-in-law of Prof. Manu Pant) for their unwavering support in this academic journey.

Contributors

Hazandy Abdul-Hamid
Department of Forestry Science and Biodiversity, Faculty of Forestry and Environment, Universiti Putra Malaysia, Serdang, Selangor DE 43400 UPM, Malaysia
Laboratory of Bioresource Management, Institute of Tropical Forestry and Forest Products (INTROP), Universiti Putra Malaysia, Serdang, Selangor DE 43400 UPM, Malaysia

Rambod Abiri
Department of Plant Agriculture, Gosling Research Institute for Plant Preservation, University of Guelph, Guelph, ON N1G2W1, Canada
Department of Forestry Science and Biodiversity, Faculty of Forestry and Environment, Universiti Putra Malaysia, Serdang, Selangor DE 43400 UPM, Malaysia

Urja Agarwal
Department of Biotechnology, Graphic Era (Deemed to be University), Dehradun-248001, Uttarakhand, India

Ashima Bagaria
School of Basic Sciences, Manipal University Jaipur, Jaipur, RJ-303007, India

Shweta Bajaj
Department of Professional Communication, Graphic Era (Deemed to be University), Dehradun, Uttarakhand, India

Yanu Bam
Department of Biotechnology, School of Science, Kathmandu University, Dhulikhel, Nepal

Durga P. Barik
Department of Botany and Biotechnology (Supported by OHEPEE and DST-FIST program), Ravenshaw University, Cuttack, Odisha 753 003, India

Biswajit Basu
Department of Pharmaceutical Technology, School of Health and Medical Sciences, Adamas University, Barasat, Kolkata-700126 (WB), India

Aryaputri Bedaprabha
Department of Botany and Biotechnology (Supported by OHEPEE and DST-FIST program), Ravenshaw University, Cuttack, Odisha 753 003, India

Alaka Bhantana
Department of Biotechnology, School of Science, Kathmandu University, Dhulikhel, Nepal

Yash Bisht
Department of Biotechnology, St. Xavier's College, Mumbai, Maharashtra 400001, India

Anupriya Chatterjee
Department of Health Sciences, University of the People, Pasadena, California, USA

Pallavi Dheer
School of Natural Sciences, Birkbeck, University of London, Malet Street, London WC1E 7HX, United Kingdom

Dhurva Prasad Gauchan
Department of Biotechnology, School of Science, Kathmandu University, Dhulikhel, Nepal

Aarushi Gautam
Department of Biotechnology, Graphic Era (Deemed to be University), Dehradun, Uttarakhand, India

Diwash Gautam
Department of Biotechnology, School of Science, Kathmandu University, Dhulikhel, Nepal

Aman Sharma
Department of Biotechnology, Graphic Era (Deemed to be University), Dehradun-248001, Uttarakhand, India

Nandini Sharma
Department of Biotechnology and Bioinformatics, DY Patil Institute, Thathawade, Maharashtra 411033, India

Tara Shrestha
Department of Biotechnology, School of Science, Kathmandu University, Dhulikhel, Nepal

Ashutosh Singh
Department of Life Science, School of Earth, Biological and Environmental Sciences, Central University of South Bihar, Gaya 824236, Bihar, India

Yashaswi Singh
Department of Biotechnology, Graphic Era (Deemed to be University), Dehradun-248001, Uttarakhand, India

Amrita Srivastava
Department of Life Science, School of Earth, Biological and Environmental Sciences, Central University of South Bihar, Gaya 824236, Bihar, India

Sarita Subedi
Department of Biotechnology, School of Science, Kathmandu University, Dhulikhel, Nepal

Sanskriti Thapliyal
Department of Biotechnology, Graphic Era (Deemed to be University), Dehradun, Uttarakhand, India

Aastha Upreti
Department of Biotechnology, School of Science, Kathmandu University, Dhulikhel, Nepal

Somya Vaishnav
Department of Bioscience and Biotechnology, Banasthali Vidyapith, Newai, Tonk, Rajasthan 304022, India

Devvret Verma
Department of Biotechnology, Graphic Era (Deemed to be University) 566/6, Bell Road, Clement Town, Dehradun, Uttarakhand 248002, India

Preface

Ecosystem disturbances, anthropogenic pressure and natural calamities are common threats to the survival of medicinal plant species in a particular niche. This has necessitated the conservation of the existing germplasm and the improvement of the adaptation and production of medicinally active plants. Besides conventional methods of conservation, it becomes important to utilize biotechnological tools to develop superior quality and tolerant varieties of medicinal plants. Genetic engineering has been realized to be effective in creating elite medicinal and herbal plants with various improved traits. Many specific genetic transformation methods have been designed and applied to a wide range of medicinal plants. These genetic engineering tools facilitate the transfer of any exogenous gene of interest and regulator elements into the host nuclear or chloroplast genome and create a genetically modified (GM) plant variety. Consequently, novel medicinal plant varieties that are better adapted to biotic and abiotic stresses, or those giving better yields in terms of production and active metabolites can be developed. These transformation methods have been crucial in circumventing the slow methods of conventional plant breeding to create improved plant varieties for sustainable utilization. The enforcement of the Convention on Biological Diversity (CBD) represents the significance of the conservation of biodiversity and sustainable use of biological components. Changing climatic conditions have emphasized the development of novel ex-situ conservation methods to ensure their preservation. Tissue culture practice has opened new vistas for germplasm conservation, and increased production and sustainability of medicinal plants. Besides, tissue culture technology offers scope for inducing genetic variability in both medicinal and herbal plants. The *in vitro* techniques allow maximal production in a limited space and time, allowing the improved or the conserved plant material to be available to the grower throughout the year. Technology has been widely applied for large-scale production of medicinal and aromatic plants. Besides, DNA banks have been realized to be another crucial means of germplasm conservation of medicinal and herbal plant species. Tissue culture and genetic engineering methods have also been used to enhance the production of plant secondary metabolites leading to bioprospecting and biofortification of medicinally active plants. Such studies also open up scope for the preparation of improved drugs against a variety of novel pathogens. Consequently, the maintenance of medicinal species diversity and the creation of better-quality plants must go hand in hand for sustainable development and utilization of medicinal and herbal plants.

With this background knowledge, the present book provides a wide spectrum of the latest information on genetic transformation and the conservation of important medicinal and herbal plant species. The book presents up-to-date information on the aforementioned topics related to conservation for sustainable utilization of important medicinal and aromatic plants, genetic engineering tools for plant transformation, and transgenics for improved traits in the medicinally active plant species.

We thank our publishers for supporting us in publishing a book on this significant topic. This book is the result of the efforts of contributing authors from across the world, who have presented valuable information in their respective research areas in well-crafted chapters.

We hope that the book will be an important source of information for students, researchers and academics working in the field of plant conservation and improvement for a sustainable future.

Manu Pant
SS Sohrab
Azamal Husen

1 Sustainable Conservation and Management Strategies of High-Value Medicinal and Aromatic Plants (MAPs)

Manu Pant, Aarushi Gautam, Poorva Saxena, Prachi Rawat, Sanskriti Thapliyal, Deepshika Kothari, Azamal Husen, Shweta Bajaj and Arasu Raman*

*Corresponding author

1.1 INTRODUCTION

Medicinal and aromatic plants (MAPs) are mostly used in pharmacy or perfumery for their therapeutic and fragrant properties. While aromatic plants are used for their flavour and aroma, medicinal plants are primarily those employed in formal and traditional medicine. MAPs specifically are natural and sustainable sources of alkaloids, essential oils, flavonoids, saponins, etc. (Agarwal et al., 2023). These plant-derived materials or products with therapeutic or other human benefits that comprise either raw or processed substances from one or more plants are classified as "herbal medicines" by the World Health Organization. Approximately 60,000 years ago, the very first evidence of humans using plants was found in a Neanderthal flower burial site in Northern Iraq, as the remains of a body were found enclosed by at least 7 medicinal plants (Jamshidi-Kia et al., 2018). More historical evidence supports the beneficial effects of aromatic plants, herbs, and their essential oils on a variety of disorders and their use as the first pharmacological substances to be used to treat illnesses or other abnormal conditions (Christaki et al., 2012; Giannenas et al., 2013; Giannenas et al., 2018; Giacometti et al., 2018; Oliveira et al., 2018; Husen 2021, 2022, 2023a, b; 2024a, b). The pursuit of new applications, fragrances, and medicines through plant discovery and botanical sourcing carries on the traditions of early people concerning medicinal and fragrant plants. In recent times, the production of aromatic and medicinal plants has emerged as a lucrative export sector. Around 1624 medicinal plant species in the Nepal Himalaya region, with almost 100 species utilized for subsistence economics export medicines, plants, and other MAPs for Rs 18,000 crores annually (Baral and Kurmi, 2006). The Himalayas serve as a repository for a wide variety of therapeutic plants because of the extreme diversity of their habitats. One diversified habitat Nanda Devi Biosphere Reserve NDBR in the western Himalayas is home to around 640 species of medicinal plants (Bajpayee, 2023) like *Taxus wallichiana* used for anti-cancer properties, *Paris polyphylla* used to treat cancer, abnormal uterine bleeding and menorrhagia, *Dioscorea deltoidei* used to treat abdominal pains, irritability, anemia, urogenital disorders and many others.

A large category of plants has also been utilized for ages to flavor and preserve food, improve and sustain health, and infuse daily life with color and fragrances. These plants have been used for a multitude of purposes throughout recorded history, improving human existence; even if they are occasionally referred to as culinary herbs, potherbs, spices, seasonings, medicinal herbs, or other titles to distinguish plant use. Since many plants classified as MAPs are also utilized cosmetically,

DOI: 10.1201/9781003482659-1

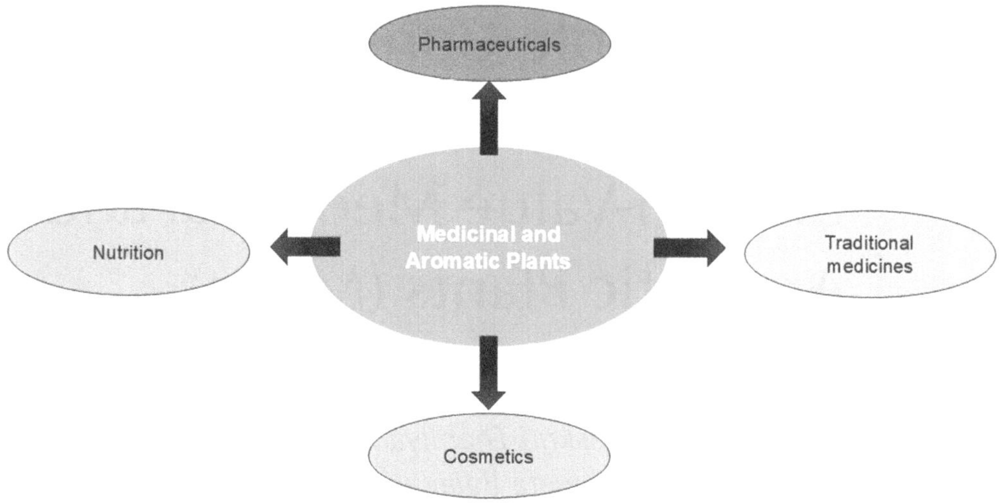

FIGURE 1.1 Uses of medicinal and aromatic plants MAPs in various industries.

another classified term medicinal, aromatic, and cosmetic MAC plants is now suggested to be used (Slikkerveer, 2006). Consequently, MAPs and MACs are of immense use in developing medicines, perfumes, dyes, colorants, products like cop protectants, etc., when manufacturing industrial goods and fine chemicals from plants (Figure 1.1).

With the increasing market demand for MAPs/ MACs, dwindling land availability and pressure on natural populations, it becomes imperative to conserve and propagate the priority plants for sustainable utilization. The conservation methods and the sustainable use of these medicinal plants are essential for maintaining the ecological balance, safeguarding health, and securing livelihoods (Pidigam et al., 2023). This chapter covers some of the major methods of MAP cultivation and conservation.

1.2 CULTIVATION OF MAPS

Medicinal and aromatic plants (MAPs) have properties that are highly valued in the pharmaceutical, cosmetic, fragrance, food and flavour industries. However, to date, the production of MAPs still relies to a large degree on wild collection. With the extinction of about 20 to 25% of MAPs, the practice of harvesting wild plants coupled with land conversion or habitat loss is currently threatening. To avoid draining wild populations, species-specific propagation and cultivation protocols are necessary. MAPs cultivation requires intensive care and management. The conditions and duration of cultivation vary depending on the quality of MAP materials required. Changing environmental conditions and unscientific harvesting practices further complicate the cultivation practices (Das et al., 2016). Nevertheless, planned cultivation of these plants is the only way to cater to the growing market demands of MAPs, while also empowering the farmers and conservation of these threatened plant species (Kala, 2015; Chandra et al., 2016). In the process of planned cultivation, the selection of geographical sites is very important and has to be tailor-made to meet the requirements of a particular botanical or a group of botanicals with ecological homogeneity. Soil moisture, nutrient availability, photoperiodic conditions, and rainfall patterns are other factors that largely influence MAP cultivation. For instance, while commercial cultivation of *Mentha,* antibacterial, antifungal and antiallergic; *Glycyrrhiza,* antimicrobial, anti-inflammatory, antioxidant and hepatoprotective activity; *Coptis,* antibacterial, antipyretic, anticancer and antidiarrheal; *Humulus,* a sedative, analgesic, antirheumatic and anti-inflammatory agent, prefer more than 14 hours of day length for high crop yield, MAPs like *Citronella*, *Pelargonium* and *Pogostemon* are short-day plants where essential oil production can be increased by nearly 25% by satisfying appropriate light requirements. The effects of photoperiodic cycles long days have also been established in the cultivation of other medicinal plants like *Solanum*, *Digitalis*, *Rauvolfia* and *Dioscorea* spp. (Ahad et al., 2021).

While seeds are the most preferred method of plant propagation, in the case of MAPs, the non-availability of viable seeds, minuscule seed size, and hindrances in seed collection and preservation limit the large-scale propagation of MAPs as found in plants like *Swertia chirata*. In these cases, other methods of cultivation like cuttings or root-based propagation are mostly used for MAP cultivation like *Cinnamomum* spp., *Centella asiatica*, *Salacia* spp., and *Pogostemon heyneanus*. While some species readily root from cuttings, others may require specific conditions or hormonal treatments. However, adventitious root development from cuttings continues to be a complex process and poor root development is a major challenge in cutting-based propagation (Waman et al., 2019). Besides, cutting-based methods are labor-intensive and very slow (Gonçalves and Romano, 2013). Likewise, domestication of uncommon or endangered plants may be difficult, especially when the accessibility of initial propagation material is limited for these plants (Grigoriadou et al., 2011). These limitations can be overcome by *in vitro* propagation methods that can increase the probability of success primarily when the endangered species are targeted (Zuzarte et al., 2010).

1.3 CONVENTIONAL METHODS OF CONSERVATION

Conventionally, MAPs can be conserved through in-situ and ex-situ methods. These methods prioritize protecting endemic species and their plant ecosystems over just internationally endangered ones (Nazari et al., 2023).

In-Situ Conservation – In-situ conservation of medicinal plants focuses on preserving the wild genetic diversity of these plants in their natural habitats. This is achieved through various protected areas, including national parks, wildlife sanctuaries, and biosphere reserves. These areas are designated to conserve and restore biodiversity, ensuring the survival of medicinal plants within their native ecosystems. The governments also develop policies to develop national parks to enhance the conservation and sustainable use of medicinal plants within these protected areas (Table 1.1). Medicinally significant plants *Saussurea obvallata, Meconopsis aculeata*, and *Asimina parviflora* have been protected through these conservation strategies (Semwal and Painuli., 2019; Bahukhandi et al., 2019; Chauhan et al., 2021).

Ex-Situ Conservation – The ex-situ conservation method aims to preserve the species facing threats by relocating them to safer environments such as botanical gardens, zoological gardens, seed banks, and gene banks. The main goal is to ensure the survival of threatened taxa and maintain genetic diversity, supporting reintroduction into their natural habitats once conditions improve. Examining research on species-recovery attempts forms part of the process of utilizing ex-situ collections to reconstruct plant genetic diversity. More than fifty percent of these attempts are aimed at seed banks and botanic gardens and are either partially or completely successful. Factors limiting the effectiveness of ex-situ material include scarcity, low viability, reduced longevity, low genetic variation, lack of evolution, and socioeconomic challenges. Modern practices, such as improved collecting protocols and storage conditions, offer opportunities for better conservation outcomes.

In India, the Botanical Survey of India BSI works extensively towards these conservation strategies. BSI has developed a network of botanic gardens across India to conserve endemic and threatened flora using seed storage, pollen conservation, low-temperature plant shoot storage, and advanced methods like plant tissue culture.

TABLE 1.1
National Parks and Biosphere Reserves in Asia

S. No.	National Park Asia Pacific	Biosphere Reserve Asia Pacific
1.	Afghanistan-2	Australia- 5
2.	Armenia-4	Cambodia-1
3.	Azerbaijan-11	China-34
4.	Bangladesh-18	Japan-12
5.	Bhutan-5	India-19
6.	Brunei-2	Indonesia-13
7.	Cambodia-13	Iran-10
8.	China-5	Kazakhstan-15
9.	Georgia-11	Korea-14
10.	India-116	Kyrgyzstan-2
11.	Indonesia-56	Malaysia-3
12.	Iran-23	Maldives-3
13.	Iraq-2	Micronesia-2
14.	Israel-49	Mongolia-9
15.	Japan-34	Myanmar-2
16.	Kazakhstan-13	Pakistan-2
17.	Kyrgyzstan-13	Palau-1
18.	Lebanon-4	Philippines-3
19.	Malaysia-29	Sri Lanka-4
20.	Mongolia-29	Thailand-5
21.	Myanmar-9	Turkmenistan-1
22.	Nepal-12	Uzbekistan-2
23.	North Korea-9	Vietnam-11
24.	Oman-1	
25.	Pakistan-26	
26.	Philippines-36	
27.	Russia-64	
28.	Saudi Arabia-3	
29.	South Korea-22	
30.	Sri Lanka-26	
31.	Taiwan-9	
32.	Tajikistan-1	
33.	Timor-Leste-1	
34.	Thailand-156	
35.	Turkey-43	
36.	United Arab Emirates-4	
37.	Uzbekistan-4	
38.	Vietnam-32	
39.	Yemen-2	

1.4 ADVANCED METHODS OF MAP CONSERVATION

The tissue culture technique for medicinal and aromatic plants is applied as a modern approach to address concerns about over-harvesting and extinction of these valuable plant species. Tissue culture, or micropropagation, involves growing plant cells or tissues under controlled conditions in a laboratory setting. This technique allows for the mass production of plants with desirable traits, including those with medicinal properties, in a relatively short period compared to traditional cultivation methods. Researchers can stimulate the growth of new plants by culturing small pieces of plant tissue, such as stem or leaf fragments, in a nutrient-rich medium. This method is beneficial for species that are slow-growing or difficult to propagate through conventional means.

Growing rare and elite varieties can solve issues like species depletion and environmental degradation in response to the growing demand for highly desirable medicinal plants. To fulfill industrial needs for clones, particularly those with high-yielding chemotypes, micropropagation becomes a key response (Moraes et al., 2021). Additionally, plant tissue culture provides disease-free plant material, improving the safety and quality of herbal medicines, and serves as a platform for research and development in medicinal plant biology (Jain et al., 2023).

Tissue culture techniques also offer a solution to the challenge of sustainable production by enabling the mass propagation of these plants in controlled environments, ensuring a continuous and reliable supply of bioactive compounds. Medicinal plants contain secondary metabolites, a heterogeneous group of natural compounds not essential for plant growth but crucial for various biological functions, including defense mechanisms. These metabolites, which include terpenoids, alkaloids, steroids, and others, are synthesized through diverted metabolic pathways and play vital roles in pharmaceuticals, flavors, fragrances, and other industries (Chandran et al., 2020).

In vitro technologies are essential for the production of secondary metabolites and pathogen-free plants in the field of medicinal plants. In general, efficient *in vitro* plant regeneration and minimizing somaclonal variations in medicinal and aromatic plants depends on an understanding of the complex interactions between genotype, explant type, nutritional medium composition, and environmental factors (Duta-Cornescu et al., 2023).

Additionally, plant tissue culture makes it easier for desirable genotypes to multiply efficiently, guaranteeing the availability of genetic resources for upcoming breeding initiatives and scientific studies (Shahzad et al., 2017). Tissue culture methods of clonal propagation like axillary bud culture, somatic embryo development or meristem tip culture, provide a rapid and efficient way to multiply desirable genotypes while preserving genetic uniformity and superior qualities. Mastering clonal propagation enables us to reduce pressure on wild populations, ensure consistent product quality, and promote the cultivation of MAPs, thereby contributing to human health and ecosystem preservation (Waman et al., 2019). In plant tissue culture, some factors like source and sterilization of explants, nutrient medium selection and preparation, selection of plant growth regulators and media supplements, standardization of physicochemical conditions, etc., are some of the factors that critically affect the success of the experiment (Mir et al., 2014; Gaurav et al., 2018; Qu et al., 2019; Lemma et al., 2020).

In culture medium, the basal nutrient media provide essential inorganic nutrients, carbon source is essential for energy.

1.5 GENETIC IMPROVEMENT OF MAPS

Genetic improvement of MAPs is required to develop more productive and tolerant crops. In most cases, these strategies are aimed at improving the secondary metabolite production in the target species as MAPs are a huge reserve of these secondary metabolites (Shafi and Zahoor, 2021). It is these secondary metabolites that impart a plethora of therapeutic properties to the plants like antimalarial (Mota et al., 2012), antifungal (Bahraminejad et al., 2016), antidiabetic (Salehi et al., 2019), antispasmodic (Heghes et al., 2019), anthelminthic (Zaman et al., 2020), antarthritic (Banik et al., 2020), antioxidant (Chrysargyris et al., 2020), anti-inflammatory (Banik et al. 2020), anticancer (Kaplan, 2022), antimicrobial (Salvi et al., 2022), antiviral (Kaplan, 2022), antitubercular (Nandikatti et al., 2023), antithyroid (Sajid-ur-Rehman et al., 2023), and antihistaminic properties (Nenni and Karahuseyin, 2024). The presence of these metabolites not only makes MAPs most sought after in the traditional and modern systems of medicine, but also in the nutrition, cosmetic, perfumery, and alcohol industries. Besides, the metabolites help them adapt to changing environmental conditions and stresses.

Improvement of MAPs can be brought about by conventional breeding methods and by advanced biotechnological methods like transgenics and tissue culture.

1.5.1 Breeding

The breeding methods commonly used for MAPs are

Combination Breeding: Crossing is done to combine the trait expressions of the parents, who are the donors of specific characteristics, into common progenies. During the resulting segregating generations, elite plants possessing the ideal combination of traits are chosen and bred for consistency. This technique is most frequently used in MAP breeding to produce novel diversity (De, 2017).

Mutation Breeding: There is a great potential to improve the alkaloid levels, oil output, and other biochemical aspects of aromatic and medicinal crops by mutation breeding. The main benefit of this is the creation of diversity and difference in a specific crop. Mutagenesis, together with in vitro selection and plant biotechnology techniques, enables plant breeders to identify traits that have proven difficult to achieve through breeding. By employing variability to construct a novel gene combination through mutagenesis, the researchers can overcome the yield plateau and improve tolerance (Agrawal and Kumar, 2021).

Hybrid Breeding: In hybrid breeding, desirable features from two or more kinds are combined via the hybridization process. Choosing ideal parent plants becomes an important element in this breeding strategy because the genetic composition of the progeny is entirely dependent on the qualities of the parent plants. Moreover, hybridization makes use of a phenomenon known as heterosis, or hybrid vigor, in which the progeny outperforms either of the parent plants in terms of performance (Fridman, 2015).

Many priority crops have been bred to improve their genetic makeup. Two novel cultivars of *Calendula officinalis*, named Nataly and Diana, were created through hybridizations and individual selections. These cultivars can produce over 1,000 kg/ha of dry inflorescences, and depending on the variety, they have a flavonoid and polyphenol content of 0.624–0.873% and 0.988–1.038%, respectively (Gonceariuc, 2014). Two interspecific *Ocimum* hybrids that resulted from a natural cross between *O. kilimandscharicum* and *O. basilicum* have been investigated and found to be distinct chemotypes through metabolite profiling using gas chromatography-mass spectrometry. These hybrids were found to possess a distinct combination of specialized metabolites from their parental species, which are either abundant in terpenes or phenylpropanoids. They are abundant in phenylpropanoids or terpenes. Therefore, these two hybrids of *Ocimum* represented new chemotypes that could be useful in commercial cultivation to yield new bioactive constituents and essential oils that can be used in healthcare and bioremediation (Gurav et al., 2020).

1.5.2 Biotechnological Interventions For Maps Improvement

Developing *in vitro* plant regeneration systems and genetically modifying plants make it easier to create secondary product metabolic pathways. Plant manipulation strategies are made efficient by developments in gene cloning related to relevant pathways, high throughput screening systems for chemical and biological activity generating resources and understanding that secondary plant metabolism is regulated at a higher level at the whole plant level (Gómez-Galera et al., 2007). The methods involved are explained in more detail below.

Somaclonal Variations – Somaclonal variations refer to the genetic variations in plants generated by *in vitro* propagation methods. These are caused by genetic and epigenetic changes that may arise from the instability of in vitro cultures. Occasionally, these modifications have positive outcomes; for instance, they can be applied in breeding initiatives to create new cultivars with desired traits. To create genetic diversity and facilitate the introduction of novel genotypes resistant to abiotic stresses, diseases, and pests, somaclonal variation has been employed in genetic improvement programs for the world's most economically significant crops (Ferreira et al., 2023). The most popular technique for creating somaclones involves applying plant growth regulators PGRs like dichlorophenoxyacetic acid 2,4-D and other media modifications. This is crucial when dealing with species that are hard to breed or plants that have little genetic diversity. Somaclonal variation can have effects that are noticeable in a short amount of time and without the need for advanced technology (Duta-Cornescu et al., 2023).

Somaclonal variations have been used to increase essential oil content in *Cymbopogon winterianus* with around 1.5 fold increase in desired components like geraniol, citronellol, citronellal and geranyl acetate (Mathur et al., 1988). *Pogostemon patchouli* is another important aromatic plant valued for its essential oil that is extensively used in cosmetic and fragrance products. Somaclonal variation technique has been used to enhance plant height, herb yield, and the amounts of patchouli alcohol, α-guaiene, α, δ-patchoulene, and α-bulnesene in the essential oil (Ravindra et al., 2012).

Transgenics for MAP improvement – Biotechnology-based breeding methods (BBBMs) become necessary when traditional breeding methods are unable to fulfill the enormous demands (Roychoudhury and Bhowmik, 2021). The two primary BBBMs that rely on plant tissue culture directly are artificial polyploidy induction and agrobacterium-mediated gene transformation. Targeted modification of secondary metabolite pathways in medicinal plants can be facilitated by controlled regulation of endogens and/or transferred genes through the use of engineered zinc-finger proteins or transcription activator-like effectors. The genome editing techniques that can result in user-designed medicinal plants are zinc-finger nucleases, sequence-specific nucleases of transcription activator-like effector nucleases TALENs, and clustered regularly interspaced short palindromic repeats-associated Cas. The targeted genome editing techniques readily accessible today can control plant synthetic biology and create new opportunities for the introduction of medicinal plants into specific industries (Niazian, 2019). Biotechnology advancements, specifically in the areas of haploid and double haploid DH production, have created new opportunities for genetics, breeding, transformation, and mapping research in these MAPs. This will make it possible to commercially exploit these rare plants and the compounds they yield for use in aromatic industries, medicinal applications, plant growth, and weed and insect control (Sharma et al., 2018).

In aromatic plants, stress conditions increase secondary metabolites. Aromatic plants inoculated with stress exhibit distinct enzymatic antioxidant responses (Chiappero et al., 2021). According to a study, when compared to its wild-type counterpart, the transgenic Geranium *Pelargonium graveolens* showed a notable increase in plant height ~20–23%, total biomass 18–60%, leaf: stem ratio 12–21%, and oil quantity. Under the control of the 35 S CaMV promoter, transgenic *P. graveolens* expressing the ACC deaminase gene from *Achromobacter xylosoxidans*, encoding 1-aminocyclopropane-1-carboxylilate, demonstrated enormous tolerance to salinity and drought stress compared to the wild-type plant (Singh et al., 2021). A more recent study revealed that the overexpression of *CiMYC2* in tobacco led to shorter plant height, darker leaf colour, higher levels of chlorophyll and carotenoid content, more long-handle and short-handle trichomes, and higher levels of terpenes. Besides, some novel terpenes were found in *CiMYC2* transgenic tobacco lines which also expressed more genes related to the terpenoid biosynthesis pathways than the wild type lines (Gao et al., 2023). *Atropa belladonna, Artemisia annua, Panax ginseng, Papaver somniferum, Dioscorea sp., Medicago sativa* and other significant medicinal crops have benefited from genetically transformed hairy root cultures and pathway engineering, which have increased production of target compounds (Bindu et al., 2018; Kaur and Ganjewala., 2019).

1.6 COMMONLY USED MEDICINAL AND AROMATIC PLANTS

Ocimum sanctum – Ocimum sanctum, also known as holy basil, is a highly prized culinary and medicinal herb belonging to the Lamiaceae family (Figure 1.2). It is native to the Indian subcontinent and has been used in Ayurvedic medicine for more than three millennia. It is widely grown in certain parts of Asia and Africa, and grows profusely in warm, tropical climates . It naturally thrives in moist soil, the type of soil, and differences in rainfall regulate the size, shape, and therapeutic qualities of the plant. Tulsi is recognized to treat a wide range of common health issues and is often referred to as a "Solution of Life" in the Ayurvedic framework because of its healing properties (Rathod et al., 2023). The leaf extract is generally used to treat a variety of ailments, including the common cold, intestinal illness, heart disease, headaches, stomach problems, kidney stones, skin disorders, fevers etc. The leaves help to enhance memory and are considered a nerve tonic. A concoction of juice from tulsi leaves can be used to treat various types of internal illnesses, skin disorders, and fevers (Bhadra and Sethi, 2020; Rao et al., 2023). The herb also serves as an excellent repellent against flies, mosquitoes, etc.

Holy basil has a very complex chemical makeup that includes a wide range of nutrients and other biologically active substances, the amounts of which can differ significantly between strains and even between plants growing in the same area. Notably, differences in growing, harvesting,

FIGURE 1.2 *Ocimum sanctum.*

processing, and storage circumstances have a substantial impact on the quantity of many of these elements, variables that are still little understood. Despite its reputation as a general vitalizer and its ability to boost athletic endurance, Tulsi doesn't contain caffeine or other stimulants. The stems and leaves of Holy basil are rich in substances such as saponins, flavonoids, triterpenoids, tannins, rosmarinic acid, caryophyllene, carvacrol, eugenol, ursolic acid, to name a few (Deshmukh et al., 2015). It is these metabolites that impart the plant a wide range of therapeutic properties. The plant is commercially propagated by seeds (Sharma et al., 2014; Srivastav et al., 2018) but traditionally, some families exchange cuttings and seeds among themselves (Shukla et al., 2021). Tulsi being a highly medicinally valuable plant still lacks large-scale production strategies and conventional propagation through seeds fails to meet the demand of industries, *in vitro* propagation techniques ensure quality plant material while also producing physiologically uniform plants (Shukla et al., 2021). Many reports on the *in vitro* propagation of tulsi have been published utilizing explants such as nodal segments (Aggarwal et al., 2020), leaf (Mishra, 2015), etc.

Trachyspermum ammi – This is an annual herbaceous plant vernacular name ajwain that is a prized member of the Apiaceae family, which has significant medicinal and aromatic use (Gersbach and Reddy, 2002). The herb is commonly grown in arid and semi-arid areas with highly-salted soil. Indigenous to Egypt, the crop is widely grown and distributed around the world, including Europe, Iran, Afghanistan, India and Pakistan (Shojaaddini et al., 2008). The fruits and seeds of the plant are much valued for their commonly considered medicinal and nutritious properties (Chauhan et al., 2012) (Figure 1.3).

The plant content includes fiber (11.9%), carbohydrates (24.6%), tannins, glycosides, moisture (8.9%), protein (17.1%), fat (21.1%), saponins, flavones, and other components 7.1% involving calcium, phosphorous, iron, cobalt, copper, iodine, manganese, thiamine, riboflavin, and nicotinic acid

FIGURE 1.3 *T. ammi* seeds.

(Qureshi et al., 2010). The phytochemicals, in particular, include alkaloids, chalcones, coumarins, flavonoids, glycosides, saponins, steroids, and tannins (Soltani et al., 2018; Kumar and Khurana, 2018). According to Minija and Thoppil (2002), the fruits of the plant can accumulate up to 5% essential oil within their compartments. Thymol is typically the primary component of Ajwain essential oil, with yields ranging from 35% to 60% (Ishikawa et al., 2001; Zarshenas et al., 2014). According to Gaba et al., (2018), the volatile oil that was steam-distilled from *T. ammi* included nine monoterpenes, including two alcohols (2.9%) and seven hydrocarbons (97.1%). All these volatiles and phenolics of *T. ammi* have strong anticholinesterase and antioxidant properties. The oil is suitable for boosting appetite, treating stomach discomfort, ensuring the smooth operation of the respiratory system, and supporting the kidneys. It has been used traditionally as a household folk remedy for cough, cold, asthma, diarrhea, influenza, and cholera (Kolbadinejad et al., 2020). According to reports, the dietary threonine levels and the ajwain seed supplement separately enhanced the broiler chickens' development performance, ileal microbial population, and jejunum morphology (Raeisi et al., 2016), while the chilled fish product had the longest shelf life when it contained 3% Ajwain extract (Piri et al., 2020).

T. ammi is generally cultivated for its therapeutic properties like hepatoprotective, analgesic, antiflatulent, antiulcer, and antihypertension (Mishra, 2020). Several products such as ajwain honey, ajwain ark, and ajwain water are available in the market.

Zingiber officinale – Commonly known as ginger, the crop originated in the Indo-Malayan region, and is now widely used as a culinary spice, condiment, and medicinal plant in many nations (Daily et al., 2015). Ginger is regarded as a strong, dry, warming yang herb in Traditional Chinese Medicine (TCM) and is used to treat conditions such as the common cold, aches, diarrhea, nausea, cholera, hemorrhages, rheumatism, and toothaches. In the Ayurvedic system of medicine, while in the Arabic traditional system, ginger is an aphrodisiac medicine (Shahrajabian et al., 2019; Mahmood, 2019).

As a potent antioxidant, ginger extract can eliminate illnesses brought on by oxidative stress. Research has demonstrated the numerous neuroprotective benefits of existing phenolic compounds and anthocyanins, such as gingerols and the sugevals, including analgesic effects, memory enhancement, and learning delayed by aging (Fadaki et al., 2017). Ginger is a versatile ingredient in cooking that may be used for a wide range of dishes, including savory ones like soups, sauces, stews, grilled meats, roasts, and apple pie, as well as sweet ones like drinks, puddings, cakes, pieces of bread, and candies (Oludoyin and Adegoke, 2014). Ginger has been shown to have antimicrobial, anti-nausea, antipyretic, anti-inflammatory, analgesic, hypoglycemic, anti-ulcer, antiemetic, cardiotonic, anti-hypertensive (Al-Amin et al., 2006), anti-platelet aggregation (Bordia et al., 1997), and anti-inflammatory effects in both laboratory animals and human subjects.

Ginger is cultivated through rhizome (Figure 1.4) and is sometimes utilized as a flavoring agent in foods and beverages, also used as fragrances in cosmetics and soaps (Alam, 2013). Ginger products like biscuits, capsules, tea, candy, ginger syrup, wine, power, etc., are available in national as well as international markets (Sang et al., 2020).

Curcuma longa – *C. longa* or turmeric is a medicinal rhizome that is widely discussed in Indian material medica Dravyaguna Sastra (Krup, 2013). Belonging to the family Zingiberaceae, turmeric is extensively grown throughout Asia. Turmeric powder comes from dried rhizomes (Figure 1.5), and because of its fragrant and pleasing yellow hue, it has special significance for humans (Labban, 2014).

The plant contains an underground stem called a rhizome, which grows thick and girdled with the roots of old leaves. To manufacture the unique brilliant yellow spice, rhizomes are cooked, dried, and powdered. The flavor of turmeric powder is spicy and pungent, and it smells mildly of ginger and orange. Due to its many health benefits, turmeric has been utilized extensively in traditional medicine in Asian countries, in addition to its culinary applications.

FIGURE 1.4 Ginger rhizome.

According to Neha et al. (2009), it helps treat a variety of conditions, including diabetes, hemorrhoids, anemia, jaundice, cough, asthma, wound healing, colic, gout, renal calculi, poisoning, freckles, skin, and neurological diseases. Assamese tribal women apply fresh rhizome paste to their skin to improve and shield their complexion from infection. Cattle are given this plant's rhizome combined with other components to treat loose stools (Krup, 2013). The paste can be applied topically to treat a variety of skin conditions, including herpes zoster, pemphigus, boils, bruising, blisters, ulcers, eczema, insect bites, parasite infections, hemorrhages, and acne (Labban, 2014). Widely considered a herbal medicine panacea, the plant with its antioxidant capabilities, can also help with cancer, ulcers, and inflammation, among other ailments (Nasri et al., 2014).

The correlation between the medicinal applications of this plant and its inclusion of numerous metabolites, including flavonoids, saponins, tannins, coumarins, terpenoids, alkaloids and quinones, phenolics, some essential amino acids, has been documented by Saragthem and Haokip (2010). The antifungal, antibacterial, antiparasitic, antimutagenic properties and curative properties against renal problems, Alzheimer's disease, arthritis, and allergies have been well studied (Negi et al., 1999; Naz et al., 2010; Akter et al., 2016). The crop is cultivated through rhizome-based propagation conventionally, while tissue culture techniques have been employed for its large-scale cultivation to meet the rising industrial demand for turmeric (Sinchana et al., 2020; Taghavi et al., 2021; Brijesh and Ajjappala, 2023).

Aloe barbadensis – Also known as Aloe vera, the plant belongs to the family Alliaceae and is the most biologically active among 400 species (Chandegara and Varshney, 2013). The plant is native

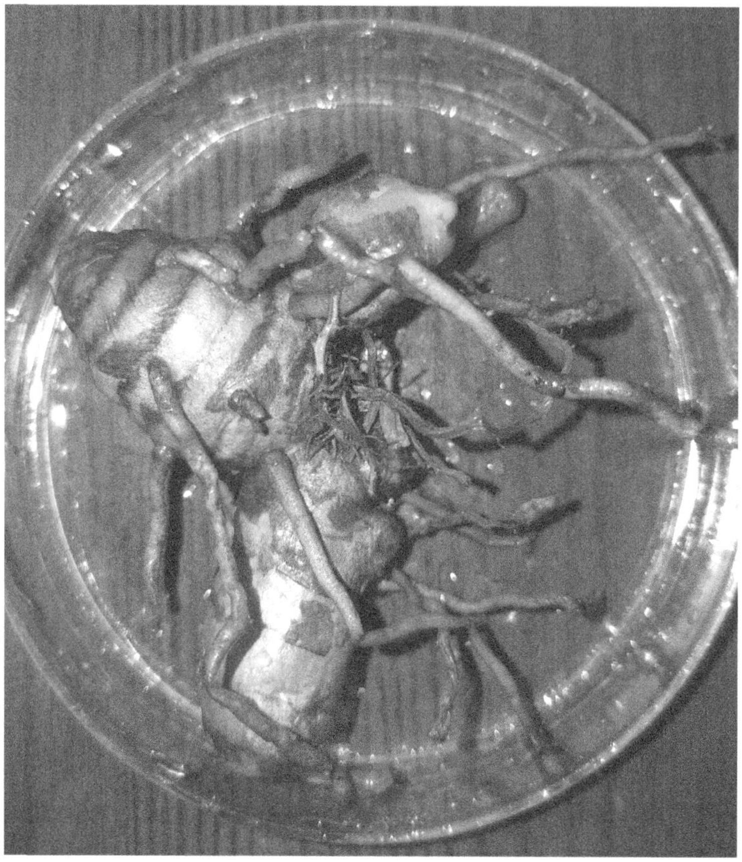

FIGURE 1.5 *Curcuma longa* rhizome.

to southern and eastern Africa along the upper Nile in the Sudan, and subsequently introduced into northern Africa and naturalized in the Mediterranean region and other countries across the globe. The plant is commercially cultivated in Aruba, Bonaire, Haiti, India, South Africa, the United States of America, and Venezuela (Manvitha and Bidya, 2014).

In Ayurveda, it has been defined as a multi-functional herb, which is an anti-inflammatory, diuretic, uterine tonic, spermatogenic, laxative, fever reliever, blood purifier. In Ayurveda, it has also been used as an appetite stimulant, and for treatment of cough, cold, piles, debility, dyspnoea, asthma and jaundice (Joseph and Raj, 2010). The pharmacological actions of aloe vera consist of anti-inflammatory and anti-arthritic activity as well as anti-bacterial and hypoglycaemic effects. Due to its wound and burn healing properties, it is known as the healing plant or the silent healer. The plant has great demand in both the cosmetics and pharmaceutical sectors.

Aloe vera constitutes likely 75 active constituents that include enzymes, minerals, sugars, lignin, saponins, salicylic acids, amino acids, anthraquinones and vitamins (Sahu et al., 2013). Vitamins E, A, B-12 and C are present in abundance bestowing the plant with antioxidant properties (Kar and Bera, 2018). Steroids like campesterol, cholesterol, lupeol and beta-sitosterol are present in the plant and have painkiller and anti-inflammatory properties (Lanka, 2018). The parenchymatous cells of fresh leaves of aloe vera are used to produce a colorless gel (Figure 1.6) (Terefe and Neges, 2017) which is much prized in the cosmetic and herbal industry due to its chemical composition.

The plant propagation strategies include cultivation of the lateral shoots suckers or via seeds of the plant, but due to male sterility and self-incompatibility conventional propagation methods fail to

FIGURE 1.6 *Aloe vera* plant.

meet the demands of pharmaceutical, food, medicinal and cosmetic industries (Ahmed et al., 2024). *In vitro* propagation ensures year-round production and high-quality plant material.

Matricaria chamomilla – Commonly known as chamomile, the plant is an essential medicinal herb indigenous to Europe and Asia belonging to the family Asteraceae formerly Compositae, and also dubbed as the "star among medicinal species" (Ali et al., 2020). Traditionally, chamomile is propagated through seeds, via raising seedlings, or by cuttings (Timothy and Mwangi, 2015; Chauhan et al., 2022). German chamomile, wild chamomile *Matricaria discoidea* DC., Valley mayweed *Matricaria occidentalis* G., *Matricaria aurea* Loefl., field or corn chamomile *Anthemis arvensis* L., stinking chamomile *Anthemis cotula* L., scentless or false chamomile *Tripleurospermum inodorum* L., dyer's chamomile *Cota tinctoria* L., etc. are several botanical species that are known as chamomile (Tsivelika et al., 2021).

Chamomile flowers have various phenolic chemicals, majorly consisting of the flavonoids patuletin, apigenin, quercetin, luteolin and their glucosides. Dicycloethers and coumarins are also present in the blooms. The main ingredients of the essential oil that is extracted from the flowers are terpenoids such as α-bisabolol and its oxides, as well as azulenes like chamazulene. When chamomile is consumed as tea other components such as flavonoids, coumarins, and phenolic acids that are water soluble exert therapeutic effects (Catani et al., 2021). Chamomile oil contains chamazulene carboxylic acid, chamazulene (1–15%) and proazulenes. For instance, the essential oil of chamomile was tested for amoebicidal activity against *Acanthamoeba castellani* (Hajaji et al., 2017). Many bacterial infections of the skin, mouth cavity, gums, and respiratory tract are treated with chamomile, as well as inflammations of the skin and mucous membranes. Chamomile extracts are popular for having anti-hypertensive effects (Awaad et al., 2018).

Aerva lanata **Mountain knotgrass** – *Aerva lanata L.* Juss. Amaranthaceae is an important medicinal plant growing in the tropical regions of Africa and Asia. The plant is generally cultivated via seed propagation (Athira and Nair, 2017; Prajapati et al., 2022). It has been demonstrated to have numerous uses in traditional medicine and has various names depending on the region in which it occurs (Sivasankari et al., 2014; Musaddiq et al., 2018). Ayurveda describes *A. lanata L. Juss.* as a diuretic with analgesic, antibacterial, anti-inflammatory, and anthelmintic properties. It is also used as a therapy for diabetes (Gunatilake et al., 2012). The anti-inflammatory and antibacterial qualities of leaf and root paste are used in the treatment of various skin infections and pustules (Nagaratna et al., 2014; Athira and Nair, 2017). It is also present as a major constituent in various cosmetic products. For around thirty years, the plant has been the focus of pharmacological and chemical research due to its widespread use in folk medicine. The presence of active substances including alkaloids, saponins, terpenoids, and other phenolic compounds (Kumar et al., 2013) is linked to its wide range of medicinal activity. Several recent researches attest to the broad range of applications for which *A. lanata L.* Juss is suitable. The antioxidant (Kumar et al., 2013; Ragavendran et al., 2012); anti-inflammatory (Balasuriya and Dharmaratne, 2007; Ramachandra et al., 2011); anti-cancer; and hepatoprotective effects of the plant have been shown by many studies. The plant's well-known diuretic, antiurolithic and nephroprotective properties have been validated by various scientific research (Balasuriya and Dharmaratne, 2007; Kumar et al., 2015). Moreover, recent reports have linked the anti-diabetic properties of *A. lanata L.* Juss. root extracts that are helpful in the inhibition of some enzymes (Rajesh et al., 2012; Riya et al., 2015; Akanji et al., 2018).

Centella asiatica – Also known as Indian pennywort, *C. asiatica* is a member of the Apiaceae and has long been used as a vegetable in Oceania, Africa, China, Southeast Asia, India, and Sri Lanka. In Southeast Asia, it has a long history of usage to cure a wide range of conditions, including skin conditions, rheumatism, inflammation, syphilis, mental illness, epilepsy, hysteria, dehydration, diarrhea, nerve illnesses and improved memory and skin conditions. A variety of neurological diseases are reportedly relieved by *C. asiatica* and its triterpenes, but the most researched and studied are Alzheimer's disease AD (Song et al., 2018) and Parkinson's disease (Nataraj et al., 2017).

Asiatic acid, asiaticoside, and madecassoside form the major constituents responsible for pharmacological value apart from being rich in flavonoids and terpenoids (Roy et al., 2013), P-cymene-44% along with other volatile compounds was found to be in a prominent amount in the essential oil of *C. asiatica.* Asiatic acid, in particular, has been shown to have a strong antiproliferative effect on cancer cells. A study on the lung cancer xenograft model, oral treatment of AA dramatically reduced weight and tumor volume (Wu et al., 2017).

The plant is commonly propagated through seeds (Kumar, 2017), however, due to the overexploitation of *Centella* from its natural habitat, it has posed a massive threat to the wild germplasm and diversity of this precious medicinal plant. *In vitro* techniques have been employed for successful rapid propagation and germplasm conservation through leaves (Kumar, 2017), axillary meristem (Siddiqui et al., 2019) and somatic embryogenesis (Biradar, 2017).

1.7 CONCLUSION

Medicinal and aromatic plants offer several advantages to human health and environmental well-being. Nature has blessed the planet with these valuable resources. However, unthoughtful anthropogenic activities, changes in climatic and ecological conditions, novel pests and pathogen attacks, dwindling land availability, and many such factors have caused a drastic decline in the availability of superior-quality medicinal and aromatic plants. The preservation of these precious resources for future generations may be protected by encouraging sustainable and ethical cultivation techniques. While the resources are dwindling, there is a contrasting rise in their demand in the pharmaceutical, cosmetic and nutritional markets. Advanced propagation techniques can increase the availability of medicinal plants, making them more accessible to communities that rely on traditional medicine. This can improve healthcare options and potentially reduce reliance on expensive conventional medications. Planned cultivation will not only protect our natural heritage and foster innovation in healthcare, but it will also provide resources for income generation among the growers. This sustainable approach ensures a future where we can continue to benefit from the remarkable potential integrated within the diverse world of medicinal plants.

REFERENCES

Agarwal, S., Chauhan, A., Anand, J., Abate, L., and Rai, N. 2023. Medicinal plant-based alkaloids: A suitable precursor for nanoparticle synthesis and their various applications. In: *Secondary Metabolites from Medicinal Plants: Nanoparticle synthesis and their applications* (Eds Bachheti, R.K. and Bachheti, A.), CRC Press, pp 111–132.

Aggarwal, D., Neeti, N., Reddy, M. S., and Kumar, A. 2020. Shoot organogenesis and assessment of clonal fidelity of regenerated plants of Ocimum tenuiflorum L.: queen of herbs. *Vegetos, 33*, 420–429.

Agrawal, L., and Kumar, M. 2021. Improvement in ornamental, medicinal, and aromatic plants through induced mutation. *Journal of Applied Biology and Biotechnology, 94*, 162–169.

Ahad, B., Shahri, W., Rasool, H., Reshi, Z. A., Rasool, S., and Hussain, T. 2021. Medicinal plants and herbal drugs: An overview. In Tariq Aftab and Khalid Rehman Hakeem (eds.), *Medicinal and Aromatic Plants: Healthcare and Industrial Applications*, 1st ed., pp. 1–40.

Ahmad, S., Tariq, M., Saniya, C. G., Belwal, V., Shah, M. H., Kumar, P., ... and Kaushik, P. 2024. Efficient in-vitro regeneration protocol for large-scale propagation of Aloe vera L. Burm. f. *Journal of Applied Research on Medicinal and Aromatic Plants, 100588*. https://doi.org/10.1016/j.jarmap.2024.100588

Akanji, M. A., Olukolu, S. O., and Kazeem, M. I. 2018. Leaf extracts of Aerva lanata inhibit the activities of type 2 diabetes-related enzymes and possess antioxidant properties. *Oxidative Medicine and Cellular Longevity, 20181*, 3439048.

Akter, K., Barnes, E. C., Brophy, J. J., Harrington, D., Community Elders, Y., Vemulpad, S. R., and Jamie, J. F. 2016. Phytochemical profile and antibacterial and antioxidant activities of medicinal plants used by aboriginal people of New South Wales, Australia. *Evidence-Based Complementary and Alternative Medicine, 20161*, 4683059.

Alam, P. 2013. Densitometric HPTLC analysis of 8–gingerol in Zingiber officinale extract and ginger–containing dietary supplements, teas and commercial creams. *Asian Pacific Journal of Tropical Biomedicine, 38*, 634–638.

Al-Amin, Z. M., Thomson, M., Al-Qattan, K. K., Peltonen-Shalaby, R., and Ali, M. 2006. Anti-diabetic and hypolipidaemic properties of ginger Zingiber officinale in streptozotocin-induced diabetic rats. *British Journal of Nutrition, 964*, 660–666.

Ali, A. E., Ali, S. M. E. H., EL-Rhman, S. A. E. A., Mohamed, M. A. E., and Khider, T. O. 2020. Brine-shrimp lethality bioassay of different extracts of the medicinal plant matricaria chamomilla flowers. *Journal of Medicine and Healthcare*, *4*(6), 1–4.

Athira, P., and Nair, S. N. 2017. Pharmacognostic review of medicinal plant Aerva lanata. *Journal of Pharmaceutical Sciences and Research, 99*, 1420.

Awaad, A. A., El-Meligy, R. M., Zain, G. M., Safhi, A. A., Al Qurain, N. A., Almoqren, S. S., ... and Al-Saikhan, F. I. 2018. Experimental and clinical antihypertensive activity of Matricaria chamomilla extracts and their angiotensin-converting enzyme inhibitory activity. *Phytotherapy Research, 328*, 1564–1573.

Bahraminejad, S., Seifolahpour, B., and Amiri, R. 2016. Antifungal effects of some medicinal and aromatic plant essential oils against Alternaria solani. *Journal of Crop Protection, 54*, 603–616.

Bahukhandi, A., Sekar, K. C., Barola, A., Bisht, M., and Mehta, P. 2019. Total phenolic content and antioxidant activity of Meconopsis aculeata Royle: a high value medicinal herb of Himalaya. *Proceedings of the National Academy of Sciences, India Section B: Biological Sciences, 89*, 1327–1334.

Bajpayee, K. K. 2023. Cultivation of medicinal plants. Cultivation of medicinal plants. *Journal of Science, Research and Teaching, 27*, 229–243.

Balasuriya, G. M. G. K., and Dharmaratne, H. R.W. 2007. Cytotoxicity and antioxidant activity studies of green leafy vegetables consumed in Sri Lanka. *Journal of the National Science Foundation of Sri Lanka, 354*, 255–258.

Banik, B., Das, S., and Das, M. K. 2020. Medicinal plants with potent anti-inflammatory and anti-arthritic properties found in Eastern parts of the Himalaya: An ethnomedicinal review. *Pharmacognosy Reviews, 14(28)*: 121–137.

Baral, S. R., and Kurmi, P. P. 2006. A compendium of medicinal plants in Nepal. IUCN, 1–534.

Bhadra, P., and Sethi, L. 2020. A review paper on the Tulsi plant Ocimum sanctum. *Indian Journal of Natural Sciences, 1060*, 20854–20860.
Bindu, K. H., Mythili, J. B., and Radhika, R. M. 2018. Genetic engineering in medicinal and aromatic plants. In *Genetic engineering of horticultural crops* pp. 249–271. Academic Press.
Biradar, S. R. 2017. Somatic embryogenesis of medicinally important herb Centella asiatica L. *Bioscience Discovery, 8*, 295–299.
Bordia, A., Verma, S. K., and Srivastava, K. C. 1997. Effect of ginger Zingiber officinale Rosc. and fenugreek Trigonella foenumgraecum L. on blood lipids, blood sugar and platelet aggregation in patients with coronary artery disease. *Prostaglandins, Leukotrienes and Essential Fatty Acids, 565*, 379–384.
Brijesh, H., and Ajjappala, B. 2023. Micropropagation strategies in medicinally important turmeric Curcuma sp: Current research and future challenges. *Journal of Applied Biology and Biotechnology, 113*, 1–8.
Catani, M. V., Rinaldi, F., Tullio, V., Gasperi, V., and Savini, I. 2021. Comparative analysis of phenolic composition of six commercially available chamomile Matricaria chamomilla L. extracts: Potential biological implications. *International Journal of Molecular Sciences, 2219*, 10601.
Chandegara, V. K., and Varshney, A. K. 2013. Aloe vera L. processing and products: a review. *International Journal of Medicinal and Aromatic Plants, 34,* 492–506.
Chandra, S., Dwivedi, P., and Shinde, L. P. 2016. An industrial cultivation of Tulsi Ocimum sanctum for medicinal use and rural empowerment. *Journal of Medicinal Plants Studies, 46*, 213–218.
Chandran, H., Meena, M., Barupal, T., and Sharma, K. 2020. Plant tissue culture as a perpetual source for production of industrially important bioactive compounds. *Biotechnology Reports, 26*, e00450.
Chauhan, B., Kumar, G., and Ali, M. A. 2012. Review on phytochemical constituents and activities of Trachyspermum ammi L. Sprague fruits. *American Journal of Pharmtech Research, 24*, 329–40.
Chauhan, R., Singh, S., Kumar, V., Kumar, A., Kumari, A., Rathore, S., Kumar, R., and Singh, S. 2021. A Comprehensive review on biology, genetic improvement, agro and process technology of German Chamomile Matricaria chamomilla L. *Plants, 111*, 29.
Chiappero, J., del Rosario Cappellari, L., Palermo, T. B., Giordano, W., Khan, N., and Banchio, E. 2021. Antioxidant status of medicinal and aromatic plants under the influence of growth-promoting rhizobacteria and osmotic stress. *Industrial Crops and Products, 167*, 113541.
Christaki, E., Bonos, E., Giannenas, I., Florou-Paneri, P. 2012. Aromatic plants as a source of bioactive compounds. *Agriculture, 2*(3), 228–243.
Chrysargyris, A., Mikallou, M., Petropoulos, S., and Tzortzakis, N. 2020. Profiling of essential oils components and polyphenols for their antioxidant activity of medicinal and aromatic plants grown in different environmental conditions. *Agronomy, 105*, 727.
Daily, J. W., Yang, M., Kim, D. S., and Park, S. 2015. Efficacy of ginger for treating Type 2 diabetes: A systematic review and meta-analysis of randomized clinical trials. *Journal of Ethnic Foods, 21*, 36–43.
Das, M., Jain, V., and Malhotra, S. K. 2016. Impact of climate change on medicinal and aromatic plants: review. *Indian Journal of Agricultural Sciences, 86* 11, 1375–1382.
de Oliveira, M. S., Almeida, M. M., Salazar, M. L. A. R., Pires, F. C. S., Bezerra, F. W. F., Cunha, V. M. B., ... and de Carvalho Junior, R. N. 2018. Potential of medicinal use of essential oils from aromatic plants. *Potential of Essential Oils*, 1, 1–21.
De, L. C. 2017. Breeding of medicinal and aromatic plants-an overview. *International Journal of Botany and Research, 72*, 25–34.
Deshmukh, A. S., Deshmukh, G. B., and Shirole, P. D. 2015. Ocimum sanctum: A medicinal gift from nature. *International Journal of Pharmacognosy, 212*, 550–559.
Duta-Cornescu, G., Constantin, N., Pojoga, D. M., Nicuta, D., and Simon-Gruita, A. 2023. Somaclonal variation—Advantage or disadvantage in micropropagation of the medicinal plants. *International Journal of Molecular Sciences, 241*, 838.
Fadaki, F., Modaresi, M., and Sajjadian, I. 2017. The effects of ginger extract and diazepam on anxiety reduction in animal model. *Indian Journal of Pharmaceutical Education and Research, 513*, 159–162.
Ferreira, M.D.S., Rocha, A.D.J., Nascimento, F.D.S., Oliveira, W.D.D.S., Soares, J.M.D.S., Rebouças, T.A., Morais Lino, L.S., Haddad, F., Ferreira, C.F., Santos-Serejo, J.A.D. et al. 2023. The role of somaclonal variation in plant genetic improvement: A systematic review. *Agronomy, 13*, 730.
Fridman, E. 2015. Consequences of hybridization and heterozygosity on plant vigor and phenotypic stability. *Plant Science, 232*, 35–40.

Gaba, J., Sharma, S., Joshi, S., and Gill, P. 2018. Gas chromatography-mass spectrometric analysis of essential oil, nutritional and phytochemical composition of ajwain seeds Trachyspermum ammi L.. *Journal of Essential Oil Bearing Plants, 214*, 1128–1137.

Gao, W., Meng, Q., Wang, X., Chen, F., Zhou, Y., and He, M. 2023. Overexpression of CiMYC2 transcription factor from Chrysanthemum indicum var. aromaticum resulted in modified trichome formation and terpenoid biosynthesis in transgenic tobacco. *Journal of Plant Growth Regulation, 427*, 4161–4175.

Gaurav, N., Juyal, P., Tyagi, M., Chauhan, N., and Kumar, A. 2018. A review on in vitro propagation of medicinal plants. *Journal of Pharmacognosy and Phytochemistry, 76*, 2228–2231.

Gersbach, P. V., and Reddy, N. 2002. Non-invasive localization of thymol accumulation in Carum copticum Apiaceae fruits by chemical shift selective magnetic resonance imaging. *Annals of Botany, 902*, 253–257.

Giacometti, J., Kovacevic, D.B., Putnik, P., Gabric, D., Bilusic, T., Kresic, G., Stulic, V., Barba, F.J., Chemat, F., BarbosaCanovas, G., Jambrak, R. 2018. Extraction of bioactive compounds and essential oils from Mediterranean herbs by conventional and green innovative techniques: a review. *Food Research International, 113*, 245–262.

Giannenas, I., Bonos, E., Christaki, E., and Florou-Paneri, P. 2018. Oregano: A feed additive with functional properties. In A.M. Holban and A.M Grumezescu (Eds.), *Therapeutic Foods* pp. 179–208. Academic Press.

Giannenas, I., Bonos, E., Christaki, E., and Florou-Paneri, P. 2013. Essential oils and their applications in animal nutrition. *Journal of Medicinal and Aromatic Plant Sciences, 2*, 140, 2167–0412.

Gómez-Galera, S., Pelacho, A. M., Gené, A., Capell, T., and Christou, P. 2007. The genetic manipulation of medicinal and aromatic plants. *Plant Cell Reports, 26*, 1689–1715.

Gonceariuc, M. 2014. Moldavian medicinal and aromatic plants varieties. *Hop and Medicinal Plants, 221/2*, 51–62.

Gonçalves, S. and Romano, A. 2013. In vitro culture of lavenders (Lavandula spp.) and the production of secondary metabolites. *Biotechnology Advances, 31*(2), 166–174.

Grigoriadou, K., Krigas, N., and Maloupa, E. 2011. GIS-facilitated in vitro propagation and ex situ conservation of Achillea occulta. *Plant Cell, Tissue and Organ Culture PCTOC, 107*, 531–540.

Gunatilake, M., Lokuhetty, M. D. S., Bartholameuz, N. A., Edirisuriye, D. T., Kularatne, M. U., and Date, A. 2012. Aerva lanata Polpala: Its effects on the structure and function of the urinary tract. *Pharmacognosy Research, 4*, 181–188.

Gurav, T. P., Jayaramaiah, R. H., Punekar, S. A., Dholakia, B. B., and Giri, A. P. 2020. Generation of novelties in the genus Ocimum as a result of natural hybridization: a morphological, genetical and chemical appraisal. *Industrial Crops and Products, 156*, 112859.

Hajaji, S., Sifaoui, I., López-Arencibia, A., Reyes-Batlle, M., Valladares, B., Pinero, J. E., ... and Akkari, H. 2017. Amoebicidal activity of α-bisabolol, the main sesquiterpene in chamomile Matricaria recutita L. essential oil against the trophozoite stage of Acanthamoeba castellani Neff. *Acta Parasitologica, 622*, 290–295.

Heghes, S. C., Vostinaru, O., Rus, L. M., Mogosan, C., Iuga, C. A., and Filip, L. 2019. Antispasmodic effect of essential oils and their constituents: A review. *Molecules, 249*, 1675.

Husen, A., 2021. Traditional herbal therapy for the human immune system. CRC Press, Boca Raton, FL. https://doi.org/10.1201/9781003137955

Husen, A., 2022. Herbs, shrubs and trees of potential medicinal benefits. Taylor & Francis Group, NW. https://doi.org/10.1201/9781003205067

Husen, A., 2023a. Exploring poisonous plants: Medicinal values, toxicity response and other potential uses. CRC Press, Boca Raton, FL. https://doi.org/10.1201/b23017

Husen, A., 2023b. Antidiabetic medicinal plants and herbal treatments. CRC Press, Boca Raton, FL. https://doi.org/10.1201/b23347

Husen, A., 2024a. Antimalarial medicinal plants. Taylor & Francis Group, NW. ISBN: 9781032457154 https://doi.org/10.1201/9781003378396

Husen, A., 2024b. Medicinal spices and condiment crops. Taylor & Francis Group, NW. ISBN: 9781032469478 https://doi.org/10.1201/9781003387046

Ishikawa, T., Sega, Y., and Kitajima, J. 2001. Water-soluble constituents of ajowan. *Chemical and Pharmaceutical Bulletin, 497*, 840–844.

Jain, S., Rathod, M., Mishra, G., Kumari, V., Baksh, H., Saxena, S., and Prasad, L. 2023. Plant tissue culture for medical therapy: Unlocking the potential of medicinal plants. *Current Journal of Applied Science and Technology, 4246*, 7–22.

Jamshidi-Kia, F., Lorigooini, Z., and Amini-Khoei, H. 2018. Medicinal plants: Past history and future perspective. *Journal of Herbmed Pharmacology, 7(1)*, 1–7.

Joseph, B., and Raj, S. J. 2010. Pharmacognostic and phytochemical properties of Aloe vera linn an overview. *International Journal of Pharmaceutical Sciences Review and Research, 42*, 106–110.

Kala, C. P. 2015. Medicinal and aromatic plants: Boon for enterprise development. *Journal of Applied Research on Medicinal and Aromatic Plants, 24*, 134–139.

Kaplan, A. 2022. The nanocomposites designs of phytomolecules from medicinal and aromatic plants: promising anticancer-antiviral applications. *Beni-Suef University Journal of Basic and Applied Sciences, 111*, 17.

Kar, S. K., and Kanti, T. B. 2018. Phytochemical constituents of Aloe vera and their multifunctional properties: a comprehensive review. *International Journal of Pharmaceutical Sciences and Research, 9*, 1416–1423.

Kaur, G., and Ganjewala, D. 2019. Stress protectant secondary metabolites and their metabolic engineering to enhance abiotic stress tolerance in plants. In Kumar, M., Muthusamy, A., Kumar, V., Sarin, N.B (Eds.), *In vitro Plant Breeding towards Novel Agronomic Traits: Biotic and Abiotic Stress Tolerance*, 197–216.

Kolbadinejad, A., and Rezaeipour, V. 2020. Efficacy of ajwain Trachyspermum ammi L. seed at graded levels of dietary threonine on growth performance, serum metabolites, intestinal morphology and microbial population in broiler chickens. *Journal of Animal Physiology and Animal Nutrition, 1045*, 1333–1342.

Krup, V., Prakash, L. H., and Harini, A. 2013. Pharmacological activities of turmeric Curcuma longa Linn: a review. *Journal of Homeopathy and Ayurvedic Medicine, 2133*, 2167–1206.

Kumar, J., Basak, B. B., Jat, R. S., and Reddy, N. 2016. Medicinal and aromatic plants making healthy and wealthy. *Indian Horticulture, 616*, 61–64.

Kumar, G., Longanathan, K. and Rao, B. 2013. Phytochemical composition and in vitro antioxidant activity of aqueous extract of Aerva lanata (L.) Juss. ex Schult. Stem (Amaranthaceae). *Asian Pacific Journal of Tropical Medicine*, 6: 180–187.

Kumar, K., Prabhu, S., Ravishankar, B., and Sahana, Y. B. 2015. Chemical analysis and in vitro evaluation of antiurolithiatic activity of Aerva lanata Linn. Juss. Ex Schult. roots. Research and reviews. *Journal of Pharmacognosy and Phytochemistry*, 33, 1–7.

Kumar, M. S. 2017. Rapid in vitro multiplication of Centella asiatica L. Urban through multiple shoots from leaf explants. *European Journal of Biotechnology and Bioscience*, *53*, 41–47.

Kumar, N., and Khurana, S. 2018. Phytochemicals and bioactive potential of Trachyspermum ammi L. *Der Pharmacia Lettre, 108*, 48–56.

Labban, L. 2014. Medicinal and pharmacological properties of Turmeric Curcuma longa: A review. *International Journal of Pharmaceutical and Biomedical Science, 51*, 17–23.

Lanka, S. 2018. A review on aloe vera-The wonder medicinal plant. *Journal of Drug Delivery and Therapeutics, 85*, 94–99.

Lemma, D. T., Banjaw, D. T., and Megersa, H. G. 2020. Micropropagation of medicinal plants. *International Journal of Plant Breeding and Crop Science, 7*, 796–802.

Mahmood, S. 2019. A critical review on pharmaceutical and medicinal importance of ginger. *Acta Scientific Nutritional Health, 3*, 78–82.

Manvitha, K., and Bidya, B. 2014. Aloe vera: a wonder plant its history, cultivation and medicinal uses. *Journal of Pharmacognosy and Phytochemistry*, 25, 85–88.

Mathur, A. K., Ahuja, P. S., Pandey, B., Kukreja, A. K., and Mandal, S. 1988. Screening and evaluation of somaclonal variations for quantitative and qualitative traits in an aromatic grass, Cymbopogon winterianus Jowitt. *Plant Breeding, 1014*, 321–334.

Minija, J., and Thoppil, J. E. 2002. Essential oil composition of Trachyspermum ammi l. sprague from South India. *Indian Journal of Pharmaceutical Sciences*, *643*, 250–251.

Mir, B. A., Mir, S. A., and Koul, S. 2014. In vitro propagation and withaferin A production in Withania ashwagandha, a rare medicinal plant of India. *Physiology and Molecular Biology of Plants, 20*, 357–364.

Mishra, S. 2020. A review of super food ajwain and its pharmacological actions. *International Journal of Research in Pharma and Pharmaceutical Science, 11*, 30–33.

Mishra, T. 2015. Protocol establishment for multiplication and regeneration of 'Holy Basil'Ocimum sanctum Linn. An important medicinal plant with high religious value in India. *Journal of Medicinal Plants Studies*, *34*, 16–19.

Moraes, R. M., Cerdeira, A. L., and Lourenço, M. V. 2021. Using micropropagation to develop medicinal plants into crops. *Molecules*, *266*, 1752.

Mota, M. L., Lobo, L. T. C., da Costa, J. M. G., Costa, L. S., Rocha, H. A., e Silva, L. F. R., Pohlit, A. M., and de Andrade Neto, V. F. 2012. In vitro and in vivo antimalarial activity of essential oils and chemical components from three medicinal plants found in northeastern Brazil. *Planta Medica*, *7807*, 658–664.

Musaddiq, S., Mustafa, K., Ahmad, S., Aslam, S., Ali, B., Khakwani, S., ... and Jabbar, A. 2018. Pharmaceutical, ethnopharmacological, phytochemical and synthetic importance of genus Aerva: A review. *Natural Product Communications*, *133*, 375–385.

Nagaratna, A., Prakash, L. H., and Harini, A. 2014. A pharmacological review on gorakha ganja Aerva lanata Linn Juss. ex. Schult. *Journal of Pharmacognosy and Phytochemistry*, *3*, 253–7.

Nandikatti, V., Nagasree, K. P., and Kumar, M. K. 2023. Antitubercular activity of essential oils extracted from some aromatic plants, flowers and seeds. *Journal of Young Pharmacists*, *152*, 380–383.

Nasri, H., Sahinfard, N., Rafieian, M., Rafieian, S., Shirzad, M., and Rafieian-Kopaei, M. 2014. Turmeric: A spice with multifunctional medicinal properties. *Journal of HerbMed Pharmacology*, *31*, 5–8.

Nataraj, J., Manivasagam, T., Justin Thenmozhi, A., and Essa, M. M. 2017. Neurotrophic effect of asiatic acid, a triterpene of Centella asiatica against chronic 1-methyl 4-phenyl 1, 2, 3, 6-tetrahydropyridine hydrochloride/probenecid mouse model of Parkinson's disease: The role of MAPK, PI3K-Akt-GSK3β and mTOR signalling pathways. *Neurochemical Research*, *42*, 1354–1365.

Naz, S., Jabeen, S., Ilyas, S., Manzoor, F., Aslam, F., and Ali, A. 2010. Antibacterial activity of Curcuma longa varieties against different strains of bacteria. *Pakistan Journal of Botany*, *421*, 455–62.

Nazari, M., Kordrostami, M., and Ghasemi-Soloklui, A. A. 2023. Conservation of medicinal plants by tissue culture techniques. In *Medicinal plants: Biodiversity, biotechnology and conservation* pp. 801–818. Springer Nature Singapore, Singapore.

Negi, P. S., Jayaprakasha, G. K., Jagan Mohan Rao, L., and Sakariah, K. K. 1999. Antibacterial activity of turmeric oil: a byproduct from curcumin manufacture. *Journal of Agricultural and Food Chemistry*, *4710*, 4297–4300.

Neha, S., Ranvir, G. D., and Jangade, C. R. 2009. Analgesic and antipyretic activities of Curcuma longa rhizome extracts in Wister Rats. *Veterinary World*, *28*.

Nenni, M., and Karahuseyin, S. 2024. Medicinal plants, secondary metabolites, and their antiallergic activities. In *Biotechnology of medicinal plants with antiallergy properties: Research trends and prospects* pp. 37–126.Springer Nature Singapore, Singapore.

Niazian, M. 2019. Application of genetics and biotechnology for improving medicinal plants. *Planta*, *249*, 953–973.

Oludoyin, A. P., and Adegoke, S. R. 2014. Effect of ginger Zingiber officinale extracts on blood glucose in normal and streptozotocin-induced diabetic rats. *International Journal of Clinical Nutrition*, *22*, 32–35.

Pandey, A., Adhikari, P., Dhyani, A., and Palni, L. M. S. 2021. Use of root microbiome in conservation of Himalayan Yew Taxus wallichiana Zucc.: An ecofriendly approach. *National Academy Science Letters*, *443*, 249–251.

Pidigam, S., Geetha, A., Pandravada, S. R., Khan, S., Thuraga, V., Sivaraj, N., and Nagaraju, K. 2023. Conservation strategies for Indian medicinal plants. *Medicinal and Aromatic Plants of India*, *2*, 279–302.

Piri, A., Sahebzadeh, N., Zibaee, A., Sendi, J. J., Shamakhi, L., and Shahriari, M. 2020. Toxicity and physiological effects of ajwain Carum copticum, Apiaceae essential oil and its major constituents against Tuta absoluta MeyrickLepidoptera: Gelechiidae. *Chemosphere*, *256*, 127103.

Prajapati, G. M., Kadam, R. S., and Ghuge, D. R. 2022. Aerva lanata: A comprehensive profile in the pharmaceutical and food Industries. *International Journal of Pharmacognosy and Chemistry*, 72–82.

Qu, Y., Safonova, O., and De Luca, V. 2019. Completion of the canonical pathway for assembly of anticancer drugs vincristine/vinblastine in Catharanthus roseus. *The Plant Journal*, *972*, 257–266.

Qureshi, A. A., and Kumar, K. E. 2010. Phytochemical constituents and pharmacological activities of Trachyspermum ammi. *Plant Archives*, *102*, 955–959.

Raeisi, S., Sharifi-Rad, M., Quek, S. Y., Shabanpour, B., and Sharifi-Rad, J. 2016. Evaluation of antioxidant and antimicrobial effects of shallot Allium ascalonicum L. fruit and ajwain Trachyspermum ammi

L. Sprague seed extracts in semi-fried coated rainbow trout Oncorhynchus mykiss fillets for shelf-life extension. *LWT-Food Science and Technology*, *65*, 112–121.

Ragavendran, P., Sophia, D., Raj, C. A., Starlin, T., and Gopalakrishnan, V. K. 2012. Phytochemical screening, antioxidant activity of Aerva lanata L–An Invitro study. *Asian Journal of Pharmaceutical and Clinical Research*, *52*, 77–81.

Rajesh, R., Chitra, K., and Paarakh, P. M. 2012. Anti hyperglycemic and antihyperlipidemic activity of aerial parts of Aerva lanata Linn Juss in streptozotocin induced diabetic rats. *Asian Pacific Journal of Tropical Biomedicine*, *22*, 924–929.

Ramachandra, Y. L., Shilali, K., Ahmed, M., Sudeep, H. V., Kavitha, B. T., Gurumurthy, H., and Rai, S. P. 2011. Hepatoprotective properties of Boerhaavia diffusa and Aerva lanata against carbon tetra chloride induced hepatic damage in rats. *Pharmacologyonline*, *3*, 435–441.

Rao, S. K., Sharma, A., and Jain, D. 2023. A review medicinal and traditional uses on Tulsi plant Ocimum sanctum L.. *World Journal of Biology Pharmacy and Health Sciences*, *131*, 450–456.

Rathod, T. P., Wable, V. K., Jadhav, S. B., and Ramteke, P. P., 2023 An overview of Tulsi and New Innovations in Tulsi Production. *International Journal of Research Publications and Reviews*, *43*, 4657–4666.

Ravindra, N. S., Ramesh, S. I., Gupta, M. K., Jhang, T., Shukla, A. K., Darokar, M. P., and Kulkarni, R. N. 2012. Evaluation of somaclonal variation for genetic improvement of patchouli Pogostemon patchouli, an exclusively vegetatively propagated aromatic plant. *Journal of Crop Science and Biotechnology*, *15*, 33–39.

Riya, M. P., Antu, K. A., Pal, S., Chandrakanth, K. C. Anilkumar, K. S., Tamrakar, A. K., Srivastava, A. K., and Raghu, K. G. 2015. Antidiabetic property of Aerva lanata L. Juss. ex Schult. is mediated by inhibition of alpha glucosidase, protein glycation and stimulation of adipogenesis. *Journal of Diabetes*, *7*(4), 548–561.

Roy, D. C., Barman, S. K., and Shaik, M. M. 2013. Current updates on Centella asiatica: phytochemistry, pharmacology and traditional uses. *Medicinal Plant Research*, *3*.

Roychoudhury, A., and Bhowmik, R. 2021. State-of-the-art technologies for improving the quality of medicinal and aromatic plants. *Medicinal and Aromatic Plants: Healthcare and Industrial Applications*, 593–627.

Sahu, P. K., Giri, D. D., Singh, R., Pandey, P., Gupta, S., Shrivastava, A. K., ... and Pandey, K. D. 2013. Therapeutic and medicinal uses of Aloe vera: a review. *Pharmacology and Pharmacy*, *408*, 599.

Sajid-ur-Rehman, M., Ishtiaq, S., Khan, M. S., Jabeen, Q., Youssef, F. S., Ahmed, S. A., Elhady S. S., and Ashour, M. L. 2023. Characterization of Sesuvium sesuvioides using high-performance liquid chromatography and validation of its in vivo anti-hyperthyroidism effect via suppressing oxidative stress and inflammatory markers. *ACS Omega*, *848*, 45896–45905.

Salehi, B., Ata, A., Kumar, N. V. A., Sharopov, F., Ramírez-Alarcón, K., Ruiz-Ortega, A., Ayatollahi, S. A., Fokou, P. V. T., Kobarfard, F., Zakaria, Z. A., Iriti, M., Taheri, Y., Martorell, M., Sureda, A., Setzer, W. N., Durazzo, A., Lucarini, M., Santini, A., Capasso, R., … Sharifi-Rad, J. 2019. Antidiabetic potential of medicinal plants and their active components. *Biomolecules*, *910*. https://doi.org/10.3390/biom9100551

Salvi, P., Kumar, G., Gandass, N., Kajal, Verma, A., Rajarammohan, S., Rai, N., and Gautam, V. 2022. Antimicrobial potential of essential oils from aromatic plant Ocimum sp.; A comparative biochemical profiling and in-silico analysis. *Agronomy*, *123*, 627.

Sang, S., Snook, H. D., Tareq, F. S., and Fasina, Y. 2020. Precision research on ginger: the type of ginger matters. *Journal of Agricultural and Food Chemistry*, *6832*, 8517–8523.

Sarangthem, K., and Haokip, M. J. 2010. Bioactive components in Curcuma caesia Roxb. grown in Manipur. *Bioscan*, *51*, 113–115.

Semwal, P., and Painuli, S. 2019. Antioxidant, antimicrobial, and GC-MS profiling of Saussurea obvallata Brahma Kamal from Uttarakhand Himalaya. *Clinical Phytoscience*, *5*, 1–11.

Shafi, A., and Zahoor, I. 2021. Metabolomics of medicinal and aromatic plants: Goldmines of secondary metabolites for herbal medicine research. In *medicinal and aromatic plants* pp. 261–287. Academic Press.

Shahrajabian, M. H., Sun, W., and Cheng, Q. 2019. Clinical aspects and health benefits of ginger Zingiber officinale in both traditional Chinese medicine and modern industry. *Acta Agriculturae Scandinavica, Section b—Soil and Plant Science*, *696*, 546–556.

Shahzad, A., Parveen, S., Sharma, S., Shaheen, A., Saeed, T., Yadav, V., ... and Upadhyay, A. 2017. Plant tissue culture: applications in plant improvement and conservation. *Plant Biotechnology: Principles and Applications*, 37–72.

Sharma, N. K., Choudhary, R. C., and Kumar, M. 2014. An improved plant regeneration system of *Ocimum sanctum* L. – An important Indian Holy Basil plant. *Journal of Cell and Tissue Research, 141*, 4143.

Sharma, S., Satardekar, K. V., and Barve, S. S. 2018. Genetic improvement of medicinal and aromatic plants through haploid and double haploid development. *Biotechnological Approaches for Medicinal and Aromatic Plants: Conservation, Genetic Improvement and Utilization*, 523–556.

Shojaaddini, M., Moharramipour, S., and Sahaf, B. 2008. Fumigant toxicity of essential oil from Carum copticum against Indian meal moth, Plodia interpunctella. *Journal of Plant Protection Research, 484*(4), 411–419.

Shukla, M. R., Kibler, A., Turi, C. E., Erland, L. A., Sullivan, J. A., Murch, S. J., and Saxena, P. K. 2021. Selection and micropropagation of an elite melatonin rich Tulsi Ocimum sanctum L. germplasm line. *Agronomy, 112*, 207.

Siddiqui, S., and Thomas, T. 2019. Standardization of an efficient protocol for in vitro micropropagation of Centella asiatica L.– an important medicinal plant. *Pharmaceutical and Biosciences Journal*, 7(4), 22–26.

Sinchana, N. S., Kattimani, K. N., Prabhuling, G., Sudesh, K., and Jagadeesha, N. 2020. Standardization of tissue culture protocol for turmeric Curcuma longa L. cv. Salem. *International Journal of Chemical Studies, 81*, 2721–2726.

Singh, P., Pandey, S. S., Dubey, B. K., Raj, R., Barnawal, D., Chandran, A., and Rahman, L. U. 2021. Salt and drought stress tolerance with increased biomass in transgenic Pelargonium graveolens through heterologous expression of ACC deaminase gene from Achromobacter xylosoxidans. *Plant Cell, Tissue and Organ Culture PCTOC, 1472*, 297–311.

Sivasankari, B., Anandharaj, M., and Gunasekaran, P. 2014. An ethnobotanical study of indigenous knowledge on medicinal plants used by the village peoples of Thoppampatti, Dindigul district, Tamilnadu, India. *Journal of Ethnopharmacology, 1532*, 408–423.

Slikkerveer, L. J. 2006. The challenge of non-experimental validation of MAC plants. In Bogers, R.J., Craker, L.E., Lange, D. (Eds.), *Medicinal and aromatic plants,* Vol. *17*, pp. 1–28. Springer, Dordrecht.

Soltani, H. M., Sadat Noori, S. A., Shariati, J. V., and Niazian, M. 2018. Essential oil chemotype of Iranian ajowan Trachyspermum ammi L.. *Journal of Essential Oil Bearing Plants, 211*, 273–276.

Song, D., Jiang, X., Liu, Y., Sun, Y., Cao, S., and Zhang, Z. 2018. Asiaticoside attenuates cell growth inhibition and apoptosis induced by Aβ1–42 via inhibiting the TLR4/NF-κB signaling pathway in human brain microvascular endothelial cells. *Frontiers in Pharmacology, 9*, 28.

Srivastava, R.K., Kumar, S., Sharma, R.S. 2018. *Ocimum* as a promising commercial crop. In: Shasany, A., Kole, C. (Eds.), *The Ocimum Genome. Compendium of Plant Genomes*. Springer, Cham.

Taghavi, T., Rahemi, A., Rafie, R., and Kering, M. K. 2021. Optimizing turmeric tissue culture, testing different media and a plant growth regulator matrix. *HortTechnology, 316*, 692–704.

Terefe, T., and Neges, T. 2017. Review on therapeutic and medicinal use of Aloe vera. *Cancer Biology, 74*, 29–38.

Timothy, K. K., and Mwangi, M. 2015. Studies on German Chamomile Matricaria recutita L. propagation and the effect of light and age on seed viability. *Journal of Animal and Plant Sciences, 242*, 3781–3786.

Tsivelika, N., Irakli, M., Mavromatis, A., Chatzopoulou, P., and Karioti, A. 2021. Phenolic profile by HPLC-PDA-MS of Greek chamomile populations and commercial varieties and their antioxidant activity. *Foods, 1010*, 2345.

Waman, A. A., Smitha, G. R., and Bohra, P. 2019. A review on clonal propagation of medicinal and aromatic plants through stem cuttings for promoting their cultivation and conservation. *Current Agriculture Research Journal, 72* (2), 122–138.

Wu, T., Geng, J., Guo, W., Gao, J., and Zhu, X. 2017. Asiatic acid inhibits lung cancer cell growth in vitro and in vivo by destroying mitochondria. *Acta Pharmaceutica Sinica B, 71*, 65–72.

Zaman, M. A., Abbas, R. Z., Qamar, W., Qamar, M. F., Mehreen, U., Shahid, Z., and Kamran, M. 2020. Role of secondary metabolites of medicinal plants against Ascaridia galli. *World's Poultry Science Journal, 763*, 639–655.

Zarshenas, M. M., Samani, S. M., Petramfar, P., and Moein, M. 2014. Analysis of the essential oil components from different Carum copticum L. samples from Iran. *Pharmacognosy Research, 61*, 62.

Zuzarte, M. R., Dinis, A. M., Cavaleiro, C., Salgueiro, L. R., and Canhoto, J. M. 2010. Trichomes, essential oils and in vitro propagation of Lavandula pedunculata Lamiaceae. *Industrial Crops and Products, 323*, 580–587.

2 *In Vitro* Clonal Propagation as an Effective Tool for The Conservation and Production of Elite Medicinal and Aromatic Plants

*Siddhartha Kafle, Yanu Bam, Diwash Gautam, Aakriti Koirala, Divya Panta, Aastha Upreti, Tara Shrestha and Dhurva Prasad Gauchan**

*Corresponding author

2.1 INTRODUCTION

Medicinal and aromatic plants (MAPs) are essential in traditional medicine and modern pharmacology due to their therapeutic bioactive compounds (Singh and Sharma, 2015). Herbs such as *Aloe vera*, *Lavandula angustifolia* (lavender), and *Curcuma longa* (turmeric) are utilized for their antioxidant, antibacterial, anti-inflammatory, and anticancer qualities (Ekor, 2014). The increased global demand for natural remedies, herbal cosmetics, and nutraceuticals has increased the usage of MAPs. Efforts in developing methods for the extraction and applicability of active constituents from these plants were furthered by biotechnological and pharmacognostic advances. On the other hand, needs assessment has to be balanced with sustainable harvesting and conservation to avoid loss of biodiversity and the availability of these resources (Barata et al., 2016).

Plants were first used in medicine 5000 years ago on a Sumerian clay slab, in the Indian Vedas, the Ebers Papyrus, and in texts from Hippocrates and Dioscorides. Other important history of MAPs includes Arab texts that influenced European medicine, new plants from the Americas, and Paracelsus, who advocated for chemically prepared drugs. The botanical classification by Linnaeus changed plant identification in the 18th century. The isolation of alkaloids and glycosides from plants in the 19th century was a giant step in scientific achievement in the histories of the pharmaceutical sciences (Bauer, 2012). Today, with concerns over synthetic drug efficacy and side effects, there is renewed interest in natural remedies. Medicinal and aromatic plants continue to address various health concerns, supported by modern research validating their benefits. One example of an anticancer medication that the FDA has approved for commercial manufacture is Taxol, which is derived from the Pacific yew tree and is widely used as the first-choice treatment for a variety of cancers, most notably breast cancer. The FDA has also approved topotecan and irinotecan, which are derived from *Camptotheca acuminata*, for the treatment of small-cell lung, metastatic colorectal, and ovarian malignancies, respectively. Etoposide and teniposide, derived from *Podophyllum peltatum*, are semisynthetic derivatives used in cancer therapy. These examples underscore the valuable contributions of plants to modern medicine, highlighting their role in combating serious illnesses like cancer (Inoue and Craker, 2014). The use of medicinal and aromatic plants has continued up to

DOI: 10.1201/9781003482659-2

the present moment, reflecting the dynamic integration of traditional wisdom, scientific progress, and cultural practice. With continued research into their bioactive compounds and mechanisms of action, their potential contribution to medicine can only keep getting bigger (Inoue et al., 2014).

2.2 GLOBAL DEMAND OF MAPS

It has been seen that medicinal and aromatic plants (MAPs) play a pivotal role in the food, cosmetic, and pharmaceutical industries. They relatedly play cultural and economic roles in these sectors (Inoue et al., 2019). In medicinal plants trade, China and India dominated with 14.8% each, with Egypt in third place at 5.9%. Spain and Poland ranked among the top five with shares of 4.4% and 4.1%, respectively. A similar ranking is observed for lac, gums, resins, saps, and extracts, where China was the leading exporter in 2020 with a 20.7% share, followed by India (9.2%) and the USA (7.7%), along with France (7.1%), and Germany (6.5%). In essential oil exports, the top five countries remained stable in 2020, with India leading at 15.7%, followed by the USA (13.4%), France (8.6%), China (8.2%), and Brazil with 6.3% (Silalahi et al., 2023).

There are four widely different major systems of herbal medicine: Western herbalism, Ayurvedic herbalism, traditional Chinese herbalism, and traditional Arabic and Islamic herbalism (Chopra and Kamal, 2009). This only goes to prove that the application of medicinal plants is part of various cultures across the world. The ethnopharmacological approach, being focused on a traditional medical system, provides the very foundation for developing phytopharmaceutical medicines (Bhattarai et al., 2023). Most of the pharmaceutical medications are prepared from the secondary metabolites that medicinal plants produce. Furthermore, MAPs possess a sound cultural basis in the cuisines of several countries. Several culinary herb species find application in many cuisines of different countries—basil, chervil, coriander, marjoram, sage, chives, fennel, mint, tarragon, oregano, dill,

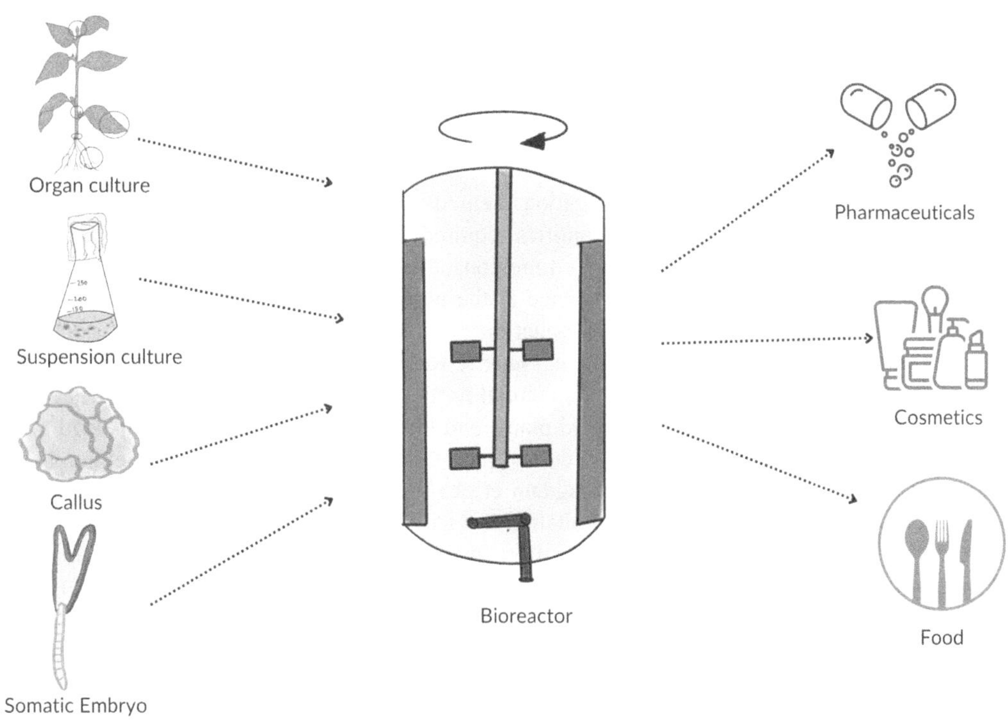

FIGURE 2.1 Plant tissue culture applications.

thyme, laurel, and rosemary. There is also a correlation evident between plant resources and goods, and culture along with its associated customs occurring within a society (Guleria et al., 2014).

MAPs have great worldwide demand as spices and foods. Controlled cultivation and export of such species can boost a country's economy. The dermatological properties of plants have been identified and used for a long time in different cultures for its use as relaxants, perfumes, anti-inflammatory and anti-scarring compounds. Typically, essential oils have been known for its use in cosmetological applications due to its pleasing and refreshing aroma. Extracts from plants have a major role in cosmetic industries because of their antiaging, anti-inflammatory, anti-oxidants and moisturizing properties. Figure 2.1 shows the industrial application of MAPs.

2.3 NEED AND SIGNIFICANCE OF CONSERVATION OF MAPS

However, high trade demand, low population density, and extremely fundamental life forms lead to overexploitation, which is the most critical threat facing MAPs. Some of the aspects that threaten the wild populations of medicinal plants thriving in mountainous regions include overgrazing, habitat degradation, early harvesting, and overharvesting. Environmental factors such as temperature and precipitation changes, degraded soils, and fragmented habitats equally pose a danger. The Himalayan highlands represent one of the richest and most endemic habitats of medicinal plants, of which the subalpine and alpine regions are especially significant. In addition, being an integral part of herbal medicine, medicinal plants also serve as a means of subsistence for the local residents. The herbs are gathered and conserved for the treatment of fever, stomachaches, burns, coughs, colds, diarrhea, and dysentery. Species like *Swertia chirayita*, *Paris polyphylla*, and *Neopicrorhiza scrophulariiflora* are most favored, while *Rheum austral*, *Dactylorhiza hatagirea*, and *Aconitum gammiei* are other preferred species (Pant et al., 2010; Khakurel et al., 2024). No threat level has ever been established for these species. Tools like the Rapid Vulnerability Assessment and the Conservation Assessment and Management Plan try to assess the magnitude of risks and accordingly suggest conservation plans. However, the threats that have been found indicate the need for their conservation and sustainable production for biodiversity conservation, subsequent plant breeding and the development of new formulations of medicinal compounds (Kulak et al., 2022; Salgotra and Chauhan, 2023).

2.4 CHALLENGES IN MAPS CONSERVATION

Major limitations to conventional propagation methods include low multiplication rates, which impede the rapid production of large quantities required, especially those plants in high demand or requiring urgent conservation. These are time-consuming, labor-intensive techniques involving cutting preparation, grafting, and maintenance in the nursery, thus increasing the cost. Added to this, seasonal dependencies allow the propagation only at certain times of the year, as specific environmental conditions are essential for it. Genetic variability, introduced by seed propagation, complicates the uniformity of plants in commercial horticulture and breeding programs. Pathogen transmission through cuttings from infected plants and low germination rates in plants lead to the inadequacy of seed propagation. The high cost of production makes these methods ineffective and less economical for mass production (Nongdam et al., 2023). A lot of medicinal plant species in the world face huge threats from overexploitation and habitat loss, resulting in dramatic ecological, cultural, and economic impacts. Indiscriminate wild harvesting thus poses threats to a number of species, with serious declines reported more precisely in Europe, South Africa, and India. Habitat loss, overharvesting, invasive species, and pollution threaten an estimated 15,000 plant species worldwide. Combined pressures thus result in the loss of essential services in ecosystems and the depletion of valuable resources (Tripathi, 2020). Somaclonal variation, epigenetic changes, chromosomal abnormalities, genetic drift, and somatic mutations can introduce genetic variability into *in vitro* cultures. Such phenomena could be responsible for some of the undesirable traits that manifest in reduced plant fitness, adaptability, or even a loss in capabilities with regard to the production of bioactive compounds (Motevali et al., 2013). Quality control in medicinal plant propagation involves exact measurements with respect to materials and reagents, temperatures, and pH conditions. The process allows for no contamination or sensible presence of contaminants on the plant materials. Sample preparation and good storage are to be followed to guarantee the material integrity and accuracy of tests (World Health Organization, 2011).

2.5 IN VITRO CLONAL PROPAGATION AS A TOOL FOR MAPS CONSERVATION

Plant tissue culture technology has been unavoidable in plant improvement for more than a century. It is crucial for the *in vitro* conservation of elite, rare, and endangered plant species' germplasm, breeding programs for a variety of crops, encapsulated seed production, and the study of plant biosynthesis via cell and root cultures (Singh et al., 2020). It also allows for the study of the relationships endophytes have with their host plants. Plant tissue culture applies the latest methods through which micropropagation rapidly reproduces elite plants. Although an extensively used process in horticulture, forestry, and agriculture, it stands first in the cultivation of medicinal plants and their transformation into economically exploitable medicinal crops (Lata et al., 2010), contributing to their conservation.

In Vitro clonal propagation, sometimes referred to as micropropagation, is used to create many genetically identical copies of a plant from a single parent. Tissue culture micropropagation was initially developed by (Morel, 1960) for the propagation of orchids, and it is currently utilized for the micropropagation of a variety of plants. Micropropagation in itself is one of the excellent modes of rapid clonal multiplication of plant species. This process of in-vitro clonal multiplication involves placing a small portion of plant tissue, such as a stem or leaf, in a nutrient-rich medium under controlled conditions. Plant tissue can be nicely cloned by allowing it to grow into an entirely new plant if placed under the right conditions. This method is frequently applied in plant tissue culture to conserve endangered species, produce large quantities of genetically identical plants, and propagate plants free from disease. The five different methods of micropropagation are used to produce plantlets artificially. In meristem culture, a leaf primordium and meristem are placed in growth media to produce elongated, rooted plantlets, which are then transferred to soil, ensuring disease-free plants and rapid multiplication of herbaceous species. Callus culture is a process where the extinction of plant tissue in the medium undergoes callus formation, followed by a transfer into a medium containing growth regulator, which would initiate the development of adventitious organs to adapt to the environment and finally become new plantlets. In suspension culture, cells will be dispersed in a sterile liquid medium for growth. Embryo culture extracts embryos for placement in nutrient-rich, sterile media. Protoplast culture isolates plant cells and cultivates them to regenerate cell walls and calluses, fostering cell division and differentiation for plant growth.

The basic approach to *in vitro* clonal propagation follows the steps mentioned hereafter. Preparation of a suitable explant and selection is a basic starting point in plant tissue culture. Primary explants may consist of sections of leaves, stem nodes, shoot ends, or even isolated cells. Sterilization methods are performed by rinsing under sterile distilled water, while a surface sterilant, like hydrogen peroxide or ethanol, will have to be used to remove contaminating agents from the surface of the explant, such as fungi or bacteria. The explant is then placed on a nutrient medium that has been prepared with 2,4-D, vitamins, minerals, and plant growth regulators like auxins and cytokinins added to it in order to promote growth. These cultures are maintained in controlled environments, typically in growth chambers with regulated humidity, light, and temperature. In the multiplication phase, new shoots grow from the explant due to cytokinin added to the medium. Then, the number is multiplied by sub-cultivation. Later, the shoots are transferred to media containing auxins, which induce rooting, and they grow to form plantlets. After dusting off any excess of medium, the plantlets are hardened against environmental conditions through an initial high relative humidity, which gradually progresses to ordinary conditions. Finally, after sufficient time, the properly rooted plantlets that have developed leaves are transferred to a soil medium or other growing substrate and grown further in a greenhouse or field (Bhojwani and Razdan, 1996). Figure 2.2 shows the visual representation of the process.

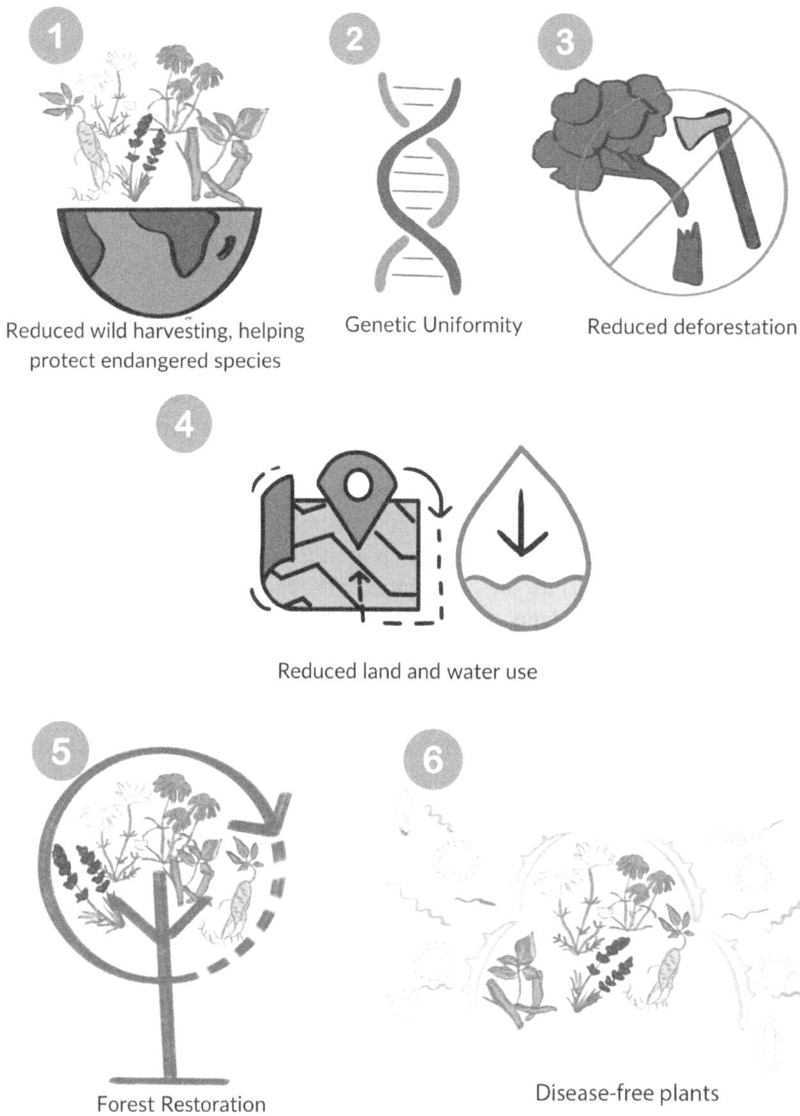

FIGURE 2.2 Applications of tissue culture-based propagation of medicinal and aromatic plants.

In vitro clonal propagation of a given MAPs is initiated by selecting a mother plant, that is true to its type. This should be free from diseases or infections; and may also be made pathogen-free through the use of certain *in vitro* techniques, which may be confirmed by routine tests as per global standards. The mother plant must also be thriving and vigorous to respond to culture conditions inducing intense cell division and regeneration. With the quality of the mother plant source, the rate of success in clonal propagation increases accordingly (Marak and Devi, 2021). Changing growth conditions for maximization of yield is called its optimization. For instance, in *Eucalyptus camaldulensis* tissue culture, maximum root and shoot regeneration was when cultured in MS medium supplemented with 1000mg/l casein hydrolysate, 1.0 mg/l BAP, and 0.1 mg/l NAA (McCOMB, 1996). The shoot crown of *Chlorophytum nepalense* gave the maximum yield of the plant when it was cultured in half-strength Gamborg's B5 and MS medium with 2% sucrose and incubated at 22°C (Basu and Jha, 2011).

In most of the case studies on in vitro plant cultivation, MS medium has been used as the growth medium, enriched with various growth regulators such as BAP, IBA, IAA, and NAA (Table 2.1).

TABLE 2.1
***In vitro* culture strategies for Medicinal and Aromatic Plants (MAPs) propagation and conservation**

Plant species	Explant used	Media/additional nutrients	Growth conditions	Growth	References
Nardostachys jatamansi	Nodal segment	Half-strength MS medium having 3% sucrose and 0.4% agar. The pH of the medium was made 5.8 by adding 0.1N HCl and 0.1N NaOH.	Temperature: 25 ± 2°C with 16 hours of light	Shoot induction: 2mg/l BAP, 1.5 mg/l kinetin Root induction: 1.5 mg/l NAA, 1mg/l BAP	Alam et al., 2016
Eucalyptus camaldulensis	Shoot tip and young leaves	MS medium supplemented with 1000mg/l casein hydrolysate, 1.0 mg/l BAP and 0.1 mg/l NAA.	Temperature: 25 ± 2°C with 16 hours of light	Shoot induction: MS medium with different concentrations of BAP and IBA. Root induction: MS medium with NAA	McCOMB, 1996
Centella asiatica	Nodal segments and axillary buds	MS media with 3% sucrose and 0.8% agar with different concentration of phytohormones like BAP, IAA, NAA and kinetin	Temperature: 25 ± 2°C with 16 hours of light	Shoot induction: Full strength Gamborg's B medium, 2% sucrose, IBA, NAA, pH – 5.7 Root induction: Half strength BS and MS basal medium, 2% sucrose, IBA (1.97–9.8 uM)	Gauchan, 2013
Chlorophytum nepalense	Shoot crown	Half Strength Gamborg's B5 and MS medium . 2% sucrose, with cytokinins (BAP, Kinetin), IBA, NAA	Temperature: 22°C Light/Dark Cycle: 16/8 hours	Shoot induction: MS media containing 22.2 µM BA along with 1.97 µM IBA Root induction: Half-strength MS media containing IBA = 1.97–9.80 µM	Basu and Jha, 2011
Aloe barbadensis	Stem disc	MS containing 7.5g/l agar and 20g/l sucrose, as well as antioxidants such as citric and ascorbic acid and pvp activated charcoal	Temperature: 25±2 °C with 16 hours light/8 hours dark photo period and 60% relative humidity and 2000-lux luminance	Shoot induction: MS media with varying concentration of BAP in range of 1.0–5.0 mg/L Root induction: MS media containing 0.2 mg/L NAA	Uikey et al., 2013
Woodfordia fruticosa	Auxillary bud	MS medium enriched with either kinetin (Kn, 4.60, 9.30, 13.90, 18.50, 23.20, and 27.20 µM) or 6-benzyladenine (BA; 4.40, 8.90, 13.30, 17.70, 22.20, and 26.50 µM) alone, or in combination with Kn (2.32, 4.60, 6.92, and 9.30 µM) and BA (2.22, 4.40, 6.62, and 8.90 µM).	Temperature: 25 ± 2°C with a 16 hours photoperiod	Shoot induction: MS medium containing 4.60 µM Kn and 4.40 µM BA Root induction: Half strength MS media containing 4.90 µM IBA	Kokkirala et al., 2012

(*continued*)

TABLE 2.1 (Continued)
***In vitro* culture strategies for Medicinal and Aromatic Plants (MAPs) propagation and conservation**

Plant species	Explant used	Media/additional nutrients	Growth conditions	Growth	References
Glycyrrhiza glabra	Nodes	MS with 3% sucrose and NB with 2% sucrose, meso-inositol 100 mg and agar 0.8% supplemented with auxins (IAA and NAA, 0.05–5.00 mg) and cytokinins (Kn and BAP, 0.05–5.00 mg)	Temperature: 25 ± 3°C with 16 hours photo period and 60–70 % relative humidity.	Shoot induction: MS media with 0.5 mg/L BAP Root induction: MS media with 1 mg/L IAA.	Kukreja, 1998
Piper longum	Leaf	Kinetin (KN) (0.25, 0.5, 1.0, 1.5, 2.0, & 3.0 mg/L) and 2,4-Dichlorophenoxyacetic acid (0.25, 0.5, 1.0, 1.5, 2.0, 3.0, 4.0, and 5.0 mg/L) alone as well as in combination with 3% (w/v) sucrose, 10% coconut water (v/v), and 0.8% (w/v) agar were added to the MS Basal media for primary culture. The MS media was supplemented with KN, BAP, and 2,4-D (0.5, 1.0, 1.5, and 2.0 mg/L) alone and in combination with BAP and NAA (0.5, 1.0, 1.5, and 2.0 mg/L) for the callus sub-culture.	Temperature: 25 ± 2°C with 12-16 h photoperiods.	The biggest calli was produced by MS media devoid of hormones. Callus was not produced by MS media containing kinetin alone. Using 1.0 mg/L 2,4-D and 2.0 mg/L KN produced the best callus induction (75%) results. When 2.0 mg/L BAP was used alone, callus proliferation peaked and fresh weight significantly increased. When it came to the proliferation of calluses, cytokinins outperformed auxins, and 10% coconut water improved induction and growth.	Thapa et al, 2022
Panax ginseng	Embryo	MS medium containing different concentrations of PGRs.	Temperature: 25+2 with 16 hours photo period	Shoot induction: MS medium with 28.9 µM GA3. Root induction: Half strength MS medium with 14.7 µM IBA.	Xu et al., 2023
Gaultheria fragrantissima	Seed	In addition to indole-3-acetic acid (IAA) and 1-naphthaleneacetic acid (NAA) (1, 2 mg/l), MRM was supplemented with TDZ (0.5, 1, 2, 3, and 4 mg/l), BA (2, 5, 10 mg/l), Kin (2, 5, 10 mg/l), and 2iP (5, 10, 15 mg/l), both individually and in combination	Temperature: 24 ± 2°C with 16 hours of light.	Shoot induction: TDZ (1–2 mg/l) containing medium. Root induction: half strength medium containing 1 mg/l IBA	Ranyaphia et al., 2011

Swertia chirata	Shoot	MS medium containing 3% sucrose and varying concentrations of BAP (4.44 to 22.20 μM) in combination with 2,4-D (0.90 to 4.52 μM)	Temperature: 25 ± 2°C with 16 hours of light	Shoot induction: MS media containing 271.45 μM adenine sulphate (Ads), 2.85 μM IAA and 8.88 μM BAP Root induction: Half-strength MS medium containing 4.90 μM IBA	Pant et al., 2015
Valeriana wallichii	Nodes	MS media with different amounts of either 6-benzyladenine (BA, 1–10 μM), kinetin (Kn, 1–10 μM), adenine sulphate (AS, 1–10 μM), or thidiazuron (TDZ, 0.1–5 μM) added to 3.0% w/v of sucrose and 0.8% w/v of agar	Temperature: 25 ± 2°C with 16 hours of light	Shoot induction: MS medium containing 3 μM BA. Root induction: MS medium without any PGRs	Singh et al.,2015
Alternanthera sessilis	Leaf	Half-strength MS medium enhanced with 1.0 mg/l BAP, 0.1 mg/l NAA, and 1000 mg/l casein hydrolysate.	Temperature: 25°C with 16 hours photo period	Shoot induction: half strength MS basal medium with 1 mg/l each of IAA and BAP Root induction: half strength MS medium with only 1 mg/l IBA	Singh et al., 2009
Zanthoxylum armatum	Leaf	WPM enhanced with varying TDZ concentrations (2.0, 4.0, 6.0, 8.0, and 10.0 μM), either separately or in conjunction with varying naphthaleneacetic acid (NAA) concentrations (0.5, 1, and 1.5 μM)	Temperature: 25 ± 2°C with 16 hours of light	Shoot induction: WPM containing varying amounts (2.0–9.0 μM) of gibberellic acid (GA3, 0.5–3.0 μM), indole-3-acetic acid (IAA, 1.0 μM), and benzylaminopurine (BA). Root induction: half-strength WPM containing 50.0 μM IBA	Purohit et al., 2019
Acorus Calamus	Rhizome	MS medium enriched with 8 g/l agar, 3% sucrose and different combinations and varying concentrations of auxins	Temperature: 25 ± 2°C with 16 hours of light	Shoot induction: MS medium enhanced with 2.4 μM IBA and 5.7 μM IAA Root induction: one quarter of MS medium containing 1 mg/L IBA	Altaf et al., 2010
Rauvolfia serpentina	Shoot tips and nodes	MS basal media enriched with 6.5% agar, 3% sucrose and varying concentrations of PGRs	Temperature: 25 ± 2°C with 16 hours of light and 50% humidity	Shoot induction: MS medium enriched with 0.54 μM NAA and 4.44 μM BA Root induction: half-strength MS medium containing 4.9 μM IBA and 5.71 μM IAA	Rajasekharan, 2010
Cephaelis ipecacuanha	Shoot tip	MS, 1/2 MS, and B5 medium with different phytohormone combinations	Temperature: 25 ± 2°C with 16 hours of light	Shoot induction: B5 medium containing 3 or 5 mg/l BA and 0.0l mg/l NAA Root induction: B5 medium containing IAA (3 mg/l)	Ideda et al.,1988

(*continued*)

TABLE 2.1 (Continued)
***In vitro* culture strategies for Medicinal and Aromatic Plants (MAPs) propagation and conservation**

Plant species	Explant used	Media/additional nutrients	Growth conditions	Growth	References
Cinnamomum verum	Embryo and axillary buds	MS basal medium at half strength with 1.5 mg/L autoclaved BAP + 0.2 mg/L IAA and 1 g/L activated charcoal added as supplements	Temperature: 25 ± 2°C with 16 hours of light	Shoot induction: Half strength MS basal medium enriched with 1.5 mg/L BAP and 0.2mg/L IAA Root induction: Half strength MS basal medium containing 4mg/L BAP with 0.1 mg/L NAA	Subasinghe et al., 2016
Chamomile recutita	Young inflorescence	MS media supplemented with 0.7% agar, 3% sucrose, 11.5 µM Kin, and 26.8 µM NAA	Temperature: 25 ± 2°C with 16 hours of light	Callus induction after 2-4 days	Ahmad & Mishra, 2015
Citronella mucronata	Nodes	4.44 µM BAP and 14.76 IBA added to the MS basal culture medium	Temperature: 25 ± 2°C with 16 hours of light	Shoot induction: MS basal media with BAP (4.44 µM) Root induction: MS basal media with 14.76 µM IBA	Guerra et al., 2022
Curcuma zedoaria	Axillary buds	Adenine sulfate (33.9–203.6 µM, Ads), indole-3-acetic acid (2.9–11.4 µM, IAA), α-naphthalene acetic acid (2.7–10.2 µM, NAA), and 6-benzyladenine (2.2–22.2 µM, BA) added to the MS basal medium in diferent combinations and quantities	Temperature: 25 ± 2°C with 16 hours of light	Shoot induction: 13.3 µM BA, 5.7 µM IAA, and 63.9 µM Ads in MS media Root induction: 7.4 µM of IBA in half strength MS medium	Jena et al.,2020
Stevia rebaudiana	Nodes	3.0% sucrose, different amounts of BAP (benzylaminopurine), Kn (kinetin), NAA, and IBA added to the MS medium	Temperature: 25 ± 2°C with 16 hours of light	Shoot induction: MS medium with 2.0 mg/L BAP Root induction: MS medium containing 0.5 mg/L NAA	Jahan et al., 2014
Dendrobium spp	Shoot tips	MS medium enhanced with 0.1 mg/L NAA and 0.5 mg/L BA	Temperature: 25 ± 2°C with 16 hours of light	Shoot induction: MS basal media containing 0.1 mg/L NAA with 0.5 mg/L BA Root induction: Same as above optimization but after 9 days	Subrahmanyeswari et al., 2022

Woodfordia floribunda	Axillary buds	MS medium enriched with either kinetin (Kn, 4.60, 9.30, 13.90, 18.50, 23.20, and 27.20 µM) or 6-benzyladenine (BA; 4.40, 8.90, 13.30, 17.70, 22.20, and 26.50 µM) alone, or in combination with Kn (2.32, 4.60, 6.92, and 9.30 µM) and BA (2.22, 4.40, 6.62, and 8.90 µM)	Temperature: 25 ± 2°C with 16 hours of light	Shoot induction: MS media enriched with 4.60 µM Kn and 4.40 µM BA Root induction: Half strength MS media containing IBA (4.90 µM)	Kokkirala et al., 2012
Satureja abyssinica	Nodes	MS medium including varying concentrations of PGRs	Temperature: 25 ± 2°C with 16 hours photo period	Shoot induction: MS medium containing 1 mg/l BAP Root induction: MS medium containing 0.5 mg/l IBA	Teshome et al., 2017
Azadirachta indica	Nodes	MS medium with PVP-40 (12.5µM) and 0.44µM BA. The carbon source was 3% sucrose. DKW medium, IBA, or NAA (with 0.22µM BA).	Temperature: 27 ± 2 °C with 12 hours photo period	Shoot induction: DKW medium (with 0.22µM BA) Root induction: 4.9µM IBA in half-strength DKW	Quraishi et al., 2004
Jatropha curcas	Nodes	In MS media containing 3% w/v sucrose and 0.8% w/v agar, BA (1.0–4.0 mg/L), NAA (0.25–2.0 mg/L), and IBA (0.25–2.0 mg/L) are added, either singly or in combination, along with L-arginine (15 mg/L) + Citric acid (50 mg/l) + Glutamine (50 mg/L) + Adenine sulfate (25 mg/L).	Temperature: 24 ± 2°C with 16 hours photo period and 50–55% relative humidity	Shoot induction: MS media containing 1.0 mg/L IBA and 3.0 mg/L BA Root induction: Half-strength MS medium with 3.0 mg/L IBA	Shrivastava and Banerjee, 2008

These growth regulators play a crucial role in having an effective rate of shoot and root induction in different species. The controlled environment in which the plants are grown contribute to effective propagation, mostly with temperatures usually kept between 25–27°C and 16 hours of light.

In almost all of the above-reviewed case studies, micropropagation was done in different medium constituents and conditions, and the optimum conditions for their substantial growth were identified. This indicates that optimization of *in vitro* clonal propagation not only enhances production efficiency but also aids in the conservation of valuable genetic resources. Such methods reduce pressure on wild stocks, provide genetically uniform and disease-free plants, and offer a continual supply for commercial purposes as well. Therefore, all these benefits with regards to the conservation and production of MAPs highlight the significance of *in vitro* propagation techniques for MAPs.

2.6 THE POTENTIAL IMPACT OF MICROPROPAGATION OF MAPS ON BIODIVERSITY CONSERVATION

In vitro clonal propagation is a crucial method for biodiversity conservation. There are two ways for the conservation of the biodiversity of plants: in situ and ex situ. In situ conservation is the preservation of plant species in their natural ecosystems or domestication and cultivation on farms and in their surroundings where they have developed unique features (UNCED, 1992). Nevertheless, numerous species, populations, and ecosystems are declining due to habitat degradation and environmental change, thereby making in situ conservation alone unable to cope with the task of conserving endangered species. Against this background, ex situ techniques for the conservation of biological material outside its natural habitat need to be resorted to (UNCED, 1992). *In vitro* clonal propagation is an ex-situ approach. Nevertheless, it should be understood that in situ and ex situ conservation measures are complementary to each other and work best in conjunction (Cruz-Cruz et al., 2013).

In vitro clonal propagation thus has a role in biodiversity conservation by remaining uniform in genetics at the level of individual species and conserving germplasm variation. Plant cultures may be stored *in vitro* for a period ranging from several months to a year or longer without subculture, depending on the plants used and the level of optimization. Modification of the culture medium can include dilution of micro- and macro elements, reduction of sugar concentration, changing concentration of growth regulators, and introduction of osmotically active chemicals (Engelmann, 2011). Typically, the culture environment is attained through the reduction in temperature, reduction of light amount, or by completely keeping the cultures in the dark (Holobiuc et al., 2007). This way, the genetic makeup of the parent plant is maintained, and the genetic diversity is preserved. Further preservation of these cultures or even explants like seeds and pollen can be done by cryopreservation (Lakshman, 2016). Besides, *in vitro* propagation occupies much less space and other resources compared to the traditional conservation methods. Thus, it is highly efficient in allowing the conservation of many species and varieties within a relatively small area. Not only that, pathogen-free and disease-resistant plants are produced by *in vitro* methods. Further, plants obtained through *in vitro* methods can be restored and rehabilitated to their ecological habitats for uplifting degraded ecosystems and enhancing biodiversity (Engelmann, 2011) (Figure 2.3).

Recent advances in *in vitro* clonal propagation have brought about a revolution in plant production, resulting in genetic uniformity and disease-free plants. Somatic embryogenesis and bioreactor systems, such as TIBs, are efficient and controlled environments for large-scale cultivation. Besides, synthetic seed and automation technologies enhance production to meet agricultural and horticultural demand. Such inventions make it possible to reproduce plants of high quality and uniformity within a very short period of time, hence making micropropagation much better than the traditional methods. Medicinal and aromatic plants are, therefore, conserved through *in vitro* clonal propagation techniques and supported by mass multiplication; continuous research in perfecting these techniques is a requisite for their sustainable cultivation.

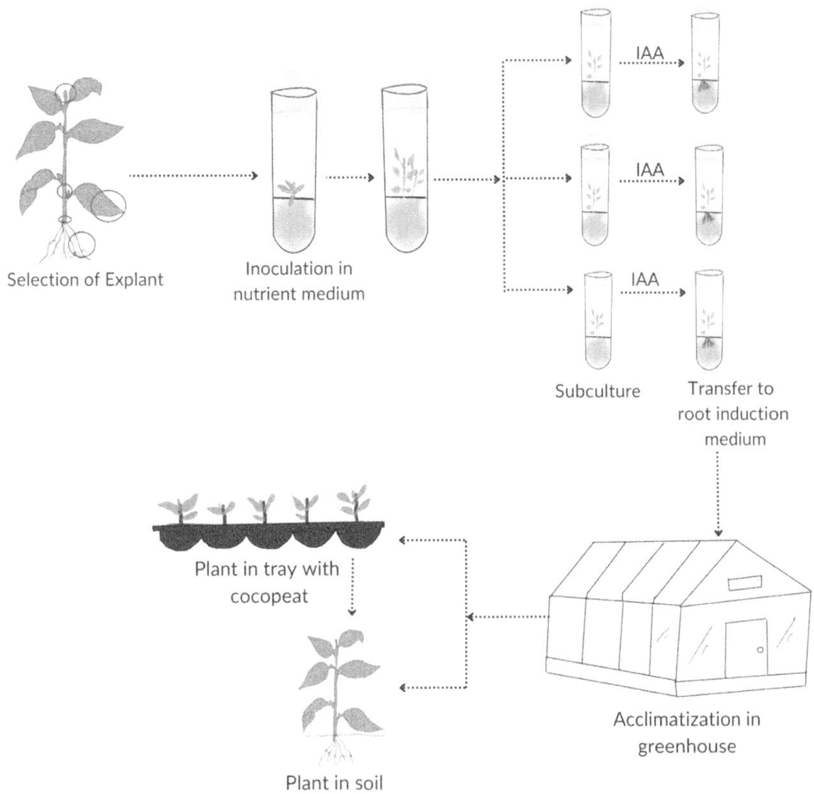

FIGURE 2.3 Plant tissue culture: Lab to land transfer of plants.

2.7 FUTURE PROSPECTS

Today, biotechnology has sophisticated medicinal and aromatic plant production through *in vitro* clonal propagation, whereby genetically identical clones are rapidly propagated in sterile environments. This method exceeds convention, producing thousands from one explant to meet the high market demand. In the case of endangered MAPs, genetic material is conserved in germplasm banks, helping in the conservation effort without depleting wild populations (Bhojwani and Dennis, 1999). Controlled cultivation ensures homogeneous, high-quality plants with controlled variables enhancing bioactive compounds, disease resistance, and growth rate. Somatic embryogenesis, synthetic seed, and genetic transformation support mass propagation and conservation. Automation and bioreactor systems enhance efficiency, scalability, and reduce production costs while minimizing contamination risks. Though challenges in the form of somaclonal variation and high initial setup costs exist, continuous research in refining protocols and the integration of molecular markers for quality control is underway. On the other hand, future developments of CRISPR/Cas9 and improvements in cryopreservation methods are bringing us closer to sustainable MAPs conservation and production (Shibli et al.,2006).

2.8 CONCLUSION

In *vitro* clonal propagation can play if integrated into issues, resulting from biodiversity loss, sustainable production and economic development, with elite medicinal and aromatic plants. Considering their therapeutic and economic value, all of these medicinal and aromatic plants become very important to be conserved in view of a sustainable supply to fulfill the growing global demand. The methods of *in vitro* clonal propagation not only provide an alternative regarding the problems that traditional propagation techniques pose but also make it possible to obtain genetic uniformity and disease-free plants with reduced dependency on wild harvesting and protection against vulnerable plant populations. The possibility of producing large numbers of elite-quality plants through micropropagation assures a constant supply for medicinal, cosmetic, and food industries to increase benefits to both local communities and global markets. However, wide-scale application of *in vitro* clonal propagation is challenging. Among these are the high production costs, requirements of specialized knowledge and infrastructure, and risk of somaclonal variation that all need to be considered and continued to be researched. Beating these challenges will be important to maximizing the gains from this technology and its long-term sustainability. *In vitro* clonal propagation, thus, remains one of the most prominent tools in elite MAPs conservation and production. The ability of this technique to provide high-quality plants while maintaining genetic uniformity aligns well with the goals of biodiversity conservation as well as sustainable development. Continued research and investment in this technology will further enhance its effectiveness in supporting the global effort on preservation and utilization of the valuable resources offered by medicinal and aromatic plants.

REFERENCES

Ahmad, R., and N. Misra. 2015. In vitro propagation of *Chamomilla recutta* from capitulum inflorescence: a medicinal plant with multiple therapeutic applications. *Era's Journal of Medical Research* 2(1):1–6.

Alam, F., Chaudhari, V.K., Biswal, P.K., and R. Kumar. 2016. A phytochemical and pharmacological review on *Nardostachys jatamansi* DC. *International Journal of Current Trends in Pharmaceutical Research* 4(2):115–8.

Altaf, A., Shashidhara, S., Rajasekharan, P.E., Hareesh, K.V., and N. Honnesh. 2010. In vitro regeneration of Acorus calamus – an important medicinal plant. *Journal of Current Pharmaceutical Research* 2(1):36–9.

Barata, A. M., Rocha, F., Lopes, V., and A. M. Carvalho. 2016. Conservation and sustainable uses of medicinal and aromatic plants: A review. *Agriculture and Agricultural Science Procedia* 8:139–46.

Basu, S., and T. B. Jha. 2011. In-vitro propagation of Chlorophytum nepalense. *Journal of Tropical Medicine* 12(1):33–9.

Bauer, P.B. 2012. Historical review of medicinal plants' usage. *Pharmacognosy Reviews* 6(11):1–5.

Bhattarai, S., Kunwar, R.M., Ojha, H. et al. 2023. Wild edible plants used by the people of Manang district, Central Nepal. *Journal of Ethnobiology and Ethnomedicine* 19:50.

Bhojwani, S.S., and M.K. Razdan. 1996. *Plant Tissue Culture Theory and Practice, A Revised Edition*. Elsevier.

Bhojwani, S.S., and T.T. Dennis. 1999. In vitro conservation of plant genetic resources. *Botanica* 49:47–52.

Chopra, I., and A. Kamal-Eldin. 2009. Plants as a source of medicines and nutraceuticals. *Evidence-Based Complementary and Alternative Medicine* 6(3):295–309.

Cruz-Cruz, C.A., González-Arnao, M.T., and F. Engelmann. 2013. Biotechnology and conservation of plant biodiversity. *Resources* 2(2):73–95. www.mdpi.com/2079-9276/2/2/73

Ekor, M. 2014. The growing use of herbal medicines: issues relating to adverse reactions and challenges in monitoring safety. *Frontiers in Pharmacology* 4:177.

Engelmann, F. 2011. Use of biotechnologies for the conservation of plant biodiversity. *In Vitro Cellular & Developmental Biology – Plant* 47 :5–16.

Gauchan, D.P. 2013. In vitro propagation of *Centella asiatica* (Linn.) Urban – an important medicinal plant. *Journal of Nepal Biotechnology Association* 3:35–8.

Guerra, F., Badilla, L., Cautin, R., and M. Castro. 2022. Micropropagation of *Citronella mucronata* D. Don, a vulnerable Chilean endemic tree species. *Plants* (Basel) 11(18):2425.

Guleria, C., Vaidya, M.K., Sharma, R., and D. Dogra. 2014. Economics of production and marketing of important medicinal and aromatic plants in mid hills of Himachal Pradesh. *Economic Affairs* 59:363–78.

Holobiuc, I., Paunescu, A., and R. Blindu 2007. Ex situ conservation using in vitro methods in some Caryophyllaceae plant species from the Red List of vascular plants in Romania. *Romanian Journal of Biology – Plant Biology* 49:3–16.

Ideda, K., Teshima, D., Aoyama, T., Satake, M., and K. Shimomura. 1988. Clonal propagation of *Cephaelis ipecacuanha*. *Plant Cell Reports* 7:288–91.

Inoue, M., and L.E. Craker. 2014. Medicinal and aromatic plants—uses and functions. *Horticulture: Plants for People and Places* 2:645–69.

Inoue, M., Hayashi, S., and L.E. Craker. 2019. Role of medicinal and aromatic plants: Past, present, and future. In *Pharmacognosy – Medicinal Plants*, ed. S Perveen, 1–13. IntechOpen.

Jahan, M.T., Islam, M.R., Roy, P.K., Mamun, A.N.K., and M.A. Islam. 2014. In vitro clonal propagation of *Stevia rebaudiana* Bertoni through node and shoot tip culture. *Nuclear Science and Applications* 23(1)(2):61–5.

Jena, S., Ray, A., and A. Sahoo. 2020. Rapid plant regeneration in industrially important Curcuma zedoaria revealing genetic and biochemical fidelity of the regenerants. *3 Biotech* 10(1):17.

Khakurel, D., Uprety, Y., Karki, S., Khadka, B., Poudel, B., Ahn, G,. Joon-Yung, C., Woe-Yeon, K., Sung-Ho, L., and S. Rajbhandary. 2024. Assessing the risks to valuable medicinal plants in Nepal from human activities and environmental factors. *Global Ecology and Conservation* 51:1–15

Kokkirala, V.R., Kota, S., Yarra, R., Bulle, M., Aileni, M., Gadidasu, K.K., da Silva, J.A., and S. Abbagani. 2012. Micropropagation via nodal explants of *Woodfordia fruticosa* (L.) Kurz. *Medicinal and Aromatic Plant Science and Biotechnology* 6(1):50–3.

Kukreja, A.K. 1998. In vitro propagation of liquorice (*Glycyrrhiza glabra* L.) through multiple shoot formation. *Journal of Spices and Aromatic Crops* 7(1):13–7.

Kulak, V., Longboat, S., Brunet, N.D., Shukla, M., and P. Saxena. 2022. In vitro technology in plant conservation: Relevance to biocultural diversity. *Plants* 11(4):503.

Lakshman, C. 2016. Biodiversity and conservation of medicinal and aromatic plants. *Advances in Plants & Agriculture Research* 5(4): 1–7.

Lata, H., Moraes, R.M., Bertoni, B., and A.M.S. Pereira. 2010. In vitro germplasm conservation of *Podophyllum peltatum* L. under slow growth conditions. *In Vitro Cellular & Developmental Biology – Plant* 46:22–7.

Marak, M.R., and A.R. Devi. 2021. Clonal propagation in horticultural crops. *The Pharma Innovation Journal* 10(8):496–9.

McComb, J.A., Bennett, I.J., and Tonkin, C. 1996. In vitro propagation of Eucalyptus species. *Tissue Culture of Australian Plants*: 112–56.

Motevali, A., Nabavi-Pelesaraei, A., Akram, A., Ghadrejani, M., and H. Hosseinzadeh-Bandbafha. 2013. Comparison of energy use pattern and greenhouse gas emissions for open-field and greenhouse strawberry production. *Turkish Journal of Agriculture and Forestry* 37(5):603–12.

Morel, G. M. 1960. Producing virus-free cymbidium orchids. *Am. Orchid Soc. Bull.*29:495–7.

Nongdam, P., Beleski, D.G., Tikendra, L., Dey, A., Varte, V.E.L., Merzougui, S., Pereira, V.M., Barros, P.R., and W.A. Vendrame. 2023. Orchid micropropagation using conventional semi-solid and temporary immersion systems: A review. *Plants* 12(5):1136.

Pant, M., Bisht, P., and Gusain, M.P. 2010. *In vitro* propagation through axillary bud culture of *Swertia chirata* Buch.-Ham ex Wall: an endangered medicinal herb. *International Journal of Integrative Biology* 10(1):48–53.

Pant, M., Bisht, P., and Gusain, M.P. 2015. Effect of temperature on tissue raised shoots of Swertia chirata Buch.-Ham. ex Wall: An endangered medicinal plant. *Journal of Chemical and Pharmaceutical Research* 7(12):542–5.

Purohit, S., Joshi, K., Rawat, V., and I.D. Bhatt. 2019. Efficient plant regeneration through callus in Zanthoxylum armatum DC: an endangered medicinal plant of the Indian Himalayan region. *Plant Biosystems – An International Journal Dealing with all Aspects of Plant Biology* 154(1):1–7.

Quraishi, A., Koche, V., Sharma, P., and S.K. Mishra. 2004. In vitro clonal propagation of neem (Azadirachta indica). *Plant Cell, Tissue and Organ Culture* 78:281–4.

Rajasekharan, P.E. 2010. Micropropagation of Rauvolfia serpentina – a potent endangered medicinal plant. *Biomed Journal* 5(2):74–9.

Ranyaphia, R.A., Maoa, A.A., and S.K. Borthakurb. 2011. Direct organogenesis from leaf and internode explants of in vitro raised wintergreen plant (*Gaultheria fragrantissima*). *Science Asia* 37:186–94.

Salgotra, R.K., and B.S. Chauhan. 2023. Genetic diversity, conservation, and utilization of plant genetic resources. *Genes (Basel)* 14(1):174.

Shibli, R.A., Shatnawi, M.A., Subaih, W.S., and M.M. Ajlouni. 2006. In vitro conservation and cryopreservation of plant genetic resources: a review. *World Journal of Agricultural Sciences* 2:372–82.

Shrivastava, S., and M. Banerjee. 2008. In vitro clonal propagation of physic nut (Jatropha curcas L.): Influence of additives. *International Journal of Integrative Biology* 3(1):73–7.

Silalahi, M., Purba, E.C., Sawitri, I.G.A.R., Wakhidah, A.Z., and E. Yuniati. 2023. International trade of medicinal and aromatic plants (MAPs). Medicinal plants: Biodiversity, biotechnology and conservation. *Sustainable Development and Biodiversity* 33:289–306.

Singh, A., Kandasamy, T., and B. Odhav. 2009 In vitro propagation of Alternanthera sessilis (sessile joyweed), a famine food plant. *African Journal of Biotechnology* 8(21):5691–5.

Singh, B., and R.A. Sharma. 2015. Plant terpenes: defense responses, phylogenetic analysis, regulation and clinical applications. *3 Biotech* 5(2):129–51.

Singh, N., Negi, A.S., and Pant, M. 2020. Tissue culture interventions in soybean production: significance, challenges and future prospects. *Ecology, Environment & Conservation.* 26 (October supplement issue):96–102.

Singh, S., Purohit, V.K., Prasad, P., and A.R. Nautiyal. 2015. Micropropagation of Valeriana wallichi DC (Indian Valerian) through nodes. *Indian Journal of Biotechnology* 14:127–30.

Subasinghe, S., Hettiarachchi, C.S., and N. Iddagoda. 2016. In-vitro propagation of cinnamon (Cinnamomum verum Presl.) using embryos and in vitro axillary buds. *Journal of Advanced Agricultural Technologies* 3(3): 164–9.

Subrahmanyeswari, T., Verma, S.K., and S. Gantait. 2022. One-step in vitro protocol for clonal propagation of Dendrobium Yuki White, a high value ornamental orchid hybrid. *South African Journal of Botany* 146:883–8.

Teshome, S., Soromessa, T., and T. Feyissa. 2017. In vitro propagation of a threatened medicinal plant Satureja abyssinica through nodal explants-Antimicrobial and antifungal herb. *International Journal of BioSciences and Technology* 10(3):20.

Thapa, C.B., Pant, K.K., Bhattarai, H.D., and B. Pant. 2022. In vitro induction and proliferation of callus in Piper longum L. through leaf culture. *Nepal Journal of Science and Technology* 21(1):13–22

Tripathi, M. 2020. Critical review of some important medicinal and aromatic plants of western Himalaya. *Research Journal of Pharmaceutical Sciences* 9(1): 7–16.

Uikey, S., Tripathi, M.K., Tiwari, G., Pandey, A., and R.P. Patel. 2013. Microcloning studies in *Aloe barbadensis. Plant Cell Biotechnology and Molecular Biology* 14:1–11.

United Nations Conference on Environment and Development (UNCED). 1992. Convention on Biological Diversity. www.cbd.int/doc/legal/cbd-en.pdf

World Health Organization. 2011. Quality control methods for medicinal plant materials. www.who.int/docs/default-source/medicines/norms-and-standards/guidelines/quality-control/quality-control-methods-for-medicinal-plant-materials.pdf?sfvrsn=b451e7c6_0

Xu, F., Valappil, A.K., Mathiyalagan, R., Tran, T.N.A., Ramadhania, Z.M., Awais, M., and Yang, D.C. 2023. In vitro cultivation and ginsenosides accumulation in Panax ginseng: A review. *Plants, 12*(17):3165.

3 Tissue Culture and Genetic Engineering of *Cannabis sativa*

Rambod Abiri

3.1 INTRODUCTION

Cannabis is a valuable pharmaceutical plant with more than 400 chemical components and secondary metabolites, belonging to the Cannabaceae family (Pain, 2015). Based on archaeological reports, the history of using these medicinal plants goes back to thousands of years ago, when this plant was used by prehistoric civilizations as an efficient remedy to reduce inflammation and pain (Ferrara, 2016). One of the early reports of the therapeutic application of cannabis as a treatment for diverse health issues was found in ancient Chinese societies as early as 2700 BCE (Ren et al., 2019). Likewise, the significant healing powers of a holy plant, cannabis, in traditional Indian medicine was reported in a religious Indian text, called the Atharva Veda (Ferrara, 2016). Next highlight of using cannabis as a natural remedy was documented in the medieval period, when various European societies and Islamic cultures widely adopted the plant, reflecting the popularity and broad application of cannabis (Ghiabi et al., 2018). More closely, historical documents confirmed the medicinal application of cannabis by European physician during the Renaissance period (Pisanti et al., 2019). The therapeutic application and adaptability potential of cannabis was confirmed when the application of this plant was shifted from traditional practice to more systematic and modern applications (Procaccia et al., 2022; Shukla et al., 2024). Interestingly, the discovery of an important and complex cell-signaling system, called the endocannabinoid system (ECS), in vertebrate's and human's body represent a notable breakthrough in the late 20th century. This finding shed light on the understanding of the mechanism of cannabis compound's interaction, such as cannabidiol (CBD) and tetrahydrocannabinol (THC), with ECS. Therefore, this information deciphered crucial insights into the physiological processes of the human and vertebrate bodies influenced by cannabis (Pacher et al., 2006). This understanding has spurred significant advancements in medicinal application of cannabis by emphasising on the possible therapeutic potential of cannabinoids in treating diverse diseases (Pacher et al., 2006).

Based on the legal status, benefits, characteristics, and unique application of this plant, cannabis, also known as marijuana, is categorised into two comprehensive class: industrial hemp and therapeutic cannabis. Notwithstanding the current protocols to categorise the cannabis plants in different regions, most importantly, the psychoactive THC level is the most critical factor which helps to determine the legal division between medicinal cannabis and industrial hemp (below 0.2% to 1% depending on jurisdiction). Nonetheless, the distinction between industrial hemp and drug-type cultivars can be straightforward using genetic standpoint (Rotherham et al., 2011; Pellati et al., 2018). Generally, hemp cultivars are mainly cultivated from seeds, while drug-type cultivars are typically propagated through cloning (Adhikary et al., 2021). Although stem cutting is one of the most common traditional methods for propagating drug-type cannabis, this method has some

DOI: 10.1201/9781003482659-3

disadvantages, including vulnerability to diseases and a potential for genetic uniformity (Caplan et al., 2018; Adhikary et al., 2021). In the cannabis industry, while conventional stem cloning method remains prevalent, there is a rapid movement towards *in vitro* propagation, which shows significant potential to surpass stem cutting. Despite both techniques demonstrating similar vegetative development and performance in cannabis industry, *in vitro* propagation methods display distinct benefits like faster multiplication rate, reduced costs, and disease-free clones. These advantages position *in vitro* propagation as the preferred method for propagation and genetic preservation in the future of the cannabis sector (Adhikary et al., 2021).

3.2 MICROPROPAGATION OF CANNABIS *IN VITRO* CULTURE

As mentioned earlier, micropropagation presents a beneficial alternative to conventional methods for propagating cannabis, enabling mass production of plants in a controlled condition (Monthony et al., 2021). Like other plants *in vitro* condition, cannabis is grown in various types of culture vessels including test tubes, magenta boxes, and dogboxes, usually located in stackable containers or multi-tiered rooms equipped with LED lighting (Shukla et al., 2017). Micropropagation enables many plants to be placed in a small area, thereby minimizing the floor space needed for maintaining mother plants, thereby optimizing the floor space necessary for mother plant maintenance. Plant tissue culture (PTC) procedures could present a wide range of approaches to maintain day-neutral genotypes and preserve the genetics for long-term. Importantly, due to the aseptic conditions during PTC, this technique can be applied to produce pathogen free plants, effectively reducing biotic pressures (Monthony et al., 2021). The initial study on the cannabis tissue culture was conducted on hemp to evaluate the stability and effectiveness of the procedure on the plant's chemical and agronomic characteristics (Richez-Dumanois et al., 1986; Michał Pławuszewski et al., 2006; Wielgus et al., 2008). Micropropagation is often classified into five stages including: Stage 0: Selection/maintenance of parent plant material; Stage 1: Initiation of cultures; Stage 2: Multiplication of shoots/embryos; Stage 3: Shoot elongation and rooting; Stage 4: Acclimatization. To produce a fully developed plant *in vitro* condition, each step requires to be optimized (Figure 3.1) (Monthony et al., 2021).

The selection of a suitable and healthy donor plant is the initial critical phase that directly and significantly impacts the efficacy of the cannabis plant *in vitro* condition (Boonsnongcheep et al., 2020). Although various explants including branch tips, axillary buds, and cotyledons have been utilized for micropropagation of cannabis, meristematic tissues are mainly favored for direct regeneration of this plant (Table 3.1). Studies emerge that callus induction, frequency, and regeneration is highly influenced by various factors including the type and age of cutting explants, type of media,

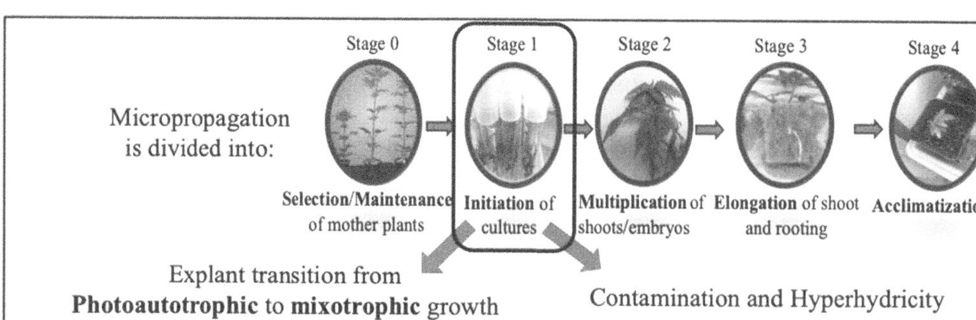

FIGURE 3.1 The schematic figure of various phases of cannabis tissue culture with some critical issues of initiation phase.

kind of species, and plant growth regulators (Table 3.1) (Wang et al., 2009; Lata et al., 2009a; Cheng ChaoHua et al., 2016; Boonsnongcheep et al., 2020). For instance (Slusarkiewicz-Jarzina et al., 2005) discovered that young leaves and petioles had the highest rates of callus production (Slusarkiewicz-Jarzina et al., 2005; Wielgus et al., 2008).

During phase 1, initiation of cultures, surface sterilization is a crucial process that greatly affect the efficacy of cannabis tissue culture. Sodium hypochlorite is a commonly used disinfectant applied during seed, nodal segments, immature leaves, and other explant cultivations *in vitro* (Boonsnongcheep et al., 2020). Various concentrations of ethanol (70–75%) are beneficial for sterilizing seeds after they have been carefully cleaned under running tap water (Slusarkiewicz-Jarzina et al., 2005; Wielgus et al., 2008; Cheng ChaoHua et al., 2016). Effective uses have also been made of other agents, such as 0.1% mercuric chloride and 5% calcium hypochlorite (Wang et al., 2009). Washing the explants after surface sterilization is the ultimate phase before culturing the explants into the vessels. The most common element resources in artificial media for *in vitro* culture include macro- and micronutrients, carbon sources, and vitamins (Boonsnongcheep et al., 2020). Depending on the type and age of the cannabis plants, as well as the aim of the tissue culture process, these elements are usually solidified with various gelling agents and supplemented with other additives such as plant growth regulators, proline, gibberellic acid, silicon, silver nanoparticles, and other substances. Although Murashige and Skoog (MS) (Murashige et al., 1962) supplemented with a carbon source of 3% (w/v), is the most commonly used artificial medium for growing *C. sativa in vitro*, other modified variants, such as half-strength MS medium containing 1% (w/v) sucrose, is used for various purposes, including seed germination (Wang et al., 2009). To induce shoot formation and root development, MS media supplemented with various type of plant growth regulators. These improvements and modifications to the MS medium help in adapting the growth condition to the unique requirements (Wang et al., 2009; Boonsnongcheep et al., 2020). Notwithstanding the frequency of applying MS media for cannabis tissue culture, other artificial media are also highly used to cultivate the plant *in vitro*. For instance, Daria medium, a variation of Knopp's solution, has proven to be an efficient medium for cannabis seed germination. To promote the regeneration of germinated seeds *in vitro* culture, various types of plant growth regulators have been added to the Daria medium (Wielgus et al., 2008). This shift from the standard MS medium to its variants highlights the advantages of exploring different culture formulations. By modifying the artificial medium to meet the unique requirements of *C. sativa*, researchers can achieve a more efficient micropropagation process, leading to better growth, development, and regeneration. Table 3.1 summarised a wide range of cannabis tissue culture protocols implementing for different cutting explanting, using various plant growth regulators.

The results of investigations on the production of *C. sativa* under *in vitro* condition confirmed the efficacy of using a 16-hour photoperiod and a temperature range of 22–26°C. However, some researches have not followed this protocol. For instance, seed germination and callus induction have been successfully implemented under total darkness (Slusarkiewicz-Jarzina et al., 2005; Wielgus et al., 2008). By comprehending diverse environmental factors that impact different phases of *C. sativa* growth and development, it may be possible to optimize the most efficient protocols for the micropropagation of the plant (Slusarkiewicz-Jarzina et al., 2005; Wielgus et al., 2008). The potential benefits of extra lighting on cannabis plants were documented, when the cotted stem form *in vivo* mother plants were mainlined under controlled light condition in the greenhouse. This study showed the positive correlation between the implementation of red-green-blue and red-blue subcanopy lighting with bud biomass and phytochemical profiles. Therefore, the results of these investigations emphasize the importance of light quality and intensity for optimal tissue culture development of *C. sativa*. As a results of this type of experiment, it can be concluded that exploring the influence of different light spectra on the growth, development, and biochemical properties of *C. sativa* could be a reliable path for future studies. However, a deeper understanding of light impact on cannabis tissue culture could improve the quality and yield of *C. sativa*, improving both

TABLE 3.1
Various optimum micropropagation procedures of cannabis. The table focuses on studies which rely on shoot formation, multiplication, rooting, and shoot regeneration

Type	Cutting explants	Tissue culture phases	Optimal media and results		References
Drug-type	Axillary nodes (Shoot multiplication and rooting)	Selection/maintenance Initiation of cultures Shoot elongation/ rooting Acclimatization	**Shoot multiplication**		(Lata et al., 2009a)
			Full MS media supplemented with TDZ (0.5μM)	12.6 shoots/ explant % Response: 100	
			Rooting		
			Half MS media supplemented with IBA (2.5μM) and activated charcoal (0.05%)	4.8 roots/ explant % Response: 95	
Fiber-type	Apical and axillary nodes (Shoot multiplication)	Selection/maintenance Initiation of cultures Shoot elongation/ rooting Acclimatization	**Shoot multiplication**		(Richez-Dumanois et al., 1986)
			MS supplemented with IBA (0.1μM) and BAP (0.5μM)	2 apical meristems (shoots) per explant % Response: data not shown	
			Rooting		
			MS media added with IBA (10 μM) and activated charcoal (0.2%)	% Response: 47.7	
Fiber-type	Shoot-tips (Shoot multiplication)	Selection/maintenance Initiation of cultures	Shoot multiplication		(Lalge et al., 2016)
			MS media supplemented with BAP (1.78μM) and NAA (0.54μM)	Shoot multiplication: Shoots/ explant N.S. Response N.S. %	
Drug-type	Immature and mature inflorescences (shoot induction and rooting)	Selection/maintenance of parent plant material; Initiation of cultures	**Shoot induction**		(Piunno et al., 2019)
			MS media added with 10μMTDZ	4 shoots/floral cluster % Response: data not shown	
			Rooting		
			MS supplemented with 1.86 μM kinetin and 0.54 μM NAA	Describes 'most' cultures as rooting	
Fiber-type	Cotyledon (callogenesis + shoot regeneration, rooting)	Initiation of cultures; Shoot elongation and rooting	**Callogenesis/shoot regeneration:**		(Chaohua et al., 2016)
			MS media supplemented with 1.82 μM TDZ and 1.07μM NAA	3 shoots/explant (3-day-old cotyledons) 51.7% regeneration	
			Rooting		
			Half MS media added with 2.46–9.84 μM IBA	% Response: 80	

Drug-type	Young leaves (callus induction and shoot regeneration)	Shoot elongation and rooting	Callogenesis: MS+0.5µMNAA+1µM TDZ Shoot regeneration: MS+0.5µMTDZ	Callogenesis: 100% response across all 10 cultivars Shoot regeneration: Not achieved	(Monthony, et al., 2021)
Drug-type	Alginate encapsulated axillary nodes (Shoot induction and rooting)	Shoot elongation and rooting	**Shoot induction**		(Lata et al., 2009b)
			MS media added with 0.5µ MTDZ and 0.075% PPM	11.8 shoots/explant (90 days; avg. 30 explants)	
			Rooting		
			(1:1) sterile fertilome: coco natural growth medium + MS media added with 0.5% PPM	100% conversion from encapsulation (90 days)	
Drug-type	Axillary nodes (SM and rooting)	Selection/maintenance of parent plant material; Initiation of cultures; Shoot elongation and rooting; Acclimatization	**Shoot induction**		(Lata et al., 2016a)
			MS media added with 0.5µM TDZ and 0.075% PPM	11.8 shoots/explant (90 days; avg. 30 explants)	
			Rooting:		
			(1:1) sterile fertilome: coco natural growth medium + MS media with 0.5% PPM	100% conversion from encapsulation (90 days)	
Drug-type	Juvenile leaves (Callogenesis, shoot induction and rooting)	Selection/maintenance of parent plant material; Initiation of cultures; Shoot elongation and rooting; Acclimatization	**Callogenesis**		(Lata et al., 2010b)
			MS media added with 0.5µM NAA and 1 µM TDZ	% Response: 93.3	
			Shoot induction		
			MS supplemented with 0.5µMTDZ	12.3 shoots/explant % Response: 96.6	
			Rooting		
			1half MS media supplemented with 2.5 µM IBA	10 roots/ explant % Response: 96.6	
Drug-type	Immature and mature inflorescences (shoot induction and rooting)	Selection/maintenance of parent plant material; Initiation of cultures	**Shoot induction**		(Piunno et al., 2019)
			MS+10µMTDZ MS+1.86µMkinetin+ 0.54µMNAAa	4shoots/ floral cluster	
			Rooting		
			Describes 'most' cultures as rooting		

(continued)

TABLE 3.1 (Continued)
Various optimum micropropagation procedures of cannabis. The table focuses on studies which rely on shoot formation, multiplication, rooting, and shoot regeneration

Type	Cutting explants	Tissue culture phases	Optimal media and results		References
Fiber-type	Shoot apex, isolated apical meristem, and cotyledonary nodes from seedlings (SM, shoot development, and rooting)	Initiation of cultures Shoot elongation and rooting	**Shoot multiplication**		(Smýkalová et al., 2019)
			IMB4 and 6.97 µM KIN and 0.81µM BAP 9 THP + 0.11mM adenine hemi sulphate	4.4shoots/explant (isolated meristems) % response~96	
			Shoot development		
			Half MS medias with no added PGRs	N.S.	
			Rooting		
			Half MS media with 0.20 µM NAA	% Response: 50	
Fiber-type	Shoot tips and nodes from axillary branches (SM and rooting)	Selection/maintenance of parent plant material; Initiation of cultures; Multiplication of shoots/embryos; Shoot elongation and rooting; Acclimatization	Shoot multiplication		(Wróbel et al., 2022)
			Half MS media added with 2.85 µM IAA	2.5shoots/explant % Response: 70	
			Rooting		
			Half MS media supplemented with 2.85 µM IAA	% Rooted: 74.6	
Drug-type	Leaves (Callus suspension cultures and somatic embryogenesis)	Selection/maintenance of parent plant material; Initiation of cultures; Shoot elongation and rooting	**Suspension culture:**		(Flores-Sanchez et al., 2009)
			MS media added with B5 vitamins and 4.52 µM 2,4-D and 4.65 µM kinetin Somatic embryogenesis:	Growth rate: data not shown	
			Somatic embryogenesis:		
			Media composition: data not shown	Number of embryos: data not shown	
Drug-type	Axillary nodes (Shoot multiplication)	Multiplication of shoots/embryos;	Shoot multiplication: DKW+0.5µMTDZ	Shoot multiplication: 2.23shoots/explant 80%response	(Page et al., 2020)
Drug-type (2/2)	Single and pairs of florets (floral reversion and rooting)	Multiplication of shoots/embryos; Shoot elongation and rooting; Acclimatization	**Floral reversion**		(Monthony, Page, et al., 2021)
			DKWw media /vitamins and 1 µM mT	Estimated 18.2 explants derived from one *InVitro* flowering plant % Response: 81	
			Rooting		
			DKWw/vitamins	% Rooted: 44	

Drug-type	Juvenile leaves (Callogenesis and shoot regeneration)	Data not shown	**Callogenesis**		(Farag, 2014)
			B5 supplemented with 2.69 µM NAA and 22.20 µM BAP and 0.11 mM adenine hemi sulfate	% Callusing response: 50	
			Shoot regeneration		
			B5 added with 1.44 µM GA3	8.5 shoots/callus % Regeneration ratio: data not shown	
			Rooting		
			B5 added with 8.56 µM IAA	2.75 roots/explant % Response: 100	
Fiber-type	Juvenile leaves, petioles, internodes, and axillary nodes (Callus induction, shoot induction, and rooting)	Selection/maintenance of parent plant material; Initiation of cultures; Shoot elongation and rooting; Acclimatization	**Callus induction:**		(Slusarkiewicz-Jarzina et al., 2005)
			MS media supplemented with dicamba (9.05 and 13.57 µM a)	52.3% (5 CV Average; petioles)	
			Rooting		
			MS media supplemented with 0.57 µMIAA+ 0.54µM NAAa	69.9% plantlets formed roots	
Drug-type (0/1)	Leaves, flowers, and seedling roots, stems, and shoots (Callogenesis, callus suspension cultures)	Initiation of cultures; Multiplication of shoots/ embryos	**Callogenesis**		(Raharjo et al., 2006)
			MS media supplemented with 0.56 mM meso-inositol and 29.65µM thiaminedi HCl + 4.86 pyridoxine HCl + 8.12 µM nicotinic acid + 4.52 µM 2,4-Da	Statistical analysis: data not shown Callusing was greatest using flowers and seedling shoots	
			Suspension culture (2 -steps)		
			Step1:MS (as above, aqueous; 2weeks) Step2:B5media+9.05µM 2,4-D+2.85µMIAA+ 2.69µMNAA+5.12µM potassiuma	Continued callus growth, no regeneration	
Fiber-type	Axillary nodes (Direct organogenesis) Stems and roots (Indirect somatic embryogenesis)	Selection/maintenance of parent plant material Initiation of cultures Shoot elongation and rooting	**Callus induction:**		(Michał Pławuszewski et al., 2006)
			DARIAind++NAA+BAP (concentrations N.S.)	% callusing data not shown	
			Shoot proliferation: DARIA pro +NAA +BAP (concentrations N.S.)		
			Somatic embryogenesis: DARIA pro++NAA+BAP (concentrations N.S.)	Shoot proliferation: Adventitious shoot formation from axillary nodes and somatic embryo formation from stem tissue reported % N.S.	
			Somatic embryogenesis: data not shown		
			Rooting: DARIA root+ IAA (concentrations N.S.)	Rooting: N.S.	

(*continued*)

TABLE 3.1 (Continued)
Various optimum micropropagation procedures of cannabis. The table focuses on studies which rely on shoot formation, multiplication, rooting, and shoot regeneration

Type	Cutting explants	Tissue culture phases	Optimal media and results		References
Drug-type	Axillarynodes (SM)	Selection/maintenance of parent plant material Initiation of cultures Shoot elongation and rooting	Shoot multiplication: Formula βH media	Shoot multiplication: Shoots/explant N.S. 58%response	(Codesido et al., 2018)
Fiber-type	Hypocotyl, cotyledon and first two true leaves (direct organogenesis and rooting)	Selection/maintenance of parent plant material; Multiplication of shoots/embryos; Shoot elongation and rooting; Acclimatization	Organogenesis: MS+1.82 µMTDZ+1.07µM NAAa Rooting: MS+1.82 µMTDZ+1.07µM NAAa	Organogenesis: 1.49 shoots/hypocotyl 54.17% response Rooting: ~18% rooted	(Galán-Ávila et al., 2020)
Fiber-type	Cotyledons, axillary nodes, androots (Callus induction, shoot induction, and rooting)	Selection/maintenance of parent plant material; Multiplication of shoots/embryos; Shoot elongation and rooting;	**Callus induction**		(Wielgus et al., 2008)
			DARIA (ind+) + 4.65µM kinetin+0.27 µM NAAa	Best morphogenic callus induction: stem explants (all cultivars) Statistic alanalysis data not shown	
			Shoot induction		
			DARIA (pro+)+ 0.89 µM BAP +0.16 µM NAAa	15.56% with cotyledon explants (cv. Beniko)	
			Rooting		
			DARIA (root+)+ 11.42µM IAAa	Statistical analysis data not shown	
Drug-type	Cotyledons and epicotyls (callogenesis+ shoot regeneration, rooting)	Selection/maintenance of parent plant material; Initiation of cultures; Multiplication of shoots/embryos; Shoot elongation and rooting; Acclimatization	**Callogenesis/shoot regeneration:**		(Movahedi et al., 2015)
			MS+8.88 µMBAP+2.46µM IBA	~2 shoots/epicotyl %response N.S.	
			Rooting		
			MS media added with 0.49 µM IBAa	%response data not shown	

Drug-type	Axillary nodes (Shoot multiplicatio)	Selection/maintenance of parent plant material; Initiation of cultures; Shoot elongation and rooting; Acclimatization	Shoot multiplication: MS+2.07μMmT	Shoot multiplication: 1.78shoots/explant 97.8%response	(Mestinšek-Mubi et al., 2020)
			Rooting		
			MS media with no PGRs	%response data not shown	
Fiber-type	Leaf, hypocotyl, cotyledon, and root (Callogenesis and shoot regeneration)	Selection/maintenance of parent plant material; Shoot elongation and rooting;	Callogenesis/ shoot regeneration: MS+B5vitamins + 13.57–45.24 μM 2,4-D + 0.04–0.44 μM BAPa	Callogenesis/shoot regeneration: One tested cultivar occasionally gave rise to organogenic callus from hypocotyl tissue. %regeneration N.S.	(Mandolino et al., 1999)

Note: SM mentions to the multiple shoots' proliferation from an existing meristem, such as floral meristems, apical nodes, and axillary. Shoot regeneration meaning the de novo shoots formation from non-meristematic tissues such as cotyledons, stems, or leaves. Cannabis type is distinct in this table as either known industrial hemp genotypes "fiber-type" (Hemp; THC<0.3%inflowering head) or psychoactive "drug-type" (Cannabis; tetrahydrocannabinol (THC)> 0.3% in flowering head).

real-world out-come and scientific progress in the fields of pharmaceutical and agriculture sectors (Slusarkiewicz-Jarzina et al., 2005; Wielgus et al., 2008; Hawley et al., 2018; Boonsnongcheep et al., 2020).

3.3 ROLE OF PLANT GROWTH REGULATORS (PGRS) IN CANNABIS TISSUE CULTURE

In vitro culture of cannabis could be influenced in terms of form, grow, and development by plant growth regulators; specially two important families of auxins and cytokinins (George et al., 2008). The applications of PGRs are mainly critical for direct organogenesis (specifically shoot formation) and indirect organogenesis (particularly callus induction and regeneration). However, the application of both PGRs treatments should be at the optimum concentrations, depending on the specific phase of cannabis *in vitro* culture. For example, shoot induction, is one of the most important phases of cannabis tissue culture, which could be influenced by the balanced application of auxin and cytokinins. To produce healthy and robust cannabis plants, adventitious root formation and development are essential once a sufficient shoot has been grown. Gaining the skill, knowledge, and understanding of utilizing plant growth regulators can highly improve the success and effectiveness of cannabis micropropagation, opening the door to the formation of superior cannabis plants for a wide range of applications (Boonsnongcheep et al., 2020).

3.3.1 Multiplication of Cannabis Shoots

As mentioned earlier, different phases of direct and indirect *in vitro* culture are mainly influenced by PGRs. For instance, Lata and colleagues (Lata et al., 2009a; Lata et al., 2009b, 2010a; Lata et al., 2010b; Lata et al., 2016a; Lata et al., 2016b; Galán-Ávila et al., 2020) studied the effect of cytokinins (0.05 to 9.0 μM), including thidiazuron (TDZ), kinetin (KN), and benzyladenine (BA), on direct organogenesis of cannabis *in vitro* culture. The target of the mentioned experiment was to establish the prime protocol for in vitro culture of cannabis clones using multiplication and elongation of cutting nodal explants containing axillary buds. The results of experiments showed that TDZ was the most effective type of cytokinin for shoot development. However, an abundant number (with an average of 13 shoots per explant and shoots growth with an average shoot length of 7.1 cm) were reported after 14 days on MS basal media. A higher shoot growth was observed when the plants were transferred into the MS basal media supplemented with gibberellic acid (7.0 μM). Although, the visual observation showed the effectives of gibberellic acid (GA3) compared to TDZ, there were no statically differences between the quantitative results. Evidence revealed that although cytokinins alone are usually efficient for shoot multiplication, supplementing the media with small concentrations of auxin may enhance shoot growth in certain conditions. Achieving superior clones can be facilitated by fine-tuning *in vitro* method for cannabis through the application of various PGRs.

3.3.2 Formation of Shoot and Buds of Cannabis *In Vitro*

To produce axillary buds, Wang et al. (Wang et al., 2009) examined the effect of MS media supplemented with thidiazuron (TDZ), kinetin (KN), and benzyladenine (BA) either alone or in combination with the auxin 1-naphthaleneacetic acid (NAA) on shoot tip explants of cannabis. The results of their study showed that MS media added with NAA (0.1 mg/L) and TDZ (0.2 mg/L) produce maximum numbers of axillary buds per shoot tip (3.22) (Wang et al., 2009). Similarly, Cheng et al. (Cheng ChaoHua et al., 2016) showed that the highest number of developed shoots (3.0 ± 0.4 shoot per explants) and regenerated shoots (51.7 ± 7.2%) were obtained by subculturing the cotyledon explants into MS media supplemented with NAA (0.2 mg/L) and TDZ (0.4 mg/L) (Cheng ChaoHua et al., 2016). In addition to the mentioned studies, recently meta-topolin (mT), which is an aromatic cytokinin, has shown an effective protentional in cannabis micropropagation (Lata et al., 2016b). For instance, cultivation of nodal culture in MS media added with 2 μM mT highly developed cannabis shoot during tissue culture (Lata et al., 2016b). Its shown that all nodal explants treated with 2 μM mT induced shoots with an average length of between 11.44 ± 0.80 cm and 13.44 ± 1.38 shoots per explant (Lata et al., 2016b). This founding highlights the importance role of PGRs' type and concentrations in order to produce optimal culture of cannabis plants *in vitro* condition. Through the best selection of cytokinins and auxins, researchers can significantly improve the formation of shoots and buds, and produce healthy cloning with effective and successful protocols (Boonsnongcheep et al., 2020).

3.3.3 Indirect Organogenesis In Cannabis Tissue Culture

In plant tissue culture, indirect organogenesis refers to regeneration of shoot from the morphogenic callus. An investigation on the effect of various PGRs, including 3,6-dichloroanisic acid (DIC), 2,4-dichlorophenoxyacetic acid (2,4-D), NAA, and kinetin (KN), showed the highest effect of MS media supplemented with DIC (2.0–3.0 mg/L) on callus formation (82.7%) (Slusarkiewicz-Jarzina et al., 2005). However, their report showed the lowest rate of plant regeneration in the mentioned media with only 1.4% (Slusarkiewicz-Jarzina et al., 2005). Interestingly, a combination of cytokinin with auxin showed highest plant regeneration of cannabis plants from callus (Chandra et al., 2010; Lata et al., 2010a; Movahedi et al., 2015). For instance, it has been reported that subculturing leaf explants into MS media supplemented with different concentrations of auxins (2,4-D, NAA, indole-3-butyric acid (IBA), and indole-3-acetic acid (IAA) in combination of the cytokinin (TDZ) produced higher volume of calli. Their results showed that combination of TDZ (1.0 μM) and NAA (0.5 μM) produced highest callus formation with 93%, and the highest shoot formation were achieved using MS media supplemented with TDZ (0.5 μM) (Lata et al., 2010a). Likewise, it has been reported that subculturing cotyledon explants into MS media added with TDZ (3 mg/L) and IBA (0.5 mg/L) produced highest callus formation and proliferation (Movahedi et al., 2015). On the other hand, it has been reported that using Daria media supplemented with KN (1 mg/L) and NAA (0.05 mg/L) produced higher rate of callus induction, whereas higher rate of shoot regeneration were successfully achieved using Daria media supplemented with BA (0.2 mg/L) and NAA (0.03 mg/L) (Wielgus et al., 2008). In order to produce callus and regeneration of the plant successfully, optimising various type of PGRs and different concentration of the plant growth regulators are the reliable method in cannabis plants. By adjusting the mentioned factors, researches can produce optimum callus formation and plant regeneration *in vitro* culture, which will enable more successful and efficient micropropagation procedures in cannabis tissue culture (Boonsnongcheep et al., 2020).

3.3.4 Establishing Roots of Cannabis *In Vitro* Culture

To produce root of cannabis *in vitro* culture, auxins family, importantly indole-3-acetic acid (IAA) and Indole-3-butyric acid (IBA), showed a reliable response since cytokinin family may prevent formation and development of roots (Wang et al., 2009; Lata et al., 2009a; Lata et al., 2009b, 2010a; Cheng ChaoHua et al., 2016). On the hand, it has been reported that cannabis plants successfully induce root in MS media supplemented with mT, comparing with an auxin-supplemented medium. These results showed that healthy roots and shoots were achieved from shoots within 4–6 weeks in the MS media supplemented with mT. These results revealed the effectiveness of optimizing and selection of various PGRs and/or concentrations through cultivation of cannabis in vitro culture (Lata et al., 2016b).

3.4 IMPORTANCE OF ACCLIMATIZATION DURING THE *IN VITRO* CULTURE OF CANNABIS

The acclimatization process plays a critical role during the transmission of plants from a synthetic media (mixophotoperiod) into the natural environments (photoautoperiod) (Pospóšilová et al., 1999). Sofa, various protocols have been optimized to have an efficient acclimatization protocol for different cannabis species. For instance, (Wang et al., 2009) found the high efficiency of acclimatization protocol, when the *in vitro* explants with the root lengths between 0.5 and 1.0 cm long were carefully taken out from the culture vessels, washed with sterile water, treated with fungicide solution, and cultured into soil (Wang et al., 2009). The rooted plants then were maintained for three to four weeks under controlled conditions in green house, and their results showed a significant survival rate (up to 95%) (Wang et al., 2009). Interestingly, another guideline showed that rooted cannabis plants can be cleaned using coco natural growth (a commercially available plant growth medium which is produced from coconut fibers), after taken out from the *in vitro* vessels. Next, the plants can be pre-conditioned in the mentioned media (coco natural) within thermocol cups, sealed with polythene bags and kept in high humidity for ten days. In the next phase, the plants were acclimated in Fertiome (a sterile, larger mixing pot) and maintained until the cannabis plants reached a height of 14–16 cm over six weeks. Finally, 8 weeks after transplanting, the survival rate of plants was almost 95%. This recommended protocol emphasizes the necessity of balancing between the initial growth condition *in vitro* condition and the transplanting phases in natural environment. By implementing the recommended protocol, cannabis producers and researchers can grow plants with higher survival rates (Lata et al., 2009a; Boonsnongcheep et al., 2020).

3.5 GENETIC TRANSFORMATION ROLE IN CANNABIS

In moder molecular breeding, genetic transformation plays a critical role to produce novel plant species, investigates the molecular mechanisms of plants and utilizes genetic diversity. In cannabis research area, forward (*e.g.*, mutagenesis screens) and reverse (*e.g.*, ectopic expression) genetics are vital approaches due to the detection and manipulation of unknown genetic mechanisms. However, reverse genetic methods have recently gained more attention in modern molecular breeding, particularly due to the invention of advanced genetic sequencing tools (Feeney and Punja, 2003). Although cannabis is a recalcitrant plant for genetic transformation, meaningful advancement has been made in applying genetic transformation methods to this species (Feeney et al., 2003; Slusarkiewicz-Jarzina et al., 2005; Sirkowski, 2012; Wahby et al., 2017; Schachtsiek et al., 2019). While, there is significant progress in applying *Agrobacterium*-mediated gene transfer systems in plants, the efficiency of this approach is still low in cannabis plants due to the low regeneration of the plants after gene transformation (Feeney et al., 2003). For instance, Feeney and Punja (Feeney et al., 2003) successfully transformed the gene at the cellular level of cannabis; however, the difficulties were reported on the regeneration of whole plants. Likewise, the bottlenecks in the regenerating whole plants were reported during the hairy root formation of cotyledonary and hypocotyl node explants using *A. rhizogenes* strains (Wahby et al., 2017). In addition to the importance of the mentioned gene transformation process in cannabis area, the creation of knockout cannabis mutants of key genes, such as CBGA-, CBDA-, and THCA- synthesis, is also important for enhancing genome edited cannabis. However, the lack of enough successful report on the genome editing of cannabis should be considered (Sirkowski, 2012). These results highlight the critical necessity for standardized procedures that guarantee uniform and repeatable regeneration and transformation results across various cannabis species. These achievements emphasize the essential needs need to optimize an efficient and reliable protocols for gene transformation and genome editing of cannabis, with a focus on the increasing regeneration ratio (Schachtsiek et al., 2019).

3.6 TRANSIENT GENETIC TRANSFORMATION IN CANNABIS PLANTS

Transient genetic transformation is a kind of modification method where genetic materials are inserted into the host genome temporarily. However, this method helps to brief expression of transferred gene in the host without medication of the host genome structure (Abiri et al., 2016). Reportedly, there are various transient methods, but virus-induced gene silencing (VIGS), which is an RNA-mediated post-transcriptional gene silencing (PTGS) method, facilitates an efficient and rapid investigation of gene function (Baulcombe, 1999; Liu et al., 2002; Senthil-Kumar et al., 2014; Adhikary et al., 2019; Schachtsiek et al., 2019). It has been reported that application of the VIGS technique could help to evaluate the function of inserted genes within 3–6 weeks, before implementing stable gene transformations (Adhikary et al., 2019). Adapting the VIGS system using Cotton leaf crumple virus (CLCrV) could significantly improves the efficiency of this transient gene transformation technique in cannabis, by impairing function phenotypes of imperative genes including phytoene desaturase (PDS) and magnesium chelatase subunit I (ChlI) (Schachtsiek et al., 2019). Stablishing this pioneering experiment could facilitate the exploration of uncharacterized genes in cannabis, despite early reports indicating only minor phenotypic changes. In addition to the application of CLCrV technique, there are some other transient gene transformation methods which can be applied in cannabis plant. For example, the Tobacco rattle virus (TRV) vector, a reliable method for transient gene transformation in over 200 dicot plant species, can be utilized in cannabis due to the plant's susceptibility to viral infections (Kumar et al., 2007; McPartland, 2018). Although, TRV is a reliable method in transient gene transformation, CLCrV is a promising technique in effectively elucidating gene functions, contributing to a deeper understanding of cannabis genetics and aiding in breeding programs to develop superior cultivars. Continued research on various viral vectors of transient gene transformation including TRV and CLCrV could considerably accelerate advancements in cannabis genetics (Adhikary et al., 2021).

3.7 STABLE GENETIC TRANSFORMATION AND GENOME EDITING IN CANNABIS PLANTS

Stable genetic transformation is an efficient method for improving plants by modifying the host genomes of the plants. This method, in contrast to transient gene transformation, can insert genes into the host genome, resulting in heritable and long-lasting changes (Abiri et al., 2016; Shew et al., 2018). There are various methods which can be used as stable gen transformation in cannabis including *Agrobacterium*-mediated, biolistic (gene gun), and protoplast transformation methods. Although, clustered regularly interspaced short palindromic repeats or CRISPR, is a type of genome editing technique (Jinek et al., 2012), this method provides accurate gene modification with the consequences for both practical and fundamental investigation in cannabis biology (Aliaga-Franco et al., 2019). As mentioned earlier, its crucial to differentiate between CRISPR-mediated gene editing and conventional genetic transformation methods (Shew et al., 2018). It must be highlighted that while GM process requires the insertion of transfer DNA (tDNA) or foreign DNA within the same species or between species, CRISPR (using enzymes and guide RNA) gene editing allows genome targeted modification without always producing GMOs (Aliaga-Franco et al., 2019). Using CRISPR-Cas9 technique in cannabis research signifies an advancement for genetic studies and molecular research of this plant. In conclusion, CRISPR-Cas9 stable genetic editing has a vital capacity to edit the gene in cannabis plants. The ability to accurately edit genome sequences can provide new chances for the improvement of particular cannabis plant species with high qualities, efficiently satisfying a wide range of social and industrial demands (Adhikary et al., 2019; Adhikary et al., 2021).

3.8 APPLICATION OF HAIRY ROOT CULTURE IN CANNABIS PLANTS

Hairy root culture (HRC) technique, which is caused by infection of a negative soil born bacteria called *Agrobacterium rhizogenes,* is an effective method for generating transgenic plants and exploring gene functions in the host plants. HRC is a rapid growth technique proceed in a hormone-free media, which is a perfect model for generating secondary metabolites comparable to those compounds in wild-type plants (Pistelli et al., 2010). Moreover, HRC offers consistent, repeatable, and microbial free production in plant's bioreactors that is not influenced by seasonal or weather conditions. Therefore, application of the system can help to produce sustainable secondary metabolites in a wide range of medicinal plant species including cannabis. Hairy roots can be developed at the infection site of hypocotyls or calli by *A. rhizogenes*, and by separating these hairy roots, elevated production of specific desired chemicals compounds can be assessed. Hairy roots can also be grown up for production of high-yield chemical compounds and can be kept as pure cultures by cryopreservation at -196°C (Engelmann, 2013). There have been some reports of producing hairy roots in cannabis by applying HRC tools (Wahby et al., 2017). A report showed that the optimal condition to produce hairy roots in cannabis is using B5 media supplemented with NAA (4 mg/l) and kept under 25°C in a dark condition (Farag et al., 2015). Nonetheless, the minimal production of CBDA and THCA under the specific mentioned condition demonstrates negligible cannabinoid synthesis. During the hairy root culture by *A. rhizogenes*, some chemical components, including muscarine, atropine, and choline can be accumulated in the transformed plants, despite the fact that the typical cannabis hairy roots may not naturally produce THC (Wahby et al., 2017). It has been shown that the level of chemical components is higher in the transformed plants using hairy root technique, compared to non-transformed plants. For instance, (Wahby et al., 2017) showed higher level of choline (203 to 510 mg/L, compared to 66–153 mg/L in the control), atropine (562 to 933 μg/L, compared to 532–553 μg/L in the control), and muscarine (231–367 μg/L, undetectable in the control) production in the transformed plants. An interesting study on cannabis hairy root cultures suggests the possibility of modifying cannabinoid biosynthesis pathways using the effective expression of the THCA synthase gene in tobacco hairy root cultures (Taura et al., 1995; Sirikantaramas et al., 2004).

3.9 ASSESSING THE ROLE OF MERISTEM CULTURE OF CANNABIS

Meristem culture, which is based on the totipotent cells ability inside the apical dome, is an efficient and quick tools to produce virus-free plants in the large scale (Mori, 1971; Wang et al., 2016). The apical dome zone of the plant is mainly effective at reducing the spread of pathogens through vascular tissue with inadequate cell-to-cell transmission due to the absence of vascular connections to axillary buds, leaf primordia, and developing procambium (Wang et al., 1991). In fact, this technique can open the new door for producing large-scale virus-free plants (Wang et al., 1991; Wu et al., 2020). Typically, the procedure of meristem culture in plants requires excising tissue, typically from lateral or apical buds, that includes the apical meristem with a length between 0.1 to 0.5 mm and a small number of leaf primordia, approximately 1–2 mm. Then, the plant tissue is cultivated in an optimal plant tissue culture medium supplemented with an ideal combination of PGRs to promote better growth and development. The cells can go through different plant tissue culture ways including direct or indirect organogenesis or somatic embryogenesis, and regenerate into whole plants (Mori, 1971; Wang et al., 1991; Mordhorst et al., 2002; Al-Taleb et al., 2011; Spanò et al., 2018). Importantly, the production of virus-free plants by meristem culture have been reported in the close relative of cannabis such as *Humulus lupulus* (Adams, 1975; Grudzińska et al., 2004; Grudzińska et al., 2006). Notwithstanding all the benefits of meristem culture in different plants, this method still faces some bottlenecks in overcoming fungi and bacteria that have been linked to cannabis-related illnesses. Therefore, scholars need to apply this method to cannabis and its close relatives to find the best way to overcome these challenges (Sallie et al., 1969; Adams, 1975; Grudzińska et al., 2004; Grudzińska et al., 2006).

3.10 UNLOCKING THE EFFECTIVENESS OF PROTOPLAST CULTURE IN CANNABIS

Protoplast culture has been introduced as a mainstay in modern genetic methods including genome editing, somatic mutation, cell fusion and gene transformation (Lei et al., 2015). Notwithstanding the great achievements of applying this method in other crop species, utilizing this method in cannabis is still in the preliminary phase, particularly for the survival of transfected protoplasts and plant regeneration. Sofa, studies on protoplast culture have shown that mesophyll protoplasts have been isolated in three different cannabis cultivars (Morimoto et al., 2007; Flaishman et al., 2022). However, the low survival rate of protoplasts after 48 hours in liquid culture media and the lack of plant regeneration have made this method ineffective in the cannabis field (Flaishman et al., 2022). Despite the mentioned obstacle in protoplast culture of cannabis, protoplast is introduced as a vital tool for effectiveness and confirming of planned guide RNA (gRNA), in advance of their application in producing gene-edited plants (Adhikary et al., 2021). Optimising the protoplast culture in cannabis species can drastically improve the efficiency of genetic manipulation methods. Nowadays, scientists have been focusing on the current obstacles of this method in cannabis including survival rate and regeneration phase. However, ongoing studies will be essential for overcoming the challenges, especially during the genome editing process (Adhikary et al., 2021).

3.11 CONCLUSION

Cannabis sativa is a significant pharmaceutical plant that has been utilized in both traditional and modern medicine since the dawn of civilization. Although stem cutting is one of the best methods for cannabis propagation, tissue culture methods have shown promising results in the cannabis area. There are various factors which may influence the tissue culture of cannabis including PGRs, tissue culture medium, nutrients, light, and temperature. The successful application of tissue culture techniques has led to increased efficiency in genome modification methods. Reportedly, both direct and indirect regeneration have been reported in the cannabis area by cultivating different explant types into various media supplemented with auxin and cytokinin hormones. The plant tissue culture vessels have been kept under optimum light and temperature conditions in the growth chambre. Despite the achievements in tissue culture and gene modification of cannabis, further research is needed to enhance the efficiency of these protocols and improve plant regeneration.

CONFLICTS OF INTEREST

The authors declare no conflict of interest.

REFERENCES

Abiri, R., Valdiani, A., Maziah, M., Shaharuddin, N. A., Sahebi, M., Yusof, Z. N. B., Atabaki, N., & Talei, D. (2016). A critical review of the concept of transgenic plants: insights into pharmaceutical biotechnology and molecular farming. *Current Issues in Molecular Biology*, *18*(1), 21–42.

Adams, A. (1975). Elimination of viruses from the hop (*Humulus lupulus*) by heat therapy and meristem culture. *Journal of Horticultural Science*, *50*(2), 151–160.

Adhikary, D., Khatri-Chhetri, U., Tymm, F. J., Murch, S. J., & Deyholos, M. K. (2019). A virus-induced gene-silencing system for functional genetics in a betalainic species, *Amaranthus tricolor* (Amaranthaceae). *Applications in Plant Sciences*, *7*(2), e01221.

Adhikary, D., Kulkarni, M., El-Mezawy, A., Mobini, S., Elhiti, M., Gjuric, R., Ray, A., Polowick, P., Slaski, J. J., & Jones, M. P. (2021). Medical cannabis and industrial hemp tissue culture: present status and future potential. *Frontiers in Plant Science*, *12*, 627240.

Al-Taleb, M. M., Hassawi, D. S., & Abu-Romman, S. M. (2011). Production of virus free potato plants using meristem culture from cultivars grown under Jordanian environment. *American-Eurasian Journal of Agricultural & Environmental Sciences*, *11*(4), 467–472.

Aliaga-Franco, N., Zhang, C., Presa, S., Srivastava, A. K., Granell, A., Alabadí, D., Sadanandom, A., Blázquez, M. A., & Minguet, E. G. (2019). Identification of transgene-free CRISPR-edited plants of rice, tomato, and *Arabidopsis* by monitoring DsRED fluorescence in dry seeds. *Frontiers in Plant Science*, *10*, 1150.

Baulcombe, D. C. (1999). Fast forward genetics based on virus-induced gene silencing. *Current Opinion in Plant Biology*, *2*(2), 109–113.

Boonsnongcheep, P., & Pongkitwitoon, B. (2020). Factors affecting micropropagation of *Cannabis sativa* L.: A review. *Pharmaceutical Sciences Asia*, *47*, 21–29.

Caplan, D., Stemeroff, J., Dixon, M., & Zheng, Y. (2018). Vegetative propagation of cannabis by stem cuttings: Effects of leaf number, cutting position, rooting hormone, and leaf tip removal. *Canadian Journal of Plant Science*, *98*(5), 1126–1132.

Chandra, S., Lata, H., Mehmedic, Z., Khan, I. A., & ElSohly, M. A. (2010). Assessment of cannabinoids content in micropropagated plants of *Cannabis sativa* and their comparison with conventionally propagated plants and mother plant during developmental stages of growth. *Planta Medica*, *76*(07), 743–750.

Chaohua, C., Gonggu, Z., Lining, Z., Chunsheng, G., Qing, T., Jianhua, C., Xinbo, G., Dingxiang, P., & Jianguang, S. (2016). A rapid shoot regeneration protocol from the cotyledons of hemp (*Cannabis sativa* L.). *Industrial Crops and Products*, *83*, 61–65.

Codesido, V., Meyer, S., & Casano, S. (2018). Influence of media composition and genotype for successful Cannabis sativa L. in vitro introduction. In *XXX International Horticultural Congress IHC2018: II International Symposium on Micropropagation and In Vitro Techniques* 1285.

Engelmann, F. (2013). Cryopreservation of clonal crops: a review of key parameters. In *II International Symposium on Plant Cryopreservation 1039*, 31–39.

Farag, S. (2014). *Cannabinoids production in Cannabis sativa L.: An in vitro approach*. Ph.D. thesis, Technical University of Dortmund.

Farag, S., & Kayser, O. (2015). Cannabinoids production by hairy root cultures of Cannabis sativa L. *American Journal of Plant Sciences*, *6*(11), 1874–1884.

Feeney, M., & Punja, Z. (2003). Tissue culture and Agrobacterium-mediated transformation of hemp (Cannabis sativa L.). *In Vitro Cellular & Developmental Biology-Plant*, *39*, 578–585.

Ferrara, M. S. (2016). *Sacred bliss: A spiritual history of cannabis*. Rowman & Littlefield.

Flaishman, M. A., PEER, R. C., Cohen, O., & Bocobza, S. (2022). Methods of regenerating and transforming cannabis. In: Google Patents.

Flores-Sanchez, I. J., Peč, J., Fei, J., Choi, Y. H., Dušek, J., & Verpoorte, R. (2009). Elicitation studies in cell suspension cultures of Cannabis sativa L. *Journal of Biotechnology*, *143*(2), 157–168.

Galán-Ávila, A., García-Fortea, E., Prohens, J., & Herraiz, F. J. (2020). Development of a direct in vitro plant regeneration protocol from Cannabis sativa L. seedling explants: developmental morphology of shoot regeneration and ploidy level of regenerated plants. *Frontiers in Plant Science*, *11*, 645.

George, E. F., Hall, M. A., & Klerk, G.-J. D. (2008). Plant growth regulators I: Introduction; auxins, their analogues and inhibitors. In E. F. George, M. A. Hall, G-J De Klerk (Eds.), *Plant Propagation by Tissue Culture: Volume 1. The Background* (pp. 175–204). Springer.

Ghiabi, M., Maarefvand, M., Bahari, H., & Alavi, Z. (2018). Islam and cannabis: Legalisation and religious debate in Iran. *International Journal of Drug Policy*, *56*, 121–127.

Grudzińska, M., & Solarska, E. (2004). The elimination of viruses and hop latent viroid from hop (Humulus lupulus L.) in Poland. *I International Humulus Symposium*, *668*, 149–152.

Grudzińska, M., Solarska, E., Czubacka, A., Przybyś, M., & Fajbuś, A. (2006). *Elimination of Hop Latent Viroid from Hop plants by cold treatment and meristem tip culture*. *Phytopathologia Polonica*, *40*, 21–30.

Hawley, D., Graham, T., Stasiak, M., & Dixon, M. (2018). Improving cannabis bud quality and yield with subcanopy lighting. *HortScience*, *53*(11), 1593–1599.

Jinek, M., Chylinski, K., Fonfara, I., Hauer, M., Doudna, J. A., & Charpentier, E. (2012). A programmable dual-RNA–guided DNA endonuclease in adaptive bacterial immunity. *Science*, *337*(6096), 816–821.

Kumar, S. P. D., Liu, Y., & Schiff, M. (2007). *Tobacco rattle virus vectors and related compositions and methods*. In: Google Patents.

Lalge, A. B., Mendel, P., Vyhnanek, T., Trojan, V., Kalousek, P., & Havel, L. (2016). Effects of different morphoregulators on growth and development of Cannabis sativa L. *Proceedings of the International Ph. D. Students Conference on MendelNet*. pp. 726–730.

Lata, H., Chandra, S., Khan, I., & ElSohly, M. A. (2009a). Thidiazuron-induced high-frequency direct shoot organogenesis of Cannabis sativa L. *In Vitro Cellular & Developmental Biology-Plant*, *45*, 12–19.

Lata, H., Chandra, S., Khan, I. A., & ElSohly, M. A. (2009b). Propagation through alginate encapsulation of axillary buds of Cannabis sativa L.—an important medicinal plant. *Physiology and Molecular Biology of Plants*, *15*, 79–86.

Lata, H., Chandra, S., Khan, I. A., & ElSohly, M. A. (2010a). High frequency plant regeneration from leaf derived callus of high Δ9-tetrahydrocannabinol yielding Cannabis sativa L. *Planta Medica*, *76*(14), 1629–1633.

Lata, H., Chandra, S., Techen, N., Khan, I. A., & ElSohly, M. A. (2010b). Assessment of the genetic stability of micropropagated plants of Cannabis sativa by ISSR markers. *Planta Medica*, *76*(01), 97–100.

Lata, H., Chandra, S., Khan, I. A., & ElSohly, M. A. (2016a). In vitro propagation of Cannabis sativa L. and evaluation of regenerated plants for genetic fidelity and cannabinoids content for quality assurance. In *Protocols for In Vitro Cultures and Secondary Metabolite Analysis of Aromatic and Medicinal Plants, Second Edition*, 275–288.

Lata, H., Chandra, S., Techen, N., Khan, I. A., & ElSohly, M. A. (2016b). In vitro mass propagation of Cannabis sativa L.: A protocol refinement using novel aromatic cytokinin meta-topolin and the assessment of eco-physiological, biochemical and genetic fidelity of micropropagated plants. *Journal of Applied Research on Medicinal and Aromatic Plants*, *3*(1), 18–26.

Lei, R., Qiao, W., Hu, F., Jiang, H., & Zhu, S. (2015). A simple and effective method to encapsulate tobacco mesophyll protoplasts to maintain cell viability. *MethodsX*, *2*, 24–32.

Liu, Y., Schiff, M., & Dinesh-Kumar, S. P. (2002). Virus-induced gene silencing in tomato. *The Plant Journal*, *31*(6), 777–786.

Mandolino, G., & Ranalli, P. (1999). Advances in biotechnological approaches for hemp breeding and industry. *Advances in Hemp Research*, *9*, 185–212.

McPartland, J. M. (2018). Cannabis systematics at the levels of family, genus, and species. *Cannabis and Cannabinoid Research*, *3*(1), 203–212.

Mestinšek-Mubi, Š., Svetik, S., Flajšman, M., & Murovec, J. (2020). In vitro tissue culture and genetic analysis of two high-CBD medical cannabis (Cannabis sativa L.) breeding lines. *Genetika*, *52*(3), 925–941.

Michał Pławuszewski, W. L., & Wielgus, K. (2006). Regeneration of Polish cultivars of monoecious hemp (Cannabis sativa L.) grown in vitro. *Renewable Resources and Plant Biotechnology*, *149*, 149–188.

Monthony, A. S., Kyne, S. T., Grainger, C. M., & Jones, A. M. P. (2021). Recalcitrance of Cannabis sativa to de novo regeneration; a multi-genotype replication study. *PloS one*, *16*(8), e0235525.

Monthony, A. S., Page, S. R., Hesami, M., & Jones, A. M. P. (2021). The past, present and future of Cannabis sativa tissue culture. *Plants*, *10*(1), 185.

Mordhorst, A. P., Hartog, M. V., El Tamer, M. K., Laux, T., & de Vries, S. C. (2002). Somatic embryogenesis from Arabidopsis shoot apical meristem mutants. *Planta*, *214*, 829–836.

Mori, K. (1971). Production of virus-free plants by means of meristem culture. *Japan Agricultural Research Quarterly*, *6*, 1–7.

Morimoto, S., Tanaka, Y., Sasaki, K., Tanaka, H., Fukamizu, T., Shoyama, Y., Shoyama, Y., & Taura, F. (2007). Identification and characterization of cannabinoids that induce cell death through mitochondrial permeability transition in Cannabis leaf cells. *Journal of Biological Chemistry*, *282*(28), 20739–20751.

Movahedi, M., Ghasemi-Omran, V., & Torabi, S. (2015). The effect of different concentrations of TDZ and BA on in vitro regeneration of Iranian cannabis (Cannabis sativa) using cotyledon and epicotyl explants. *Journal of Plant Molecular Breeding*, *3*(2), 20–27.

Murashige, T., & Skoog, F. (1962). A revised medium for rapid growth and bio assays with tobacco tissue cultures. *Physiologia Plantarum*, *15*(3), 473.

Pacher, P., Bátkai, S., & Kunos, G. (2006). The endocannabinoid system as an emerging target of pharmacotherapy. *Pharmacological Reviews*, *58*(3), 389–462.

Page, S., Monthony, A., & Jones, A. (2020). Basal media optimization for the micropropagation and callogenesis of Cannabis sativa L. *BioRxiv*, *1*, 1–23.

Pain, S. (2015). A potted history. *Nature, 525*(7570), S10–S11.

Pellati, F., Borgonetti, V., Brighenti, V., Biagi, M., Benvenuti, S., & Corsi, L. (2018). Cannabis sativa L. and nonpsychoactive cannabinoids: their chemistry and role against oxidative stress, inflammation, and cancer. *BioMed Research International, 2018*(1), 1691428.

Pisanti, S., & Bifulco, M. (2019). Medical Cannabis: A plurimillennial history of an evergreen. *Journal of Cellular Physiology, 234*(6), 8342–8351.

Pistelli, L., Giovannini, A., Ruffoni, B., Bertoli, A., & Pistelli, L. (2010). Hairy root cultures for secondary metabolites production. *Bio-farms for Nutraceuticals: Functional Food and Safety Control by Biosensors, 698*, 167–184.

Piunno, K. F., Golenia, G., Boudko, E. A., Downey, C., & Jones, A. M. P. (2019). Regeneration of shoots from immature and mature inflorescences of Cannabis sativa. *Canadian Journal of Plant Science, 99*(4), 556–559.

Pospóšilová, J., Tichá, I., Kadleček, P., Haisel, D., & Plzáková, Š. (1999). Acclimatization of micropropagated plants to ex vitro conditions. *Biologia Plantarum, 42*, 481–497.

Procaccia, S., Lewitus, G. M., Lipson Feder, C., Shapira, A., Berman, P., & Meiri, D. (2022). Cannabis for medical use: versatile plant rather than a single drug. *Frontiers in Pharmacology, 13*, 894960.

Raharjo, T. J., Eucharia, O., Chang, W.-T., & Verpoorte, R. (2006). Callus induction and phytochemical characterization of Cannabis sativa cell suspension cultures. *Indonesian Journal of Chemistry, 6*(1), 70–74.

Ren, M., Tang, Z., Wu, X., Spengler, R., Jiang, H., Yang, Y., & Boivin, N. (2019). The origins of cannabis smoking: Chemical residue evidence from the first millennium BCE in the Pamirs. *Science Advances, 5*(6), eaaw1391.

Richez-Dumanois, C., Braut-Boucher, F., Cosson, L., & Paris, M. (1986). Multiplication végétative in vitro du chanvre (Cannabis sativa L.). Application à la conserva-tion des clones sélectionnés. *Agronomie, 6*(5), 487–495.

Rotherham, D., & Harbison, S. (2011). Differentiation of drug and non-drug Cannabis using a single nucleotide polymorphism (SNP) assay. *Forensic Science International, 207*(1–3), 193–197.

Sallie, J. V., & Jones, O. (1969). The culture of shoot tips of hop (Humulus lupulus L.) to eliminate viruses. *Journal of Horticultural Science, 44*(3), 281–284.

Schachtsiek, J., Hussain, T., Azzouhri, K., Kayser, O., & Stehle, F. (2019). Virus-induced gene silencing (VIGS) in Cannabis sativa L. *Plant Methods, 15*, 1–9.

Senthil-Kumar, M., & Mysore, K. S. (2014). Tobacco rattle virus–based virus-induced gene silencing in Nicotiana benthamiana. *Nature Protocols, 9*(7), 1549–1562.

Shew, A. M., Nalley, L. L., Snell, H. A., Nayga Jr, R. M., & Dixon, B. L. (2018). CRISPR versus GMOs: Public acceptance and valuation. *Global Food Security, 19*, 71–80.

Shukla, M. R., Singh, A. S., Piunno, K., Saxena, P. K., & Jones, A. M. P. (2017). Application of 3D printing to prototype and develop novel plant tissue culture systems. *Plant Methods, 13*, 1–10.

Shukla, R., Handa, M., Singh, D. P., & Dhir, A. (2024). *Cannabis and its Derivatives: Guide to Medical Application and Regulatory Challenges*. Elsevier.

Sirikantaramas, S., Morimoto, S., Shoyama, Y., Ishikawa, Y., Wada, Y., Shoyama, Y., & Taura, F. (2004). The gene controlling marijuana psychoactivity: molecular cloning and heterologous expression of Δ1-tetrahydrocannabinolic acid synthase from Cannabis sativa L. *Journal of Biological Chemistry, 279*(38), 39767–39774.

Sirkowski, E. E. (2012). Marked Cannabis for indicating medical Marijuana. In: Google Patents.

Slusarkiewicz-Jarzina, A., Ponitka, A., & Kaczmarek, Z. (2005). Influence of cultivar, explant source and plant growth regulator on callus induction and plant regeneration of Cannabis sativa L. *Acta Biologica Cracoviensia: Series Botanica, 47*(2005), 145–151.

Smýkalová, I., Vrbová, M., Cvečková, M., Plačková, L., Žukauskaitė, A., Zatloukal, M., Hrdlička, J., Plíhalová, L., Doležal, K., & Griga, M. (2019). The effects of novel synthetic cytokinin derivatives and endogenous cytokinins on the in vitro growth responses of hemp (Cannabis sativa L.) explants. *Plant Cell, Tissue and Organ Culture (PCTOC), 139*, 381–394.

Spanò, R., Bottalico, G., Corrado, A., Campanale, A., Di Franco, A., & Mascia, T. (2018). A protocol for producing virus-free artichoke genetic resources for conservation, breeding, and production. *Agriculture, 8*(3), 36.

Taura, F., Morimoto, S., Shoyama, Y., & Mechoulam, R. (1995). First direct evidence for the mechanism of. DELTA. 1-tetrahydrocannabinolic acid biosynthesis. *Journal of the American Chemical Society*, *117*(38), 9766–9767.

Wahby, I., Caba, J. M., & Ligero, F. (2017). Hairy root culture as a biotechnological tool in C. sativa. *Cannabis sativa L.-Botany and Biotechnology*, 299–317.

Wang, M., Wang, Y.-H., Avula, B., Radwan, M. M., Wanas, A. S., van Antwerp, J., Parcher, J. F., ElSohly, M. A., & Khan, I. A. (2016). Decarboxylation study of acidic cannabinoids: a novel approach using ultra-high-performance supercritical fluid chromatography/photodiode array-mass spectrometry. *Cannabis and Cannabinoid Research*, *1*(1), 262–271.

Wang, P.-J., & Charles, A. (1991). Micropropagation through meristem culture. In P.-J. Wang & A. Charles (Eds.), *High-Tech and Micropropagation I* (pp. 32–52). Springer.

Wang, R., He, L.-S., Xia, B., Tong, J.-F., Li, N., & Peng, F. (2009). A micropropagation system for cloning of hemp (*Cannabis sativa* L.) by shoot tip culture. *Pakistan Journal of Botany*, *41*(2), 603–608.

Wielgus, K., Luwanska, A., Lassocinski, W., & Kaczmarek, Z. (2008). Estimation of *Cannabis sativa* L. tissue culture conditions essential for callus induction and plant regeneration. *Journal of Natural Fibers*, *5*(3), 199–207.

Wróbel, T., Dreger, M., Wielgus, K., & Słomski, R. (2022). Modified nodal cuttings and shoot tips protocol for rapid regeneration of *Cannabis sativa* L. *Journal of Natural Fibers*, *19*(2), 536–545.

Wu, H., Qu, X., Dong, Z., Luo, L., Shao, C., Forner, J., Lohmann, J. U., Su, M., Xu, M., & Liu, X. (2020). Wuschel triggers innate antiviral immunity in plant stem cells. *Science*, *370*(6513), 227–231.

4 Pharmaceutical Properties and Genetic Engineering for *Eucalyptus*

Status, Current Enchancements and Prospects

Rambod Abiri and Hazandy Abdul-Hamid**

*Correspondence authors

4.1 INTRODUCTION

Forests are complex ecological bio-systems beneficial for economic growth and sustainable development in various parts of the world (Oliveira, 2015). Forests absorb carbon dioxide (CO_2) from below and above ground, contribute to the gas in photosynthesis, and supply oxygen (Sampaio Filho et al., 2018). On the other hand, this natural gift provides timber, food and, medicines for millions globally while serving as a habitat for countless wildlife populations (Husen et al., 2021). Furthermore, forests reduce the risk of natural disasters, including drought, wildfire, landslide, and floods (Piao et al., 2022). Notwithstanding all the benefits that forests offer to human life and wildlife ecosystems, the increasing demand for shelter, timber, food, natural medicines, and wood endangers forest resources globally (Liu et al., 2018).

Eucalyptus (family *Myrtaceae*) species are evergreen, fast-growing, high-wood quality, and very large trees native to Australia. However, *Eucalyptus* species are also found in Indonesia, Papua New Guinea, China, India, South America, and the Philippines (Figure 4.1) (Calviño-Cancela et al., 2018).

Despite enduring drought stress, *Eucalyptus* has demonstrated its versatility by flourishing in various climates worldwide. Research has shown that these trees can pass their resilience to stress on to their offspring (Pritzkow et al., 2020). However, problems arise from climate change, which puts some *Eucalyptus* species at risk of harm and increasing mortality due to circumstances beyond their natural capability (Matusick et al., 2018). Stress responses vary among *Eucalyptus* species, highlighting the importance of genetic variety. It is critical to grasp these nuanced concerns to mitigate the impact of climate change on the health of eucalyptus trees (Pritzkow et al., 2020).

Because of specific characteristics like short fiber, low flexible strength, brightness, and appropriate surface appearances, *Eucalyptus* timber plays a vital role in furniture manufacturing, the particle board industry, and plywood products (Lahr et al., 2018; Popa et al., 2018; Morimoto et al., 2019). High-yielding and fast-growing *Eucalyptus* plantations (Amazonas et al., 2018), coupled with their strong adaptability to various environmental conditions and short rotation periods, suggest the considerable potential of this tree as a reliable alternative to meet the demand for wood and contribute to forest conservation (Girijashankar, 2011).

 DOI: 10.1201/9781003482659-4

FIGURE 4.1 Schematic graph of *Eucalyptus* genus distribution all over the world. The schematic graph is duplicated from: www.discoverlife.org/mp/20m?kind=*Eucalyptus*

There are limitations to conventional breeding procedures that cause a low success rate in *Eucalyptus* (Acquaah, 2015). Gene editing and gene transformation are examples of modern biotechnology that have solved these problems (Sharma et al., 2023). These innovative techniques empower scientists to manipulate *Eucalyptus* genes, enabling the creation of desired traits through the transfer of specific genes across kingdoms and genera, a capability not afforded by traditional methods (Cao et al., 2015; Emamalipour et al., 2020). These benefits are enhanced stress tolerance, larger fiber yields, herbicide resistance, and longer tree life. Compared to conventional *Eucalyptus* breeding programs, transgenic approaches can increase productivity, decrease resource extraction, and increase trees' time to reach maturity (Chang et al., 2018; Lebedev et al., 2020; Abiri et al., 2022).

Considering the numerous advantages of modern biotechnology, the primary aim of this chapter is to provide a comprehensive overview of the recent advancements in gene transformation and genome editing techniques, specifically within the context of *Eucalyptus* species. However, we first discussed the pharmaceutical aspects of Eucalyptus, highlighting its medicinal properties and applications. By reviewing this exciting research, we aim to shed light on cutting-edge developments and their potential applications in the genetic modification and genome editing of *Eucalyptus* trees.

4.2 MORPHOLOGICAL DESCRIPTION

As mentioned, *Eucalyptus* is a taxonomic genus belonging to the family *Myrtaceae*, encompassing a total of 900 distinct species and subspecies. This woody perennial with an evergreen canopy can quickly develop to enormous proportions, spanning the height spectrum from shrubs to tall trees (Coppen, 2002). *Eucalyptus* is indigenous to Australia and Tasmania and is recognized as the second most extensive genus following *Acacia* (Davis, 2002). Nearly all 700 recognized species of *Eucalyptus* trees are found in Australia. *Eucalyptus globulus*, a species native to Australia, has been widely introduced and farmed in Mediterranean and subtropical climates, as well as in Nigeria. It is now the most commonly cultivated species among the several Australian *Eucalyptus* species. *Eucalyptus* naturally can be found across a latitude range that spans from 7° N to 43° 39' S. Their native habitat is exclusively confined to the eastern region of the putative 'Wallace's line', which separates Indo-Malayan and Austro-Malayan living forms (Williams et al., 1997). *Eucalyptus* species exhibit a wide range of heights, varying from 30 to 200 feet. The Mountain Ash, which is the highest species of *Eucalyptus* tree, can grow to a height exceeding 301 feet. *Eucalyptus* oil, which is derived from the leaves and twigs of various *Eucalyptus* species, serves as a crucial and primary ingredient in pharmaceuticals, fragrances, soaps, and detergents. The majority of *Eucalyptus* species display leaf dimorphism, with distinct mature and juvenile leaves. During the juvenile phase, the leaves are positioned in pairs, without a stalk, and have an oval to spherical shape with a bluish-green color. As they reach maturity, the leaves typically take on a petiolate, pendulous, or lanceolate shape; they are also thick, stiff, extremely coriaceous, and have a cutinized appearance. The arrangement of veins in the leaves of *Eucalyptus* species is a significant factor in their identification. Typically, these veins exhibit a pinnate, oblique, or reticulate pattern (Hardel et al., 2011). The bark of the *Eucalyptus* tree can be classified into many varieties, such as smooth or rough, seasonal or permanent, or a combination of both. The age and species of a *Eucalyptus* tree determines its bark's shedding pattern, fiber length, color, thickness, hardness, and degree of furrowing (Williams et al., 1997). Annually, all species of *Eucalyptus* undergo the process of bark shedding, promptly after the demise of the outermost coat. Species such as *Eucalyptus sheathiana* Maiden (*E. sheathiana*) and *Eucalyptus cosmophylla* F. Muell. release the decaying outer layer by shedding it in the shape of sizable slabs, ribbons, or little flakes. Some species, such as *Eucalyptus leptophleba*, F. Muell. (*E. leptophleba*), *E. radiata*, and *Eucalyptus jensenii* Maiden (*E. jensenii*) possess the ability to store and accumulate desiccated dead bark. *Eucalyptus* species such as *Eucalyptus brachycalyx* and some more are commonly referred to as 'blackbutts' or 'half-barks' due to their tendency to retain dead bark on the lower portion of their trunks or stems (Khan et al., 2020). *E. globulus* exhibits a distinctive trait where the cells in its bark are capable of photosynthesis even when there is no foliage present. This ability enhances its capability to reabsorb and utilize internal carbon dioxide after experiencing partial defoliation (Eyles et al., 2009). This trait not only makes it possible for trees to thrive in less-than-ideal environments, but it also improves their resilience to leaf damage caused by factors such as grazing, forest fires, insect infestations, and leaf fungal pathogens (Saveyn et al., 2010).

4.3 CHEMICAL COMPOSITION

The leaves of *Eucalyptus* are the primary source of essential oil (EO), with the fruits, flowers, and stems being less frequently used. This oil is typically extracted through steam distillation or hydro-distillation. *Eucalyptus* essential oil (EEO) can also be extracted using a number of alternative methods, including Soxhlet extraction, solid-phase microextraction, microwave-assisted extraction, supercritical CO_2 extraction (Zhao et al., 2014). A significant amount of volatile organic compounds characterizes the EO profile of *Eucalyptus* (Bhuyan et al., 2018) EO are a blend of naturally occurring volatile bioactive substances that contain hundreds of distinct chemical elements, primarily sesquiterpenoids, monoterpenoids, phenylpropanoids, etc. The concentration of 1,8-cineole, also known as cineole or eucalyptol, is the primary factor determining the medicinal significance of *Eucalyptus*. Eucalyptus oil is mostly composed of 1,8-cineole, which makes up over 70% (v/v) of the overall oil. Other significant ingredients found with eucalyptol include limonene and α-terpineol. The extraction of *Eucalyptus* oil for medicinal use is typically done from *E. globulus*, *E. polybractea*, or *E. smithii*, as these species have a eucalyptol concentration over 70%. This high eucalyptol content has led to its inclusion in the pharmacopoeias of numerous countries such as United Kingdom, United States, France, Australia, Belgium, Germany, Japan, Netherlands, and China. In certain species such as, *E. polybractea* and *Eucalyptus bakeri* Maiden, eucalyptol makes up over 90% of the composition (Mishra et al., 2010). Extensive research has been conducted on *E. globulus*, *Eucalyptus* species. There are noticeable variations in the therapeutic capabilities of different species of *Eucalyptus*, as shown by phytochemical examination of their monoterpene profiles. According to the phytochemical examination of *E. grandis* oil, 95.95% of the essential oil's principal components are eucalyptol and 22 other chemicals (Sewanu, 2012). The main oil components and overall proportion of important Eucalyptus species are shown in Table 4.1.

TABLE 4.1
Some of the important *Eucalyptus* species with major constituent and total percentage

Species	Major component	Total Percentage	References
Eucalyptus smithii	1,8-cineole	84.27	(Chalchat et al., 1997)
Eucalyptus globulus	1,8-cineole	83.89	(Maciel et al., 2010)
Eucalyptus maideni	1,8-cineole	83.59	(Sebei et al., 2015)
Eucalyptus bicostata	1,8-cineole	81.29	
Eucalyptus sideroxylon	1,8-cineole	80.75	
Eucalyptus cineria	1,8-cineole	79.18	
Eucalyptus leucoxylon	1,8-cineole	77.76	
Eucalyptus camaldulensis	γ-terpinene	72.50	(Mubarak et al., 2015)
Eucalyptus tereticornis	β-pinene	39.00	(Ogunwande et al., 2003)
Eucalyptus grandis	α-pinene	30.40	
Eucalyptus dives Schauer	Piperitone	53.00	(Ebadollahi, 2013)
Eucalyptus elata Dehnh.	α-Phellandrene	35.20	
Eucalyptus citriodora	β-citronellal	71.77	(Maciel et al., 2010)

4.4 THERAPEUTIC APPLICATIONS OF *EUCALYPTUS* OIL

Since ancient times, the bark and leaves of several *Eucalyptus* species have been utilized as traditional medicines for treating various illnesses. The therapeutic benefits of *Eucalyptus* spp. are mostly attributed to the presence of 1,8-cineole, which is its primary ingredient. The following section explores the diverse medicinal uses of *Eucalyptus* in treating different diseases and disorders (Abiri et al., 2022). The several medicinal uses of *Eucalyptus* spp. are illustrated in Figure 4.2.

4.5 RESPIRATORY DISORDERS

Herbal medicines like cineole capsules and Myrtol standardised have gained significant attention in the last two decades because of their possible therapeutic advantages in treating different respiratory disorders. Multiple *in vivo* and *ex vivo* investigations indicate that Myrtol standardised has notable secretolytic and secretomotoric effect, leading to enhanced patency of the upper and lower airways. Consequently, this results in prompt alleviation of symptoms experienced by patients. This medication has been utilized to treat various acute and chronic infections of the upper and lower respiratory system, such as acute and chronic rhinosinusitis, acute and chronic bronchitis, and chronic obstructive pulmonary disease (COPD) (Rantzsch et al., 2009), because of its supplementary antioxidative, anti-inflammatory, and antibacterial characteristics. In Germany, Soledum™ capsules, which are soluble in the small intestine, contain 100 mg of 1,8-cineole and are utilized for the treatment of acute and chronic bronchitis, sinusitis, and other respiratory diseases. Eucalyptol's pharmacological effects on respiratory illnesses are due to its anti-inflammatory, anti-spasmodic, and mucus secretion inhibitory properties. These actions lead to the reduction of tumor necrosis factor-α (TNFα) and inflammation cytokines such interleukin-1β (IL-1 β) (Yu et al., 2018). In addition to a rise in the anti-inflammatory cytokine IL-10, the lung tissues also showed suppression of TNFβ and IL-1β and a reduction in the expression of p65, a component of NF-kB. Caceres (Caceres et al., 2017) elucidated that the heightened effectiveness of eucalyptol in humans can be ascribed to its exceptional receptiveness to Human TRPM8 channels compared to rodents. In conclusion, 1,8-cineol has the potential to be highly effective in the management of inflammatory diseases, including asthma, (COPD, sinusitis, and bronchitis.

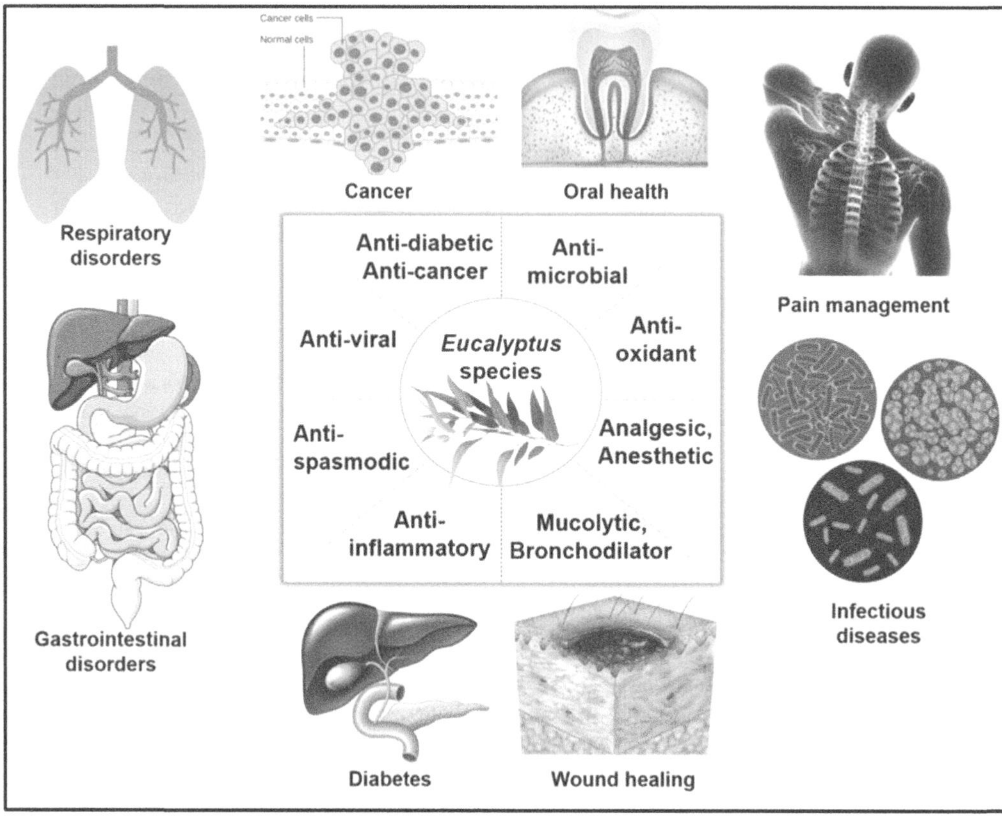

FIGURE 4.2 Various therapeutic applications of the genus Eucalyptus. (Chakraborty et al., 2020.)

4.5.1 Sinusitis

Sinusitis is the inflammation of one or more of the paranasal sinuses. This inflammation impairs the operation of the mucociliary transport, which is responsible for moving mucus. Sinusitis occurs when there is an excessive buildup of unusually thick mucus and inflammatory substances (Leung et al., 2008).. When culture results are lacking, medicines targeting the most common bacterial infections, such as *Streptococcus pneumoniae*, *Haemophilus influenzae*, *Staphylococcus aureus*, and *Moraxella catarrhalis*, have been typically recommended to 85–98% of sinusitis patients (Leung et al., 2008). The rapid and significant increase in multidrug resistance (MDR) among these prevalent pathogens is a significant public health issue. Therefore, there is an urgent need for alternative antimicrobial medicines to treat resistant pathogenic microorganisms. Research conducted by Pinheiro and his team, Pinheiro (Pinheiro et al., 2020) examined the effectiveness of *E. citriodora* EO in inhibiting the growth of several significant MDR bacteria, including *S. aureus* and strains that produce β-lactamase. According to the tests, *E. citriodora* EO successfully inhibited resistant strains, suggesting that it could be used as a substitute for these medications at a lower dose compared to what is recommended for resistant bacteria. Further research on *E. globulus* EO's effects on respiratory viruses and bacteria was conducted by Cermelli et al, 2008. The study showed that *Eucalyptus* oil was successful in combating *H. influ*enzae, *Haemophilus parainfluenzae*, and *Stenotrophomonas maltophilia*. However, *S. pneumoniae*, *Streptococcus agalactiae*, and *Klebsiella pneumoniae* were found to be resistant to EEO. The evidence suggests that EEO is an effective adjuvant therapy for sinusitis that can reduce the possibility of multidrug-resistant infections. Sudhoff et al. (Sudhoff et al., 2015) showed that eucalyptol decreases the excessive production of mucus caused by bacteria in sinusitis. This is achieved by reducing the number of goblet cells filled with mucin and the expression of the MUC2 gene in human nasal slice cultures. The authors' conclusion is that eucalyptol therapy can be remarkably effective in treating sinusitis.

4.5.2 Bronchitis

Bronchitis is a prevalent inflammation of the bronchi, which are the tubes responsible for transporting air to and from the lungs. Although viral infections are the most common cause of inflammation, antibiotics are frequently given to patients even though they don't need them (Singh et al., 2018). 1,8-cineole, the primary component of EEO, has been historically employed from ancient times for the treatment of bronchitis. Cineole functions by increasing the rate at which the cilia in the mucosal membrane beat. Additionally, it functions as a bronchodilator and anti-inflammatory medication (Fischer et al., 2013). *Eucalyptus globulus* oil was investigated by Lu et al. (2004) for its effects on bronchiolitis and mucin hypersecretion in rats with chronic bronchitis caused by lipopolysaccharides (LPS). The result indicated that the mucin content in bronchoalveolar lavage fluid (BALF) and the expression of MUC5ac in the trachea and bronchiole epithelium decreased significantly. This suggests that *E. globulus* EO has the potential to be used as a treatment for bronchitis.

4.5.3 Asthma and Chronic Obstructive Pulmonary Disease (Copd)

Asthma is defined by an inflammatory process that leads to increased sensitivity of the bronchial tubes and the ability to block the airways, which can usually be reversed. In order to study the effects of eucalyptol in bronchial asthma patients, Juergens et al.isolated monocytes from human blood (Juergens et al., 1998). According to the study, eucalyptol reduces inflammation by blocking two arachidonic acid metabolic pathways: LTB4 (leukotriene B4) and PGE2 (prostaglandin E2). This discovery marked the initial indication of the anti-inflammatory properties of 1,8-cineole, which generated significant enthusiasm among other researchers (Juergens et al., 1998). The following investigation focused on the possibility that 1,8-cineole could decrease the generation of polyclonal triggered cytokines by human unselected lymphocytes and monocytes activated with LPS (Juergens et al., 2020). According to the research, 1,8-cineol considerably reduced the production of cytokines in both monocytes and lymphocytes. In an effort to learn more about how eucalyptol affects COPD, Feitosa et al. (Kennedy-Feitosa et al., 2016) used a mouse model that had oxidative stress and inflammation brought on by cigarette smoke. It was demonstrated that eucalyptol decreased levels of cytokines and oxidative stress, suggesting that it could be used to treat COPD. The preventive impacts of EEO in LPS and *K. pneumoniae*-induced COPD were investigated by Wang et al. (2017). The result indicated that, as superoxide dismutase (SOD) activity increased, there was a marked decrease in the production of pro-inflammatory cytokines and the creation of malondialdehyde (MDA).

4.5.4 Sars-Cov-2 (Covid-19)

It is generally agreed upon that the COVID-19 pandemic is the most catastrophic global health disaster of the century and the most significant threat to humankind since the end of World War II. The new coronavirus known as SARS-CoV-2, which stands for severe acute respiratory syndrome, is responsible for this health crisis (Chakraborty et al., 2020). One of the most important enzymes in coronavirus (CoV) replication is Main protease, which transforms polypeptides into functional proteins (M^{pro}). Previous research has identified COVID-19's major protease (M^{pro}) as a potential and crucial target for inhibiting CoV replication. According to a recent in-silico study by Panikar et al. (2021), the effects of bio-compounds of *E. globulus* on M^{pro} were investigated using molecular docking. The authors found that eucalyptol had the lowest binding energy to M^{pro} and was non-toxic, suggesting that it may be useful as an inhibitor of COVID-19 and a therapy for the virus. Given the results presented, together with the previously established anti-inflammatory (Merad et al., 2020), antiviral (Usachev et al., 2013), mucolytic, and bronchodilatory (Juergens et al., 2020) benefits of *Eucalyptus*, it is crucial that additional studies be conducted on this matter.

4.5.5 Gastrointestinal Disorders

A prominent factor contributing to gastrointestinal (GIT) problems, such as peptic ulcers, is a spiral-shaped microaerophilic gram-negative bacillus, known as *Helicobacter pylori* (*H. pylori*). Infecting over 10% of the global population, this bacterium is responsible for about 95% of duodenal ulcers and up to 75% of stomach ulcers (Malfertheiner et al., 2009). *Eucalyptus* species, specifically *Eucalyptus camaldulensis* and *Eucalyptus torelliana* F. Muell, have been documented as being utilized in Nigeria for the treatment of GIT diseases (Gill, 1992). An animal model showed that the leaves extract of *E. torelliana* and *E. camaldulensis* significantly reduced stomach acid intake, making them highly effective for the treatment of gastric ulcers (Adeniyi et al., 2006). In accordance with the evidence, the *in vitro* susceptibility of six *H. pylori* strains to *E. camaldulensis* and *E. torelliana* extracts was investigated by Adeniyi et al. (Adeniyi et al., 2009). The researchers found that both extracts had effective anti-H. pyroli properties, with minimum inhibitory concentrations (MIC) ranging from 12.5 to 400 µg/mL. Additionally, a rat model with acetic acid produced ulcer was used to study the ulcer healing ability of methanolic extracts of *E. camaldulensis* and *E. torelliana*. The researchers showed that compared to the control group (distilled water and ranitidine), extracts from *E. camaldulensis* and *E. torelliana* effectively decreased the size of the ulcers by 50%. Based on their research, the authors have determined that some species of *Eucalyptus* can be utilised to effectively treat existing peptic ulcers (Lawal, 2014).

4.5.6 Wound Healing

For centuries, Indigenous people have utilised *Eucalyptus* oil-infused topical ointments in their traditional medicine practices to aid wound healing. In order to confirm the effectiveness of *Eucalyptus* in promoting wound healing, Velmurugan et al. (2014) conducted a study using Wister albino rats. They examined the wound healing properties of ethyl acetate and ethanolic extracts obtained from *Eucalyptus citriodora*. In their study, the authors found that scar epithelization was reduced when wound contraction was increased. The wounds in the group that received the ointment extracts healed completely, and the collagen and reticulin architecture returned to normal. In addition, the increased cross-linking of collagen fibers led to a noticeably higher collagen content, which in turn raised the skin breaking strength. Moreover, due to the higher protein content, an increase in the weight of dry granulation tissue were observed. According to the authors, during the proliferative stage, EEO demonstrates wound- healing activity through processes such as angiogenesis, collagen deposition, granulation tissue development, epithelization, and wound contraction. In a more recent investigation, laser Doppler flowmetry and histopathology data were used by Kadhim and Amer (Chandorkar et al., 2021) to assess the therapeutic potential of EEO on rats. The results showed that compared to the control rats, the rats treated with *Eucalyptus* had much smaller wounds and considerably higher blood perfusion. Furthermore, the *Eucalyptus* group showed significantly less inflammation and faster epithelialization.

There are a lot of variables that can slow down the complex wound healing process and one of the main factors contributing to chronic wounds is wound infection (Robson, 1997). Scientific investigations into the antimicrobial properties of EEO were conducted by Bachir and Benali (2012) in relation to two different pathogens: *S. aureus*, which causes toxic shock syndrome, post-operative wound infections, and food poisoning, and *Escherichia coli*, which causes urinary tract infections (UTI). The study found that EEO inhibited the growth of both bacterial strains in a dose-dependent manner. The antibacterial properties of *Eucalyptus* make it a highly efficient medicine for treating small cuts and wounds. Evidence suggests that EEO can be a valuable tool in wound management, particularly for chronic wounds where addressing inflammation and infection is crucial.

4.5.7 Oral Health

Dental plaque contains biofilms composed of cariogenic and periodontopathic microorganisms. Plaque bacteria, specifically *Streptococcus mutans* and *Streptococcus sobrinus*, are recognized for their ability to inhabit the surface of teeth and lead to the development of dental caries (Signat et al., 2011). The antibacterial properties of EEO have been widely acknowledged and utilized to combat various bacteria and fungus, including those that cause mouth infections. Rasooli et al. (2009) conducted a study on the antibacterial properties of *E. camaldulensis* EO against oral infections, both *in vivo* and *in vitro*. They also investigated its capacity to inhibit the production of biofilms. Researchers discovered that compared to chlorhexidine, eucalyptus oil has significantly stronger antibacterial and in vivo biofilm prevention effects. The findings provide support for the use of EEO as a viable option for dental caries therapy. Additionally, Goldbeck et al. (2014) tested two species of *Eucalyptus E. globulus* and *E. urograndis* for their antibacterial efficacy against *Streptococcus mutans* biofilm cells and planktonic cells. According to the researchers, *E. urograndis* oil took 50 minutes to have the same bactericidal effect as *E. globulus* oil, which only needed 15 minutes of contact to kill bacteria. The variation in antibacterial activity between the two species can be attributed to the larger concentration of 1,8-cineole in *E. globulus* compared to *E. urograndis*. Due to *E. globulus*'s higher concentration of 1,8-cineole content compared to *E. urograndis*, the two species' antibacterial activity differs. The findings support the idea that EEO is an effective adjuvant treatment for periodontal disorders. It has been demonstrated that a different species of *Eucalyptus*, called *E. galbie*, has antibacterial activity against the root canal infection-causing bacteria, *Enterococcus faecalis*. The researchers stated that *E. galbie* effectively eradicated over 90% of the bacterial cells in the root canal by a flushing mechanism, indicating its potential in the field of endodontics (Nourzadeh et al., 2017).

4.5.8 Cancer

Cancer is a broad term that includes more than 277 distinct types of cancer and is a significant contributor to global mortality (Hassanpour et al., 2017). East Asian countries including Thailand, India, China, and Japan have a long history of pharmaceutical and dietary therapy based on traditional medicinal herbs. Due to their high structural variety of phytochemicals, which can be attributed to their antioxidant, anti-carcinogenic, anti-mutagenic, and antiproliferative properties, they are frequently employed in cancer therapy. Research by Takasaki (Takasaki et al., 2000) examined the impact of euglobal-G1 derived from *E. grandis* leaves on a mouse model of two-stage skin carcinogenesis. Results showed that euglobal-G1 suppressed pulmonary tumorigenesis and the two-stage carcinogenesis promotion stages. Taheri et al. (Taheri et al., 2020) evaluated the antioxidant and anticancer effects of *E. camaldulensis* EO on a colorectal cancer cell line Caco-2. The study found that when the Caco-2 cell line was treated with *E. camaldulensis* EO, there was a clear relationship between the dosage and duration of treatment and the antioxidant and anticancer effects. These effects are likely related to the activation of apoptosis and the reduction of free radicals in colorectal cancer cells. Furthermore, the anticancer and antiproliferative properties of *E. globulus* EO were studied in relation to colon cancer cell lines. Based on the results of their study, *E. globulus* EO has the potential to be an effective adjuvant treatment for colon cancer since it inhibits cell proliferation by inducing apoptosis (Khazraei et al., 2021). Additionally, through the apoptotic pathway, *Eucalyptus benthamii* Maiden & Cambage EO has showed cytotoxic and anticancer properties (Döll-Boscardin et al., 2012). Additionally, a new light-based treatment called photodynamic therapy has shown great potential in treating cancer. This method includes light, molecular oxygen, and photosensitizers such aloe-emodin, berberine, hypericin, curcumin, and others. By triggering photochemical reactions within cancer cells, this therapeutic method produces reactive oxygen species (ROS), which ultimately kills the cells (Foresto et al., 2021). The utilization of *Eucalyptus* in conjunction with photodynamic therapy exhibits promising potential in reducing adverse effects and enhancing the effectiveness and specificity of cancer treatment.

4.5.9 Immunomodulation

The primary cellular components responsible for the immune response are the monocytic/granulocytic system and differentiated macrophages. These cells are crucial in recognizing and eliminating foreign entities, such as harmful bacteria. These cells are responsible for recognizing foreign microorganisms which triggers phagocytosis and leeds to lysosomal enzymes destroying the infections (Stafford et al., 2002). Yadav and Chandra (Yadav et al., 2017) conducted a study which found that EEO suppressed inflammatory mediators, as well as enhancing the process of phagocytosis and the elimination of mycobacteria in lung macrophages. Shao et al., (Shao et al., 2020) conducted a study utilizing a rat model to gain a more comprehensive understanding of how *Eucalyptus* affects respiratory immunomodulation. There was a substantial reduction in the ratio of natural killer (NK) cells and B cells/ B lymphocytes in the bloodstream. In addition, the researchers found that *Eucalyptus* extract at doses of 30 and 100 mg/kg increased CD8 (cluster of differentiation 8), which could enhance the immune system's ability to combat inflammation. The function of CD8 T cells in immunological memory is vital during initial viral infections of the respiratory system (Zhang et al., 2011). The stimulation of CD8 at a modest EO dose presented compelling evidence for the potential of eucalyptol to enhance immune cell memory. Additional research by Nakamura et al. (Nakamura et al., 2020) confirmed that EEO can inhibit mast cell degranulation and reduce allergic responses. In addition to reducing mast cell degranulation, EEO was found to inhibit cytokine production, lipid mediator synthesis, and leukotriene release. According to these results, the authors have verified that *Eucalyptus* oil can be significantly useful in the treatment of allergic dermatitis.

4.5.10 Diabetes Mellitus

Diabetes mellitus is a group of metabolic illnesses characterized by high blood sugar due to either the body's resistance to insulin, inadequate production of insulin, or excessive secretion of glucagon (Blair, 2016). Iran, Africa, and South America have all historically utilized *Eucalyptus* as a traditional medication to treat diabetes mellitus. Ahlem et al. (Ahlem et al., 2009) explored the efficacy of *E. globulus* in protecting rats from alloxan-induced oxidative stress. The Alloxan induces diabetes is attributed to the accumulation of reactive oxygen species (ROS), leading to toxicity in pancreatic cells, resulting in a decrease in insulin production and secretion (Sakurai et al., 2001). Researchers observed that, similar to the insulin-treated group, eucalyptol might normalize blood glucose levels. In addition, there was a notable decrease in the number of proteins in the liver, as well as a reduction in the level of TBARS (thiobarbituric acid reactive substances), which is an indicator of lipid peroxidation. Furthermore, the antioxidant activity of SOD (superoxide dismutase) was also reported to have decreased. Sugimoto (Sugimoto et al., 2003) reported that rats on a high-sugar diet had their intestinal fructose absorption inhibited and their adiposity reduced following administration of *E. globulus* leaf extract. Additionally, there was a dose-dependent reduction in the amounts of triacylglycerol in both the plasma and the liver. Guillen et al. (Guillén et al., 2015) demonstrated that in a nutritional mouse model, *E. tereticornis* extracts significantly reduced fasting glucose levels and increased glucose tolerance. The potential cause may be its involvement in liver pathways that promote the suppression of glucose production, such as the insulin pathway or the activation of proteins like AMPK (5′ adenosine monophosphate-activated protein kinase) that hinder the expression of crucial gluconeogenesis genes. *Eucalyptus camaldulensis*, a different type of *Eucalyptus*, has demonstrated its anti-diabetes properties by inhibiting the enzymes α-glucosidase and α-amylase through a process known as non-competitive inhibition. These enzymes, which are responsible for breaking down carbohydrates, are now being studied to create new anti-diabetic medication. they have the ability to reduce the rise in blood sugar levels by slowing down the absorption of glucose (Bastos et al., 2009).

4.6 IMPORTANCE OF GENE TRANSFORMATION FOR *EUCALYPTUS*

Due to the increasing demand for organic resources, *Eucalyptus* has become a popular choice for the food industry and pharmaceutical companies. In general, the extracted essential oil of *Eucalyptus* can be used in three groups: medicinal, industrial, and perfumery (Abiri et al., 2022). *Eucalyptus* is a rich source of essential oils (EOs), including terpenoids, phenolic, flavonoids and alkaloids with various biological activities such as anti-cancer, anti-viral, anti-bacterial, anti-microbial, and anti-fungal activities (Ishfaq et al., 2018).

4.6.1 Monoterpene Engineering In Eucalyptus

Monoterpenes, a group of terpenes with molecular formula $C_{10}H_{16}$ that possess two isoprene units, are among essential oils' most important chemical components in fragrant botanical specimens (Al-Maharik et al., 2022). These hydrophobic constituents often possess distinct and recognizable aromas, which play a significant role in determining the market value of ornamental plants. Furthermore, some monoterpenes serve as natural sources of fragrances and flavors in food and cosmetic products. Additionally, monoterpenes display a wide range of physiological performance, including antimicrobial, anti-cancer, antioxidant, and anti-inflammatory properties (Abiri et al., 2022). The main terpenoids observed in *Eucalyptus* are 1,8-cineole, α-pinene, β-pinene, and myrcene. Reportedly, methylerythritol 4-phosphate (MEP) and mevalonate (MVA) are two mainstream pathways of terpenoid biosynthesis in plants. Isoprene compounds (C5-isoprene), like isoprene phosphate (IPP) and its dimethylallyl diphosphate (DMAPP), are the initial precursors for the synthesizing of terpenoids in plants. DMAPP is catalyzed by 1-deoxy-D-xylulose-5-phosphate synthase (DXS) enzyme in the MEP pathway. Terpenoid synthases are vital enzymes responsible for the formation of monoterpene, sesquiterpene, and diterpene synthase (Zhan et al., 2022).

Recently, metabolic engineering, which aims to enhance monoterpene biosynthesis, has emerged as an intriguing research focus for modifying the composition of volatile products (Dudareva et al., 2008). The investigations mentioned above into monoterpene engineering have been applied to various plants, such as tomato, carnation, petunia, tobacco, and peppermint (Aharoni et al., 2005). Nonetheless, few reports have been focused on the engineering of monoterpene biosynthesis in woody plants like *Eucalyptus* (Ohara et al., 2010). The first report of engineered monoterpene biosynthesis in *Eucalyptus* plants was documented when Ohara and colleagues (Ohara et al., 2010) over-expressed the *Perilla frutescens limonene synthase* (*PFLS*) gene in *Eucalyptus camaldulensis* (Figure 4.3). In the transgenic *E. camaldulensis*, the limonene, α-pinene, and 1,8-cineole levels were significantly increased compared to the wild-type tree (Ohara et al., 2010). On the contrary, there were non-significant differences in transgenic *E. camaldulensis* carrying *the mangrin* gene in terms of α-pinene and 1,8-cineole concentrations compared to control plants (Gilani et al., 2017).

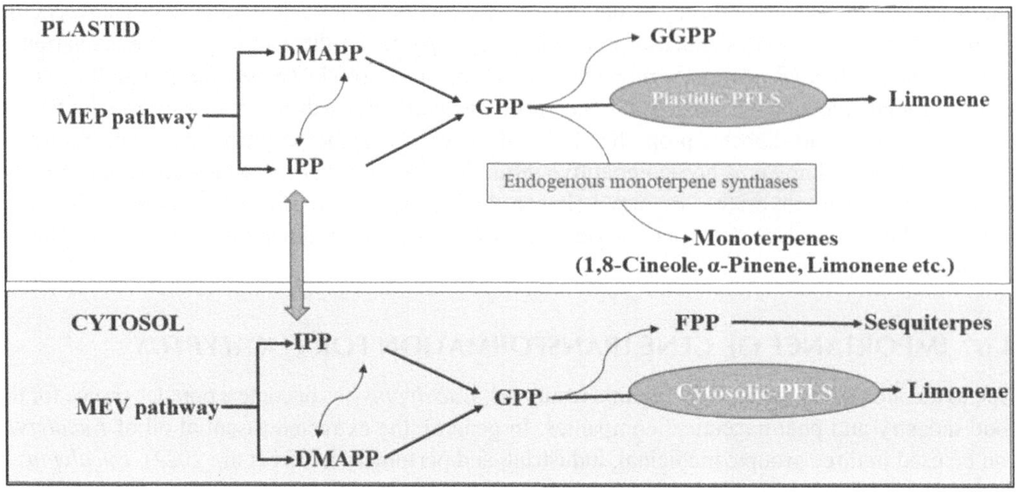

FIGURE 4.3 Schematic graph of overexpression of *Perilla frutescens limonene synthase* in the monoterpene mechanism of *E. camaldulensis*.

IMPROVING WOOD QUALITY AND QUANTITY

Biomass enhancement and wood quality improvement in forestry are among the most important characteristics of wood-based industries (Jarre et al., 2020). In the timber and paper industries, different traits such as lignin content, flexibility and strength of the wood can determine the wood quality (Wang et al., 2018). Similarly, the structure of the secondary cell wall is another vital trait that could be regulated by lignin, cellulose, and other cellulose polysaccharides. The mentioned secondary cell wall composition supports the cells by representing 90% of the wood-dry weight (Zhang et al., 2014). Besides, other compounds of the cell wall supplemented with proteins can directly affect the function and structure of the cell wall (Savidge, 2000). It has been shown that the composition changes can directly or indirectly influence the quantity and quality of wood. The mentioned changes are responsible for dissimilarities in wood properties across and within tree species (Spokevicius et al., 2006).

4.6.2 Enhancing Biomass Yield In Transgenic Eucalyptus

Biomass enrichment could be reached by preparing cellulose biosynthesis precursors and manipulating the *cellulose synthesis* (*CesA*) gene (Phitsuwan et al., 2013). It has been reported that cellulose fiber angle (Brennan et al., 2012) and lignin content are inversely associated with the content of cellulose (Shen et al., 2019). To increase the content of cellulose, different genes play essential roles in a series of chemical reactions (Wei et al., 2016). For example, *Aspergillus nuclease S(1) – Apium graveolens Dulce Group 1* (*cel1*) and Chitin-binding domain *(cbd)* genes increase cellulose content by encoding endo-1,4-β-glucanase and cellulose- binding, respectively (Levy et al., 2002). It has been confirmed that the *cel1* gene was involved in the enlargement of the cell wall, while the *cbd* gene was implicated in plant cell elongation *in vitro* conditions (Girijashankar, 2011). Similarly, native endo-glucanase genes are associated with the elongation of cell walls and tissues through the hydrolysis process of cellulose-xyloglucan links, permitting the cellulose chains to be transported easily (Girijashankar, 2011). Overexpression of the *Arabidopsis thaliana endo-glucanase cel1* and *cbd* genes in transgenic *E. grandis*, *E. camaldulensis*, and *E. hybrid* (*E. grandis* × *E. camaldulensis*) could increase the growth and development speed through the rapid elongation of cell walls (Shani et al., 2003). Additionally, there are a few reports on developing the *Eucalyptus* species by carrying different genes affecting the biomass enhancement pathway. For instance, the overexpression of *Enoyl-CoA Hydratase, Short Chain 1* (*EcHB1*) gene in *E. camaldulensis* positively functions in xylem development (Sonoda et al., 2009). Compared to wild-type plants, the results of developing transgenic plants carrying *EcHB1* showed an increasing trend in height (50%) and a greater length of fiber (50%). Likewise, the growth of stems, roots, and leaves has been significantly increased in transgenic plants. Furthermore, the lower content of hemicellulose and acid-soluble lignin has been observed in transgenic plants compared to wild-type plants. The above results highlighted that metabolic flexibility could be implemented in the transgenic plant improvements to the biosynthesis of the xylem cell wall. It may help make some advantages and modifications in the process of growth that can present prolonged structural integrity to the woody perennials (Sonoda et al., 2009). Gibberellins (GAs) are crucial in regulating plant development, wood formation, and stress responses in perennial woody plants. In a study conducted by Wu et al. (Wu et al., 2023), an extensive and methodical investigation was carried out on the genes associated with GA metabolism and signaling. The study successfully identified the regulatory functions of *GA20ox* and *GA2ox* in *Eucalyptus* plant growth, root initiation and growth, stress tolerance, and xylem development. The findings of this study have significant implications for molecular breeding strategies aimed at producing *Eucalyptus* cultivars with high yield and enhanced resistance to stress (Wu et al., 2023).

4.6.3 Modification of Lignin Pathway

Lignin is a complex and irregular phenolic heterogeneous group with a polymeric constituent (Austin et al., 2010). The wood component can generate the rigidity structure of cell walls (Weng et al., 2010), produce a physio-chemical obstacle layer against pathogens (Rajendran Ramjegathesh et al., 2015), and increase water permeability potential to xylem vessels (De Boer et al., 2003). However, high lignin content is an adverse trait for paper manufacturing due to the residual lignin in the wood fibers, which decreases and discolors the pulp brightness (Baucher et al., 2003). To overcome the mentioned obstacle in paper industries, the lignin contents should be removed during the pulping process (Huang et al., 2019). This method should be a non-destructive protocol without negatively influencing the cellulose polysaccharides using bleaching agents such as sodium sulfide and sodium hydroxide. However, using the mentioned chemical agents reduces the strength of cellulose fiber because of their chemical action, and the agents also pose a hazardous effect on the environment (De Rosa et al., 2011).

Genetic engineering of *Eucalyptus* can help researchers and industries overcome the mentioned barriers by modifying the level of lignin composition or reducing lignin content (Chang et al., 2018). It has been shown that sinapyl alcohols, coniferyl, and *p*-coumaryl can develop different lignin propyl phenol units such as syringyl (S), guaiacyl (G), and *p*-hydroxyphenyl (H), respectively. The lignin polymer units are connected by a chain of carbon-carbon and ether links (Lapierre et al., 1999). The presence of β-O-4-linkages in the lignin-rich S unit confers that this type of lignin is more appropriate than lignin-rich G units due to carbon-carbon bonds. Indeed, carbon-carbon linkage units are more tolerant of chemical delignification procedures than susceptible β-O-4-linkages (Lapierre et al., 1999). In the tree improvement field using the gene transformation method, gene coding and regulatory sequences for various monolignol biosynthetic enzyme pathways are of more interest (Wang et al., 2018). The mentioned enzymes in the pathways of monolignol biosynthetic are phenylalanine ammonia-lyase (PAL), 4-coumarate: coenzyme A ligase (4CL), ferulic acid 5-hydroxylase (F5H), caffeic acid *O*-methyltransferase (COMT), cinnamoyl-coenzyme A reductase (CCR), caffeoyl-coenzyme A *O*-methyltransferase (CCoAOMT), coumaroyl-coenzyme A 3-hydroxylase (CCoA3H), cinnamate 4-hydroxylase (C4H), coumarate 3-hydroxylase (C3H), and cinnamyl alcohol dehydrogenase (CAD) (Tu et al., 2010). In addition to the above, the transcription regulatory/ factor sequences such as *lim* and *myb* genes are applied in tree modification projects (Wu et al., 2019).

So far, a few transgenic types of research have been reported to improve the lignin pathway in *Eucalyptus*. For example, overexpression of *Escherichia coli beta-glucuronidase* (*uidA*), *Neomycin phosphotransferase* (*nptII*), and *harboring aspen C4H* (*cinnamate-4-hydroxylase*) altered lignin compositions in transgenic *E. camaldulensis* in the orientation of anti-sense and sense (Chen et al., 2001). Similarly, the overexpression of the *cad* enzyme (*cinnamyl alcohol dehydrogenase*) gene in *E. urophylla* × *E. grandis* (*Eucalyptus* hybrid) altered the lignin composition (Tournier et al., 2003). In the genome-engineered *Eucalyptus*, a significant inhibition has been observed for the activity of the *cad* gene. Likewise, the mentioned results confirmed the residual cad activity (roughly 22% and 26%) in the favorable lines under the greenhouse. Similar to the mentioned results, a reduction in the activity of the *cad* gene has been observed in 32% of engineered *E. camaldulensis* (Valério et al., 2003).

Notwithstanding the mentioned changes, and compared with the control *Eucalyptus* plant, no changes were reported in the activity *cad* or profiles lignin in engineered *Eucalyptus* after ten months (Valério et al., 2003). Up until now, the activity of two different transcription factors has been reported to modify lignin in *Eucalyptus* plants. For example, in an attempt to modify the lignin level, the anti-sense construct of tobacco transcription factor *Ntlim*1 was transformed into the genome of *E. camaldulensis.* This transcription factor is able to bind to phenylalanine ammonia lyase (PAL box sequence). As a consequence, *Ntlim*1 suppressed the transcription activities of a few genes active in phenylpropanoid mechanisms, such as *cinnamyl alcohol dehydrogenase* (*CAD*), *hydroxycinnamate-CoA ligase* (*4CL*), and *phenylalanine ammonia lyase* (*PAL*) (Valério et al., 2003). The stated changes in the pathway caused a down-regulating trend in the gene (s) active in lignin biosynthesis. As a result, the lignin content was reduced between 20–29% in cell wall residues of stem xylem tissues (Girijashankar, 2011). Likewise, a lignin content reduction was observed when *E. camaldulensis* was engineered with *LIM domain transcription factor*, *Ecliml,* extracted from the same *Eucalyptus* (Kawaoka et al., 2006).

In the lignin biosynthesis pathways, *Eclim1* is able to bind to the *cis*-acting constituent of the PAL box sequence. As a result of the mentioned gene transformation, a reduction was observed in the lignin biosynthesis genes, such as 4-hydroxycinnamate CoA ligase (*4CL*), *C4H*, and *PAL* (Kawaoka et al., 2006). Investigation into the structure of *Eucalyptus* identified that this tree can be transferred by *Agrobacterium rhizogenes* (a soilborne Gram-negative bacterium). Generally, transferring plants with *A. rhizogenes*, called a hairy root culture (HRC) system, could develop the plants' traits, focusing on accumulating secondary metabolites and other chemical compounds. It has been observed that using *A. rhizogenes* and *Eucalyptus cinnamoyl-CoA reductase1* (*EgCCR1*) gene in *E. grandis* increased the of lignin biosynthesis enzymes and led to the appearance of a thinner cell wall (Plasencia et al., 2016).

Interestingly, using a *Eucalyptus* transcription factor (synthetic and standardized) and a promoter panel could regulate secondary cell wall regulation in *Eucalyptus* (Hussey et al., 2019). Recent research has centered on altering the lignin pathway in *eucalyptus* trees. The effectiveness of enzymatic hydrolysis of lignocellulosic biomass can be enhanced by modifying *Eucalyptus* lignin for dynamic adsorption with pH-dependent lignocellulose-degradation enzymes, as demonstrated by Mou et al. (Mou et al., 2021).

4.6.4 Improvement of Micro Fibrillar Angle (MFA)

It has been reported that one of the most important characteristics of wood quality is the microfibrillary angle (MFA). MFA trait determines the wood and fiber elasticity and/or stiffness in the microfibrillar orientation of secondary fiber cells (Long et al., 2000). Interestingly, the higher mechanical strength of fibers has been observed with lesser cellulose MFA and vice versa. So far, the function of different genes in the MFA mechanism has been observed. For example, the role of *E. grandis β-tubulin* gene (*EgrTUB1*) was highlighted in the orientation of cellulose microfibrils in *E. grandis* secondary fiber cell walls (Spokevicius et al., 2006). The mentioned findings showed that *EgrTUB1* expression is correlated with MFA and causes changes due to stable gene engineering with *EgrTUB1* (Spokevicius et al., 2006). Other reports also confirmed the function of other genes, which can affect the microfibrillar orientation, like *CCR* and *FRA1. The FRA1* gene has an N-terminal microtubule-binding motor domain, and this gene can code a kinesin-like protein.

On the other hand, the *CCR* gene, which plays a vital role in the lignin biosynthetic mechanism, code cinnamoyl CoA reductase protein (Zhong et al., 2002). In *E. nitens*, the allelic variation of *CCR* has been linked with MFA variation (Thumma et al., 2005). It has been reported that *Eucalyptus* genetic modification with *CCR* and *FRA1* genes could improve the fiber's mechanical strength by reducing cellulose MFA. In addition, Fasciclin-like arabinogalactan (FLAs) proteins are important and specialized *Eucalyptus* proteins in the cell wall structure and stem biomechanics (MacMillan et al., 2010). According to a recent study, it has been found that the genetic enhancement of *E. nitens* with regards to the selection of characteristics for pulpwood does not appear to have had any negative impact on MFA (microfibril angle) and consequently, the stiffness of the timber. These findings indicate the potential for enhanced MFA in solid wood products due to selection for greater kraft pulp yield (KPY) (Rocha-Sepúlveda et al., 2021).

4.7 ADVANCES IN FERTILITY

Although there are no adequate studies on the advances in the fertility of *Eucalyptus* using genetic engineering, recent observations confirmed the effect of some genes on the plant's fertility. Accordingly, the transgenic *E. occidentalis* (an early flowering *Eucalyptus* model) and tobacco showed 100% efficacy in preventing viable pollen formation by overexpression of the *uracil phosphoribosyltransferase (barnase)* gene. The experiment also demonstrated complete pollen ablation in 90% of transgenic lines (Girijashankar, 2011). In another observation, overexpression of the *FLOWERING LOCUS T (FT)* gene induced flowering in the *Eucalyptus grandis* × *urophylla* hybrid (SP7). This result may positively affect the acceleration of genetic studies and breeding programs since the transgene can be easily segregated away to restore standard form, growth, and progeny (Klocko et al., 2016). In recent research, the authors examined potential alternatives to the existing procedures for acquiring new clones of *eucalyptus* (*Eucalyptus* spp.) based on a genetic improvement strategy that offers improved efficiency. Additionally, the feasibility of obtaining hybrid seeds was explored. The study focused on utilizing inbred progenies to enhance the species. Self-fertilization increases the number of undesired alleles exposed from heterozygotes, improving selection efficiency (Ramalho et al., 2022).

4.8 ROOT DEVELOPMENT

Root architecture development can be one of the most important characteristics of producing commercial *Eucalyptus*. The poor root system could limit *Eucalyptus* clonal production. In a hairy root culture system with *Agrobacterium rhizogenes,* the root architecture of *E. nitens*, *E. dunnii*, and *E. grandis* has been improved significantly (Macrae et al., 1993). The shoot of transformed *Eucalyptus* raised into the free hormone media and improved extensive lateral branching and growth rate. Nonetheless, the typical hairy root did not grow well in the engineered plants in the soil, but the secondary root structure was established alongside a stable root plug (Macrae et al., 1993). Similar to the above results, the developed root was observed when *E. grandis* × *E. urophylla* hybrids were engineered carrying the *Agrobacterium* strain. It has been shown that the mentioned hybrid is a hard-to-root genotype. Therefore, their root-ability was increased (Machado et al., 1997). In a recent study, researchers investigated the root traits of *E. pellita* seedlings derived from two distinct propagation methods, namely seed and stem cutting. The study aimed to compare the root development of these two propagation types under three varying nitrogen concentrations. The researchers discovered that *E. pellita* seedlings obtained through stem cuttings exhibited better growth, both above and below ground, compared to seedlings obtained from seed propagation (Hassan et al., 2021).

4.9 IMPROVEMENT OF ABIOTIC RESISTANCE IN EUCALYPTUS

So far, different attempts have been made to improve *Eucalyptus'* tolerance against various abiotic stresses such as cold, drought, acidity, and salinity. For example, the tolerance of *Eucalyptus* has been increased against salinity by introducing the *codA* gene (Yamada-Watanabe et al., 2003). It has been reported that carrying the *codA* gene in *Eucalyptus* causes the coding of choline oxidase, which leads to converting choline (an essential nutrient) to glycine betaine (organic osmolytes). Accumulation of this amphoteric compound in engineered *Eucalyptus* displayed tolerance to low and high temperatures and salinity. Similarly, modifying the *Eucalyptus hybrid* (*E. urophylla* × *E. grandis*) with a mitochondrial citrate synthetase gene caused improved phosphate uptake in acidic soils. In the mentioned modified *Eucalyptus*, a five-fold increase in the activity of enzymes caused a similar growth with colloidal aluminum phosphate compared to sodium phosphate (Girijashankar, 2011).

On the other hand, expression of four *bacterial choline oxidase A* (*codA*) genes, i.e., *codAH-1*, *codAH-2*, *codAN-1*, and *codAN-2*, showed no significant biodiversity in transgenic *E. camaldulensis* clones, compared to non-transgenic plants (Tran et al., 2018). Using the *dreb1* gene to modify *Eucalyptus* resulted in plants that could withstand drought and salinity stress. Functionally, this transcription factor regulates downstream pathways active in abiotic stresses in the genome of *Eucalyptus* (Kondo, 2002; Ishige, 2004). It has been reported that the expression of the *RNA-Binding-Pro*tein (*McRBP*) gene, which was isolated from *Mesembryanthemum crystallinum*, improved the tolerance of *Eucalyptus camaldulensis* Dehnh under salinity stress, compared to control plants (Tran et al., 2019). The expression of the *Mangrin* gene in *E. camaldulensis* harboring leads to abiotic stress tolerance under moderate and severe salinity levels (Yu et al., 2013). The overexpression of two transcription factor genes (*EguCBF1a/b*) increased the cold stress tolerance in transgenic *E. gunni* by binding to C-repeat binding factor- (CBF-) dependent signaling pathway (Navarro et al., 2011).

Furthermore, the overexpression of mentioned genes retarded growth and better water retention capacity, increased the accumulation of anthocyanins, accumulated wax deposition on the cuticle, diminished leaf area and cell size and reduced the level of stomatal density under cold stress (Navarro et al., 2011). In another investigation, cold stress tolerance was confirmed using the *CRT-repeat binding factors* (*CBFs*) gene in the *E. urophylla* x *E. grandis* hybrid. In addition to the above improvement, the ratio of syringyl/guaiacyl (S/G) and lignin content significantly increased in transgenic plants.

The recent investigation conducted by Chen et al. (2021) examined the impact of cold stress on the regulatory network of long non-coding RNA in *E. urophylla*. The research revealed that some long non-coding RNAs exhibited differential expression in response to cold stress. These identified RNAs hold promise as prospective targets for enhancing cold tolerance in *Eucalyptus* (Chen et al., 2021). Several researchers conducted a thorough and organized investigation into the genes associated with GA metabolism and signaling. They successfully determined the involvement of *GA20ox* and *GA2ox* in regulating plant development and adaptation to abiotic stress in *Eucalyptus*. This discovery has the potential to significantly assist molecular breeding efforts aimed at producing *Eucalyptus* cultivars that are both high-yielding and resistant to stress (Wu et al., 2023). Chaín et al. (2020) examined the effects of beneficial bacteria incorporated into a superabsorbent polymer on the growth promotion and drought protection of *E. grandis* seedlings. The researchers discovered that the chosen bacteria had the potential to enhance the early growth of *E. grandis* and may be conveniently introduced during transplantation by utilizing an acrylic-hydrocellulosic SAP (Chaín et al., 2020). A recent study has elucidated the significant contribution of taxol in enhancing the drought tolerance of *E. grandis* and its medicinal oil characteristics. Promoting sustainable agricultural and forestry methods necessitates emphasizing the inherent tolerance of trees during their vulnerable early stages. The results emphasize the need to conduct sophisticated scientific investigations to reveal the hidden potential of resilient trees such as *E. grandis* (Xu et al., 2023).

4.10 IMPROVEMENT OF BIOTIC RESISTANCE IN EUCALYPTUS

The utilization of phenotypic selection in conventional tree breeding has proven to be an effective strategy for achieving resistance to pests and pathogens. One instance observed in Africa involves crossbreeding *E. grandis* and *E. urophylla*, resulting in enhanced resistance against the fungal disease *Chrysoporthe austroafricana* (Wingfield, 2003). The regional heritability mapping (RHM) method identifies and quantifies the unexplained variability in genome-wide association studies (GWAS). This approach involves the analysis of small segments of the genome, which combines the influence of both common and rare single-nucleotide polymorphism (SNP) variants across individuals. By doing so, RHM aims to estimate the extent to which these specific genomic regions contribute to the observed variance in traits (Quezada et al., 2023). Resende et al. (Resende et al., 2017) conducted a comparative analysis of RHM and GWAS methodologies in the context of *Eucalyptus*. The objective was to evaluate the effectiveness of both approaches in mapping traits related to wood characteristics and the *Puccinia psidii* disease symptoms. The RHM method was deemed superior due to its ability to uncover more genomic areas to examine for potential candidate defense genes. Furthermore, the authors propose that the genomic architecture uncovered by the RHM-QTLs can potentially improve genomic selection models (Resende et al., 2017).

When a lack of inherent genetic diversity confers resistance to a specific biotic stress, the utilization of genetic engineering techniques may become imperative. Developing resistance in engineering involves several key stages, namely gene identification, candidate gene prioritization, genetic modification, and subsequent evaluation. The identification of genes in forest trees has been facilitated by the utilization of omics methodologies (Naidoo et al., 2019). Several research have been conducted on *E. grandis*, Chrysoporthe austroafricana, *E. nitens,* and *Phytophthora cinnamomi* (Mangwanda et al., 2015). Proteomics provides an additional layer of functional significance to genes implicated in plant defense mechanisms. In the study conducted by Chen et al. (2015) and Zwart et al. (2017), the iTRAQ (Isobaric Tags for Relative and Absolute Quantitation) technique was employed to investigate the response of *Eucalyptus* plants when exposed to the pathogens *C. austroafricana* and *Calonectria pseudoreteaudii* (Chen et al., 2015; Zwart et al., 2017). The study conducted by Mangwanda et al. (2015) revealed that while interacting with *C. austroafricana*, there was a more significant presence of proteins related to cell death, salicylic acid signaling, and systemic resistance, as contrasted to the findings of the transcriptome analysis. Therefore, proteomics has the potential to offer new perspectives on plant-microbe interactions due to the absence of a direct correspondence between transcript levels and protein abundance.

In addition, interactions between plants are of significant importance in the context of defense signaling against pathogens (Subrahmaniam et al., 2018). *Eucalyptus* trees also employ similar signals to prepare neighboring plants for potential threats. For instance, when *Ctenarytaina eucalypti* insects feed on *Eucalyptus* trees, it triggers the release of volatile chemicals that enhance the production of defense-related substances in nearby plants that have not been physically damaged (Christian et al., 2012). The availability of diverse omics data enables the emergence of systems biology perspectives on organisms. Mizrachi et al. (Mizrachi et al., 2017) propose a framework that connects molecular interactions and complicated features (Mizrachi et al., 2017). Tree breeding programs are crucial in establishing structured populations for various investigations (Mphahlele et al., 2021). The molecular pathways that underlie wood characteristics in *Eucalyptu*s were characterized in this manner (Mizrachi et al., 2017), with subsequent studies in systems genetics focusing on the interactions between trees and biotic stress. The increasing abundance of omics data about trees in the context of pest and disease challenges necessitates the simultaneous establishment of resources to facilitate accessibility for the scientific community. The TreeGenes database, as described by Wegrzyn et al. (Wegrzyn et al., 2008), serves as a comprehensive genomics resource for forest trees. It facilitates the integration and analysis of various genetic information, including genome sequences, transcriptomes, genetic maps, molecular markers, and phenotypic data. This resource encompasses a wide range of 1,749 species. The Plant Genome Integrative Explorer (Sundell et al., 2015) allows researchers to acquire genomic and transcriptome data of many plant species, including *Eucalyptus* (Wegrzyn et al., 2008).

To increase the tolerance/resistance of *Eucalyptus* against pests, the *Cryptochrome Circadian Regulator 1* (*cry1*) gene, which was extracted from *Bacillus thuringiensis,* showed an important function by coding crystalline toxin proteins. Additionally, the expression of *insecticidal cry3A* and *bar* (the herbicide marker gene in the construction of vector against ammonium glufosinate) genes increased the resistance of *E. camaldulensis* under the damage caused by *Chrysomelid beetle* attack (Harcourt et al., 2000). Herbicide and pest tolerance in transgenic *Eucalyptus* could provide better weed and insect control management in the plantation of this tree. In parallel with the above results, biotic stress tolerance was observed in transgenic *E. urophylla* against the bacterial wilt disease caused by the *Pseudomonas solanacearum* (Shao, 2002). Reportedly, the expression of the *N-acyl-homoserine lactones* (*AHLs*) gene in *E. urophylla* × *E. grandis* significantly increased biotic (*Ralstonia solanacearum*) resistance, compared to non-transgenic plants (Ouyang et al., 2016). Similar to the above experiments, an efficient protocol was established to transfer the *Raphanus sativus antifungal peptide* 2 (*Rs-AFP2*) gene into the genome of *E. urophylla* using the *Agrobacterium tumefaciens*-mediated transformation system. The mentioned experiment showed that transgenic *Eucalyptus* was resistant against *Phytophthora capsici* (Ouyang et al., 2012).

Companies in forest biotechnology have been cultivating genetically modified forest tree species, specifically *Eucalyptus*, to augment biomass production and fortify resistance against various environmental pressures (Du Toit et al., 2020; Teshome et al., 2020; Bhoi et al., 2022). ArborGen Inc, has achieved the development of transgenic freeze-tolerant Eucalyptus. According to Avisar et al. (2013), FuturaGene Ltd. has successfully created and brought to market the initial genetically modified *Eucalyptus* tree, which exhibits a 20% increase in biomass. Furthermore, the company is advancing efforts to commercialize disease-resistant variants of these trees (Avisar et al., 2013). Furthermore, the application of genome editing techniques exhibits significant potential for enhancing the efficiency of breeding practices in forestry, as highlighted by Bewg and Tsai (2018). Hence, upon identifying genes or metabolic pathways that are pivotal for disease resistance, it will become feasible to modify these processes efficiently and generate plants that may be readily utilized by the industry in a more expeditious manner than previously possible (Bao et al., 2019).

4.11 FACTORS AFFECTING EUCALYPTUS MICROPROPAGATION EFFECTIVENESS

Like other plant's genetic engineering platforms, the initial phase of *Eucalyptus* gene transformation is tissue culture. Generally, plant tissue culture media, vitamins, plant growth regulators, nutrients (macro- and micro-nutrients), carbohydrate resources, and solidified agents are the most important factors in the *in vitro* culture of *Eucalyptus*. To have an efficient process of *Eucalyptus* tissue culture, different protocols have been optimized (Abiri et al., 2020).

There are a number of variables that can affect the efficiency of micropropagation. In tissue culture studies, chemical substances known as growth regulators or plant growth hormones are essential in facilitating shoot proliferation and root formation. Numerous studies have indicated that shoot production rate and quality can be considerably influenced by the kind and concentration of cytokinins and auxins present in the culture media (de Oliveira et al., 2022; Tannure Faria et al., 2022). Indole-3-acetic acid (IAA) is an endogenous plant hormone that functions as a natural auxin, while indole-3-butyric acid (IBA) serves as an analog of this plant hormone. Indole-3-acetic acid

(IAA) is an endogenous plant hormone classified as natural auxin, while indole-3-butyric acid (IBA) serves as an analog of this plant hormone. The synthesis of both IAA and IBA can occur either by a tryptophan-dependent or tryptophan-independent mechanism (Teale et al., 2006; Zhao, 2014), and both can be swiftly digested in *Eucalyptus* plant tissues. Plants store auxins as either IAA or IBA, which can be converted into IAA when needed due to auxin's critical role in regulating the process of cell division and several stages of plant development, including plant elongation (Strader et al., 2011; Frick et al., 2018).

Moreover, the development of the culture medium, including the composition and concentrations of minerals, vitamins, and carbon sources, influences the proliferation and differentiation of explants. In order to initiate shoot growth, explants are typically cultured on a semi-solid medium containing gelling agents such as gelrite, agar, or phytagel. The pH of this medium is usually adjusted to fall within the range of 5 to 6. Furthermore, the utilization of liquid media for the induction and propagation of nodal segments and shoots has been documented (Bunn, 2005; Kaur, 2017). Nevertheless, there is a variance in the type of media used and the plant's response among different species of *Eucalyptus* and their hybrids. Murashige and Skoog media (varying strengths ½, ¼), JADS (Juan Antonio Diva and Silvian) medium, WPM (Woody Plant Medium), and DKW (Driver and Kuniyaki Walnut) medium have all been utilized as underlying media in earlier attempts at *Eucalyptus* micropropagation studies (Woodward et al., 2005). Moreover, the growth of plantlets in *vitro* is affected by the type of medium used, which can be either a liquid suspension medium or a semi-solid medium (Gomes et al., 2003; Brondani et al., 2012). It has been reported that *Eucalyptus* shoot multiplications are more efficient in liquid media compared to semi-solid media (Chen et al., 2001).

Organic elements like amino acids and complex organic compounds can stimulate shoot regeneration and plantlet growth. In contrast, inorganic elements like macronutrients and micronutrients are essential to a plant's health and growth. Macro- and micronutrients, including calcium (Ca), nitrogen (N), phosphorus (P), and boron (B), are crucial for plant growth; in order to encourage cell multiplication and organ development from the shoot, they have been used in *in vitro* cultures. The efficient proliferation and growth of plants in *vitro* necessitate a careful equilibrium and concentration of organic and inorganic elements (Ngomuo et al., 2014). Depending on the particular requirements of various plant species and their developmental stages, the medium's composition can be modified. Micronutrients are plant-essential elements with concentrations lower than 0.5 mM/L, whereas macronutrients are those with concentrations more than 0.5 mM/L (Pérez-Tornero et al., 2000; George et al., 2008; Al-Aizari et al., 2020).

In addition, light is a crucial environmental component that significantly impacts the efficacy of micropropagation. The appropriate light intensity and photoperiod levels can potentially influence the shoot initiation, elongation, and rooting (Shen et al., 2020). Another factor influencing plant development and differentiation is the light spectrum and, more specifically, the red to far-red light ratio. In order to maximize the efficiency of shoot proliferation and the creation of healthy plantlets, it is essential to optimize these aspects in micropropagation protocols. Numerous hybrids have been investigated to determine the effects of light and radiation, including fluorescent lamps, white, red, blue, green, yellow, purple, and red–blue LEDs, on the micropropagation of *Eucalyptus* (Souza et al., 2022; do Prado Frade et al., 2023). In addition, carbohydrates play a crucial role as biomolecules by serving as a source of energy and carbon for cellular development. Both reducing and nonreducing sugars, like glucose, fructose, sucrose, and galactose, have been used in micropropagation. Due to its simplicity in translocating through plant tissues, sucrose remains the sugar of choice for propagating *Eucalyptus* plants (Joshi et al., 2003; Gago et al., 2021; Gago et al., 2022).

Notwithstanding all the above factors in the tissue culture of *Eucalyptus*, different types of explants are still the most important targets in the genetic transformation studies (Abiri et al., 2020). Reportedly, hypocotyls, cotyledons, *in vivo* wood-producing stem segments, young seedlings, leaves from *in vitro* germinated seedlings, zygotic embryos, protoplasts, nodal stem segments of mature trees have been applied in the genetic transformation of plants (Aggarwal et al., 2016). Plant

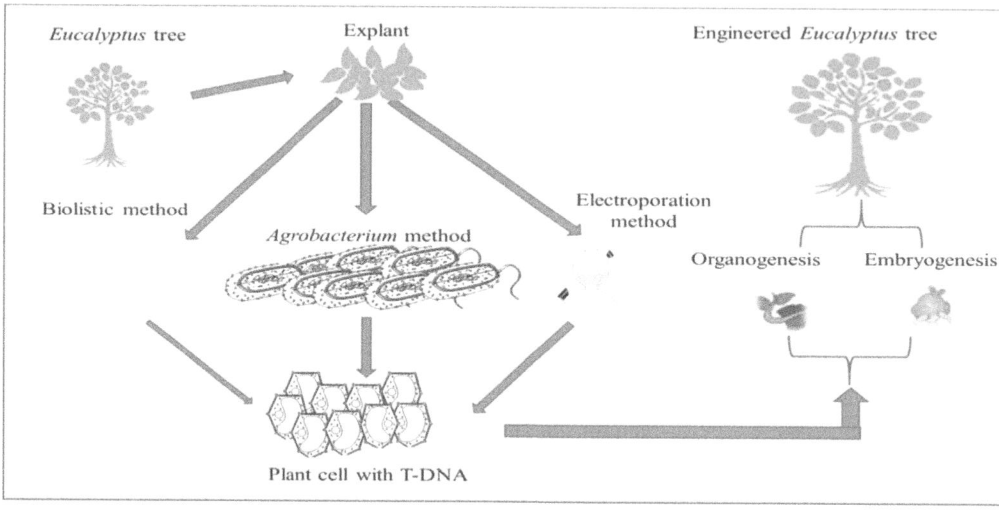

FIGURE 4.4 Schematic diagram of gene transformation using *Agrobacterium*, biolistic and electroporation method in *Eucalyptus*. Each method has different advantages and disadvantages; however, the regeneration of explants is the most important phase, determining the method's efficacy.

regeneration and transformation of *Eucalyptus* is mainly more efficient with juvenile explants, such as leaf discs, cotyledons, and hypocotyls (Corredoira et al., 2018). Nonetheless, due to the availability of less meristematic tissue in the clonally derived explants, the materials have a low percentage of regeneration (Grattapaglia, 2008). Genetic transformation advancement was observed when juvenile tissues of *E. camaldulensis* were applied to the technology (Chen et al., 2001). The most common way to have young juvenile material is the cultivation of the nodal segments *in vitro* culture. Besides the application of explants in the genetic transformation of *Eucalyptus*, the explants have been used to investigate the function of genes and promoters in wood formation (Girijashankar, 2011).

Generally, the process of gene transformation in *Eucalyptus* starts with using suitable explants *in vitro* condition. Different methods can be applied to insert T-DNA (transfer DNA) into the genome of the host tree. The explants will be regenerated directly (organogenesis) or indirectly (embryogenesis) (Figure 4.4). In the following sections, different important factors active in the gene transformation of *Eucalyptus,* such as the role of promoters, reported genes, and markers will be discussed.

4.12 PROMOTERS; THE INITIAL FACTORS DURING THE GENE TRANSFORMATION IN EUCALYPTUS

Generally speaking, achieving high-value genetic transformation procedures can be dictated by the expression of transgenes in host plants. To have a high level of gene expression, promoters can play an important function in the engineering of *Eucalyptus* (Arcuri et al., 2020). Promoters can regulate the expression of a gene by polynucleotide initiating host cells.

Exogenous and endogenous promoters are two different types of regulators (Kolachevskaya et al., 2017). Exogenous promoters are gene promoters and termination sequences derived from other plant species accomplished with functioning in the object plant (Voytas, 2013). Nonetheless, endogenous promoters are gene promoters and termination sequences obtained from the same plant species of the target plant (Porto et al., 2014). On the other hand, some promoters are constitutive, while others are a part of the non-constitutive group, including inducible promoters, cell type-specific, tissue preferred, and tissue-specific (Suqin et al., 2011). On the contrary, the constitutive promoters are able to be active permanently (Porto et al., 2014). Reportedly, the repeatedly applied promoter for *Eucalyptus* transformation investigations is the constitutive Cauliflower Mosaic Virus 35S promoter (CaMV 35S) (Yahiaoui et al., 1998). For example, the CaMV 35S promoter has been used to transfer *uidA* and *cat* genes into the genome of *E. gunni* using electroporation and polyethylene glycol (PEG) transformation methods (Teulieres et al., 1991). Similarly, *uidA* and *cat* genes were transferred into the genome of *E. citriodera* by employing CaMV 35S promoters in electroporation and PEG techniques (Manders et al., 1992). In another example, root-inducing genes of Ri plasmids were transferred into the genomes of *E. nitens*, *E. dunni*, and *E. grandis* using CaMV 35S in the *Agrobacterium rhizogenes* method (Macrae et al., 1993). The chimeric neomycin phosphotransferase II (NPT II) selection marker and the *DsRed2* reporter gene were driven by the CaMV 35S promoter in another investigation published by Zhou et al. (2022). Furthermore, certain researchers conducted experiments to assess the capability of the CRISPR/Cas9 system in facilitating gene knock-out in *E. grandis* hairy roots. In the context of CRISPR/Cas9 editing, a single vector was employed to incorporate two single guide RNAs (sgRNAs) regulated by the *Arabidopsis U6* promoter, along with *Cas9* regulated by *CaMV 35S* promoter (Nagle et al., 2023). Reportedly, genes estimated to be active under adverse conditions or in specific plant tissues or organs, when regulated by CaMV 35S (the constitutive promoter), can cause adverse modifications in engineered *Eucalyptus* (Abiri et al., 2016). For example, a reduction was reported in the roots and leaves growth of some transgenic *E. saligna* that carried the abiotic stress tolerance *Pyrroline-5-carboxylate synthase* (*P5CS*) gene under CaMV 35S (Dibax et al., 2010). Similarly, abnormal phenotypic alterations (dropped upper leaves, changed leaf size and shape) occurred when *E. camaldulensis* carried the *anti-sense Eclim1* gene controlled by CaMV 35S. In this scenario, a 29% reduction was reported in the lignin content transgenic *E. camaldulensis* compared to control plants (Kawaoka et al., 2006). Accordingly, a significant growth reduction was reported in transgenic *Eucalyptus* (*EguCBF1a*-OE A1 line) when the gene was linked to the CaMV 35S promoter (Navarro et al., 2011).

To avoid such an undesirable phenomenon, inducible and tissue-specific promoters would be a promising solution for the engineering of *Eucalyptus* (Ghanem et al., 2011). For example, *EgCAD2* is a tissue- and cell-specific expression promoter that activates the lignin biosynthetic mechanism. This promoter, isolated from *E. gunni,* is a heterologous genome in poplar. From a biotechnological points of view, the *EgCAD2* promoter is a reliable candidate that can goal transgene expression in specific tissues like vascular, dissimilar to the constitutive promoter application (Lauvergeat et al., 2002). Likewise, using the biolistic method, the *CAD* promoter derived from *Eucalyptus* controls the gene and transfers it into the *E. globulus* (Serrano et al., 1996).

Interestingly, a comparison of the function of two different promoters isolated from *E. nitens* (*EniPFLA1* and *EniPTUB1*) showed that the mentioned promoters played an important role during stress wood formation (Taylor et al., 2007). The cinnamoyl CoA reductase (CCR) is a reliable tissue-specific promoter that can be expressed in vascular tissues in woody and herbaceous transgenic plants. The *EgCCR* transformed plantlets showed high expression of GUS linked to the vascular tissues. This promoter is able to target different vascular-associated traits, such as vascular pathogen resistance or lignin content in *Eucalyptus*. Additionally, an applied *EgCCR* promoter can also inhibit the adverse pleiotropic impacts triggered by CaMV 35S (strong constitutive promoters) (de la Torre et al., 2014). As mentioned in the biotic stress section, the *Rs-AFP2* gene increases the resistance of *E. urophylla* against *P. capsici*. The gene was regulated in the construct using the inducible *prp1*–1 promoter (Ouyang et al., 2012).

Previous research has indicated that promoters with multiple types of *cis*-elements tend to demonstrate more favorable expression properties than promoters with a singular type of element (Rushton et al., 2002). The enhancement of promoter strength can be achieved through the augmentation of *cis*-elements (Rushton et al., 2002), a phenomenon associated with the augmentation of transcription factor binding sites (Rushton, 2016). Nevertheless, the proper spacing of *cis*-elements is of utmost importance for the activity of synthetic promoters. It has been observed that if these *cis*-elements are positioned too closely to each other, their activity is compromised (Rushton, 2016). Hence, it is imperative to conduct experimental investigations to ascertain the precise spatial separation between the *cis*-elements. Given the broadly conserved nature of the same promoter region across several plant species (Schlabach et al., 2010), the combination of diverse dimers with the *CaMVmini35S promoter* (Benfey et al., 1990) holds the potential to create an optimal Synthetic Pathogen-Inducible Promoter (SPIP). This would enable the precise and timely expression of disease-resistant genes (Liu et al., 2016). The investigation conducted by Huang et al. (2022) focused on the expression activity of artificial promoters for disease resistance in transgenic *E. urophylla*. The researchers designed three promoters, namely VGWSF, GWVSF, and GWSFV, positioned in the upstream, mid-stream, and downstream regions of the GWSF promoter sequence. These promoters were inserted by a V-element dimer that exhibits virus-inducing ability. The results of this study will provide valuable insights into the development and enhancement of artificial promoters in transgenic plant systems.

4.13 SELECTION MARKERS AND REPORTER GENES FUNCTION IN THE ENGINEERING OF EUCALYPTUS

Generally, reporter genes are employed to detect the transformation success in the genome of plants (Esland et al., 2018). In the process of gene transformation, the expression of reporter gene analysis can determine the efficiency of the transformation method, tissue specificity transgene expression, and promoter strength (Mortensen et al., 2019). The most common marker in Eucalyptus is the *uidA* gene coding for the enzyme β-glucuronidase (GUS). However, this marker relies on histochemical staining that caused destructive sampling. So far, different markers have been tested in *Eucalyptus* transformation, such as Yellow Fluorescent Protein (*yfp*), Cyan Fluorescent Protein (*cfp*), and Green Fluorescent Protein (*gfp*) (Nowak et al., 2004). The present investigation employed the chimeric neomycin phosphotransferase II (NPT II) selection marker and the reporter gene DsRed2 (Discosoma red fluorescent protein 2) to assess the transformation efficiency of *E. camaldulensis* and *E. robusta* genotypes (Zhou et al., 2022). Antibiotics or herbicides resistance markers have been used to select the transformed explant in Eucalyptus's gene transformation process. Accordingly, bar gene Streptomyces hygroscopicus-tolerant to ammonium glufosinate herbicide (Chauhan et al., 2014), antibiotic kanamycin resistant *nptII* gene coding for neomycin phosphotransferase-II (Gul et al., 2015), *cat* (Kendurkar et al., 2018), and acetolactone synthetase (ALS) have been frequently used in *Eucalyptus* transformation (Velini et al., 2005).

4.14 GENETIC MANIPULATION OF EUCALYPTUS

So far, different strategies of gene transformation have been applied to improve *Eucalyptus* (Grattapaglia et al., 2008). The methods include *in vivo* mature stem, *Agrobacterium-mediated*, and biolistic transformation approaches (Girijashankar, 2011). The first history of using electroporation for gene transformation into the genome of *Eucalyptus* dates back to 1990 when Kawasu and colleagues transferred a gene into the genome of *E. saligna* (Kawasu et al., 1990). In Eucalyptus's next step of gene transformation, *gus* and *cat* genes were transferred into the protoplasts *E. citriodera* and *E. gunni* using electroporation ± polyethylene glycol (PEG) method (Manders et al., 1992). However, *Eucalyptus* is a recalcitrant plant against genetic transformation technology; therefore, some techniques, like the biolistic method, were less preferred than cereal plants due to low efficacy (Girijashankar, 2011).

The particle bombardment method is another gene transformation technique that is applied to introduce DNA into the genome of *E. globulus* (Rochange et al., 1995). The result of *gus* gene expression (transient) showed the same efficacy of compressed-helium gas and gun powder device. Additionally, the results also highlighted that 6-day-old cultured embryos were the best explanation for *Eucalyptus* genetic transformation (Rochange et al., 1995). Notwithstanding the mentioned progress in the genetic engineering of *Eucalyptus*, the first history of successful regeneration was reported when Serrano and colleagues (Serrano et al., 1996) applied the biolistic transformation of zygotic embryos. Later, the biolistic transformation method was applied to introduce a gene into the genome of *E. grandis* × *E. urophylla* hybrids using cotyledons and hypocotyls explants (Sartoretto et al., 2002). Compared to other gene transformation methods, which have been applied in *Eucalyptus*, *Agrobacterium*-mediated transformation has been introduced as the most reliable method with the high efficacy of gene transformation in *Eucalyptus* plants. The first report of using *Agrobacterium*-mediated protocol went back to 1992 when Chirqui and colleagues introduced the gene into the genomes of *E. gunni* and *E. globulus* (Chriqui et al., 1991). Reportedly, gene transformation with an indirect method like *Agrobacterium*-mediated transformation is preferred over other direct methods like biolistic. In the indirect method, the insertion of multiple copies of the transgene is reduced and caused gene silencing in long-living species of trees (Mullins et al., 1997). It has been observed that different *A. tumefaciens* wild strains have various levels of pathogenicity in *E. grandis* × *E. urophylla* hybrids. This possibility helps transfer genes into Eucalyptus plants' genomes using *Agrobacterium* derived vectors (Machado et al., 1997). However, it has been shown that different *Eucalyptus* species have various responses to *Agrobacterium*. For example, compared to *E. cladocalyx* and *E. camaldulensis*, *E. occidentalis* showed more susceptibility to *Agrobacterium* (Krimi et al., 2006). As mentioned previously, the types of explants play an essential role in increasing the efficiency of gene transformation protocol. The *Agrobacterium*-mediated protocol is an effective method to transfer a gene into the callus of *E. camaldulensis*. However, the agrobiolistics transformation methods are reported as an efficient way to transfer a gene into the leaves of *E. camaldulensis* (Mendonça et al., 2013).

In addition to *Agrobacterium*-mediated transformation, sonication assisted *Agrobacterium* transformation (SAAT) method was applied in the genetic modification of the *E. urophylla* × *E. grandis* hybrid. In the seedlings stage of engineered *Eucalyptus* with the SAAT system, the transient expression of the *gus* gene showed high efficacy compared to other genetic transformations protocols (González et al., 2002). A recent study successfully conducted and refined a genetic transformation system using *Agrobacterium*-mediated techniques, employing leaves as explants in the hybrid species *E. urophylla* × *E. grandis* DH32-29. The observed frequency of genetic transformation was approximately 1.9%, enabling the identification of some pivotal genes associated with the advantageous qualities exhibited by *Eucalyptus*, such as rapid growth and extensive adaptability. Additionally, this technique has the potential to generate transgenic lines with altered characteristics, particularly in enhancing the resilience of insects and herbicides, thereby facilitating advancements in afforestation efforts (Wang et al., 2022; Su et al., 2023).

Previous studies have successfully developed hairy root transformation systems in woody species, including *E. grandis* (Plasencia et al., 2016). The recent unveiling of the genome sequence of *E. grandis* has led to the identification of numerous potential genes associated with wood formation, hence facilitating further investigations into their functional properties. The utilization of the hypervirulent A4RS strain was required to achieve effective transformation (with an average success rate of 62%) due to the resistance exhibited by *E. grandis* against *Agrobacterium*. The microscopic analysis results indicated no discernible differences in xylem development between hairy and wild-type roots. A proof of concept for the composite plant strategy was achieved by downregulating the expression of the *Eucalyptus cinnamoyl-CoA reductase1* (*EgCCR1*) gene in *Eucalyptus grandis*. This gene encodes an important enzyme involved in the biosynthesis of lignin. Consistent with expectations, the expression of an *EgCCR1* antisense construct resulted in a reduction in lignin content. In addition, the authors provided evidence that composite plants were appropriate for examining the expression pattern mediated by promoters derived from genes associated with the biosynthesis of lignin (Plasencia et al., 2016).

The availability of the different *Eucalyptus* genomes (for instance, *E. garndis*) (Myburg et al., 2014) has facilitated the comprehensive analysis of various gene families, particularly those associated with the synthesis of lignin (Carocha et al., 2015) and transcription factors that play a role in the formation of secondary cell wall (SCW), such as R2R3-MYB (Soler et al., 2015), NAC (Hussey et al., 2015), ARF (Yu et al., 2014), and Aux/IAA (Yu et al., 2015), among others. These investigations have highlighted numerous novel candidates that may potentially regulate wood production and require functional characterization. Stable transformation has consistently posed a limitation in the functional characterization of genes in *Eucalyptus*. Despite establishing stable transformation techniques for various *Eucalyptus* species (Tournier et al., 2003; Girijashankar, 2011; de la Torre et al., 2014; Cao et al., 2020), these methods are characterized by their time-consuming nature and low efficiency. Consequently, these factors can be attributed to the limited number of functional investigations conducted on transgenic eucalypts (Girijashankar, 2011). In order to address these constraints, an alternative and reliable transformation technique was established for *Eucalyptus grandis*. This system involved the utilization of *Agrobacterium rhizogenes*, enabling the generation of composite plants that possess wild-type shoots and genetically modified roots that can be readily identified through fluorescent markers (Plasencia et al., 2016). This approach is well-suited for elucidating the functional role of genes associated with xylem or SCW development, which is of particular significance for woody plant species. The researchers employed an antisense method to demonstrate the feasibility of down-regulating *Cinnamoyl CoA reductase1* (*EgrCCR1*), which is involved in the penultimate stage of the lignin branch route (Plasencia et al., 2016).

Over the past thirty years, RNA interference, antisense RNA and virus-induced gene silencing have emerged as prominent techniques for gene silencing in plants. These strategies have proven invaluable in elucidating several genes' functions, particularly in plant species without a mutant collection. Nevertheless, the efficacy of silence was diminished in mutants due to the indirect interruption of gene function through repression of the corresponding mRNA. This often resulted in a partly repressive impact and, in certain instances, unexpected consequences. The Clustered Regularly Interspaced Short Palindromic Repeats/CRISPR-associated protein 9 (CRISPR/Cas9) is a recently created and naturally occurring tool for genome editing. It has gained popularity as the preferred system for targeted mutagenesis in an increasing number of plant species, including *Eucalyptus*. In recent years, there has been a rapid adoption and refinement of techniques for precise genome sequence editing in plants (Koonin et al., 2017). The advent of genome editing has presented novel opportunities in the field of forestry, particularly with the utilization of CRISPR/Cas9 technology. The methodology employed is a highly efficient and effective method of genome editing, which has successfully implemented targeted modifications at specific loci within the genome of a forest tree (Fan et al., 2015; Demirci et al., 2018). The CRISPR/Cas9 technique relies on the activity of Cas9, an RNA-guided endonuclease derived from *Streptococcus pyogenes*, to precisely cleave a targeted DNA double strand at a specific genomic locus (Bortesi et al., 2015). The formation of a complex between Cas9 and a complementary single-guide RNA (sgRNA) enables the particular recognition of a target DNA region by base pairing. The Cas9 will cleave the DNA sequence immediately preceding a specific DNA motif known as the Protospacer Adjacent Motif (PAM), which typically consists of 2 to 6 base pairs. The sequence of PAM is determined by the source of the Cas9 endonuclease (Ma et al., 2016).

The occurrence of null mutations in the T0 generation holds particular significance for tree species with extended reproductive cycles, such as poplar or eucalypts, primarily transmitted through vegetative means (Fan et al., 2015; Tsai et al., 2015). One further trait exhibited by outcrossing trees is their notable level of genome heterozygosity. This property, along with sequence polymorphisms at the specific regions of interest, can potentially impede the effectiveness of CRISPR editing (Bewg et al., 2018). Significant advancements have been achieved in genetic modification using CRISPR technology, particularly in the poplar case, which has emerged as the pioneering tree species to undergo genome editing with remarkable efficacy (Fan et al., 2015; Zhou et al., 2015). Furthermore, the development of bioinformatics tools sensitive to allelic variations has greatly facilitated the genome editing process in species that exhibit heterozygosity (Tsai et al., 2015; Xu et al., 2015). Given the significance of wood characteristics in industrial contexts, most CRISPR gene editing investigations in poplar have focused on modifying the composition of the SCW and phenylpropanoid metabolism, specifically lignin (Wan et al., 2017; Wang et al., 2017; Xu et al., 2017; Yang et al., 2017; Shen et al., 2018). A comprehensive investigation involving over 500 transgenic events has documented the successful modification of crucial flowering genes in order to inhibit bisexual fertility (Elorriaga et al., 2018). The objective was to generate infertile trees to address the issue of potential seed or pollen dispersion, which poses a significant challenge for trees that are vegetatively propagated for industrial plantations.

Eucalypts are characterized as diploid organisms with a chromosomal count of 2n= 22. In this context, each homologous chromosome harbors a single allele, resulting in a total of two alleles per gene. In theory, the CRISPR/Cas9 system has the potential to induce two distinct types of mutations in the targeted gene for each single guide RNA (sgRNA), except for the chimeric mutation. Firstly, if both alleles undergo mutation, it results in a biallelic mutation. This biallelic mutation can manifest as either a homozygous mutation, where both alleles possess identical mutations, or as a biallelic mutation, where the mutations differ between the two alleles. Secondly, if only one of the alleles undergoes mutation, it leads to a monoallelic mutation. It is important to note that monoallelic mutations do not theoretically result in complete knock-out mutations, as a wild-type allele remains. However, they may still produce knock-down mutations. In the instance of chimera, it is observed that many mutations occur concurrently within the same plant. There exist two distinct categories of chimera. The first category comprises chimera possessing two or more mutant alleles alongside a wild-type allele. These chimeras can potentially induce knock-down effects but not complete knock-out effects. The second category encompasses chimera, in which all alleles are mutated. These chimeras can activate complete knock-out effects (Dai et al., 2020; Wang et al., 2021).

Dai et al. (Dai et al., 2020) conducted a study to investigate whether the CRISPR/Cas9 system could be used to generate gene knock-out in the hairy roots of *E. grandis*. The initial focus of their investigation was on the *EgrCCR1* (*CCR1*) gene due to the well-established understanding of the phenotypic effects resulting from the downregulation of this gene in *E. grandis* hairy roots (Plasencia et al., 2016). In addition, the researchers chose the Aux/IAA gene, *EgrIAA9A* (*IAA9A*), recognized for its preferential expression in the vascular cambium and during the development of secondary xylem (Yu et al., 2015). In the context of CRISPR/Cas9 editing, a single vector was employed to incorporate two single guide RNAs (sgRNAs) regulated by the *Arabidopsis U6* promoter, along with *Cas9* regulated by the *CaMV 35S* promoter (Brooks et al., 2014). According to their findings, the CRISPR/Cas9 system demonstrated efficacy in inducing mutations in both *CCR1* and *IAA9A* genes within *E. grandis* hairy roots. However, it is essential to note that the editing effectiveness rates varied significantly between the two target genes. The phenotypic characterization of *CCR1*-edited lines was conducted using FTIR spectroscopy and histochemical analysis, which confirmed the reduction in lignin phenotypes due to *CCR*-downregulation (Dai, 2020).

4.15 PROSPECTS FOR GENETIC ENGINEERING FOR EUCALYPTUS IMPROVEMENT

A number of promising outcomes could result from using genetic engineering to improve *Eucalyptus* trees. These outcomes could include increased growth rate, wood quality, and resistance to pests and diseases (Table 5.1) (Girijashankar, 2011). Genetic engineering holds significant promise in revolutionizing Eucalyptus tree improvement efforts across multiple fronts. Firstly, by deploying genetic modification techniques, researchers can target and enhance the tree's innate resistance mechanisms against prevalent pests, diseases, and adverse environmental conditions. This augmentation in resilience not only safeguards the health and vitality of Eucalyptus populations but also ensures sustained productivity in forestry practices. Moreover, genetic engineering presents an opportunity to fine-tune wood quality traits in Eucalyptus, including density, strength, and pulp properties, thereby catering to diverse industrial needs ranging from construction materials to paper production. Additionally, the acceleration of growth rates through genetic interventions holds immense potential in shortening the lifecycle of Eucalyptus trees, leading to quicker turnover, and increased overall biomass output. Furthermore, the adaptability of genetically engineered Eucalyptus varieties to thrive in challenging environmental settings, such as marginal lands or regions prone to drought, holds promise in expanding cultivation areas and maximizing land utility. Beyond traditional forestry applications, engineered Eucalyptus strains can serve as sustainable feedstock for bioenergy and biofuel production, contributing to renewable energy initiatives and mitigating carbon emissions. Despite these promising advancements, the responsible and ethical deployment of genetically modified organisms necessitates rigorous consideration of regulatory frameworks, environmental impacts, and societal acceptance to ensure the long-term sustainability and coexistence of genetically engineered Eucalyptus with natural ecosystems and communities. Responsible and long-term use of genetic engineering requires careful consideration of the many regulatory, ethical, and environmental factors involved. It is important to thoroughly consider public opinion, regulatory frameworks, and possible environmental implications while creating and implementing genetically modified *Eucalyptus* varieties (Aggarwal et al., 2016; Kendurkar et al., 2018).

4.16 CONCLUSION

The multifaceted advancements in Eucalyptus gene transformation and genetic engineering collectively herald a promising era for the optimization and sustainable utilization of this versatile tree species. Precision in monoterpene engineering not only broadens the applications of Eucalyptus essential oils but exemplifies the versatility of genetic engineering. Simultaneously, interventions to improve wood quality and quantity promise to revolutionize forestry and associated industries, catering to diverse end-users while fostering economically viable and sustainable practices. The augmentation of biomass yield in transgenic Eucalyptus aligns with the increasing global demand for renewable resources, supporting industrial needs and sustainable land use practices. Advances in fertility-related genetic engineering streamline cultivation and seed production, ensuring the robustness of genetically modified Eucalyptus in commercial settings. Strategic modifications in the lignin pathway offer a powerful tool for tailoring wood properties to industrial requirements, promoting environmentally friendly forestry practices.

Furthermore, improvements in the micro fibrillar angle enhance wood strength and quality, with implications for construction and pulp industries. Understanding and manipulating root development underscores the potential for optimizing growth and resource utilization. However, the responsible and sustainable deployment of genetically engineered Eucalyptus varieties necessitates a holistic approach, considering ecological implications, regulatory frameworks, and societal acceptance. Continued collaboration among researchers, industry stakeholders, and policymakers will be instrumental in unlocking the full potential of genetic engineering to shape the future of Eucalyptus cultivation and utilization.

After poplar, Eucalyptus is the most important forest tree with high economic value. New biotechnological advancements show a promising function to promote advanced mechanisms in the economical traits of Eucalyptus species. Notwithstanding all the above positive sides of the application of modern technology in Eucalyptus improvement, some bottlenecks have been observed. Optimization of an efficient in vitro culture protocol for Eucalyptus is the initial step of gene transformation. The use of inducible and tissue-specific promoters in construct structures are more appropriate, particularly in the circumstance of Eucalyptus where wood quality and yield are stated as the second-highest concerns. Nonetheless, selecting suitable promoters in the gene transformation process of trees is problematic due to the complex developmental stages that happen with increasing tissue complexity and size across years. On the other hand, applying the best marker or reporter genes is a matter of concern during the Eucalyptus genome engineering process. Overexpression of different genes in transgenic Eucalyptus species was identified to have an unfavorable consequence on the growth and morphology of the tree. In contrast, lower expression may not assist the aim of intent. As mentioned, different gene transformation strategies have been applied to transfer T-DNA into the genome of Eucalyptus. However, to avoid transgene-silencing, different strategies should be applied. All in all, the application of gene transformation has the high potential to advance the Eucalyptus tree by considering valuable traits.

REFERENCES

Abiri, R., Atabaki, N., Abdul-Hamid, H., Sanusi, R., Ab Shukor, N. A., Shaharuddin, N. A., Ahmad, S. A., & Malik, S. (2020). The prospect of physiological events associated with the micropropagation of Eucalyptus sp. *Forests*, *11*(11), 1211.

Abiri, R., Atabaki, N., Sanusi, R., Malik, S., Abiri, R., Safa, P., Shukor, N. A. A., & Abdul-Hamid, H. (2022). New insights into the biological properties of eucalyptus-derived essential oil: A promising green anti-cancer drug. *Food Reviews International*, *38*(sup1), 598–633.

Abiri, R., Valdiani, A., Maziah, M., Shaharuddin, N. A., Sahebi, M., Yusof, Z. N. B., Atabaki, N., & Talei, D. (2016). A critical review of the concept of transgenic plants: insights into pharmaceutical biotechnology and molecular farming. *Current Issues in Molecular Biology*, *18*(1), 21–42.

Acquaah, G. (2015). Conventional plant breeding principles and techniques. *Advances in Plant Breeding Strategies: Breeding, Biotechnology and Molecular Tools*, 115–158.

Adeniyi, B., Odufowoke, R., & Olaleye, S. (2006). Antibacterial and gastroprotective properties of Eucalyptus torelliana [Myrtaceae] crude extracts *International Journal of Pharmacology*, *2*(3), 362–365.

Adeniyi, C. B. A., Lawal, T. O., & Mahady, G. B. (2009). In vitro susceptibility of Helicobacter pylori to extracts of Eucalyptus camaldulensis and Eucalyptus torelliana. *Pharmaceutical Biology*, *47*(1), 99–102.

Aggarwal, D., Reddy, M. S., & Kumar, A. (2016). Biotechnological approaches for the improvement of Eucalyptus. *Plant Tissue Culture: Propagation, Conservation and Crop Improvement*, 219–244.

Aharoni, A., Jongsma, M. A., & Bouwmeester, H. J. (2005). Volatile science? Metabolic engineering of terpenoids in plants. *Trends in Plant Science*, *10*(12), 594–602.

Ahlem, S., Khaled, H., Wafa, M., Sofiane, B., Mohamed, D., & Jean-Claude, M. (2009). Oral administration of Eucalyptus globulus extract reduces the alloxan-induced oxidative stress in rats. *Chemico-biological Interactions*, *181*(1), 71–76.

Al-Aizari, A. A., Al-Obeed, R. S., & Mohamed, M. A. (2020). Improving micropropagation of some grape cultivars via boron, calcium and phosphate. *Electronic Journal of Biotechnology*, *48*, 95–100.

Al-Maharik, N., Jaradat, N., Hawash, M., Al-Lahham, S., Qadi, M., Shoman, I., Jaber, S., Rahem, R. A., Hussein, F., & Issa, L. (2022). Chemical composition, antioxidant, antimicrobial and anti-proliferative activities of essential oils of Rosmarinus officinalis from five different sites in Palestine. *Separations*, *9*(11), 339.

Amazonas, N. T., Forrester, D. I., Silva, C. C., Almeida, D. R. A., Rodrigues, R. R., & Brancalion, P. H. (2018). High diversity mixed plantations of Eucalyptus and native trees: An interface between production and restoration for the tropics. *Forest Ecology and Management*, *417*, 247–256.

Arcuri, M. L., Fialho, L. C., Vasconcellos Nunes-Laitz, A., Fuchs-Ferraz, M. C. P., Wolf, I. R., Valente, G. T., Marino, C. L., & Maia, I. G. (2020). Genome-wide identification of multifunctional laccase gene family in Eucalyptus grandis: potential targets for lignin engineering and stress tolerance. *Trees*, *34*, 745–758.

Austin, A. T., & Ballaré, C. L. (2010). Dual role of lignin in plant litter decomposition in terrestrial ecosystems. *Proceedings of the National Academy of Sciences*, *107*(10), 4618–4622.

Avisar, I., McNab, A. A., Dolman, P. J., Patel, B., Desousa, J.-L., Selva, D., & Malhotra, R. (2013). Endonasal dacryocystorhinostomy for nasolacrimal duct obstruction in patients with sarcoidosis. *Orbit*, *32*(4), 225–230.

Bachir, R. G., & Benali, M. (2012). Antibacterial activity of the essential oils from the leaves of Eucalyptus globulus against Escherichia coli and Staphylococcus aureus. *Asian Pacific Journal of Tropical Biomedicine*, 2(9), 739–742.

Bao, A., Burritt, D. J., Chen, H., Zhou, X., Cao, D., & Tran, L.-S. P. (2019). The CRISPR/Cas9 system and its applications in crop genome editing. *Critical Reviews in Biotechnology*, *39*(3), 321–336.

Bastos, V. P., Brito, T. S., Lima, F. J., Pinho, J. P., Lahlou, S., Abreu Matos, F. J., Santos, A. A., & Caldas Magalhaes, P. J. (2009). Inhibitory effect of 1, 8-cineole on guinea-pig airway challenged with ovalbumin involves a preferential action on electromechanical coupling. *Clinical and Experimental Pharmacology and Physiology*, *36*(11), 1120–1126.

Baucher, M., Halpin, C., Petit-Conil, M., & Boerjan, W. (2003). Lignin: genetic engineering and impact on pulping. *Critical Reviews in Biochemistry and Molecular Biology*, *38*(4), 305–350.

Benfey, P. N., Ren, L., & Chua, N.-H. (1990). Tissue-specific expression from CaMV 35S enhancer subdomains in early stages of plant development. *The EMBO Journal*, *9*(6), 1677–1684.

Bewg, W. P., Ci, D., & Tsai, C.-J. (2018). Genome editing in trees: from multiple repair pathways to long-term stability. *Frontiers in Plant Science*, *9*, 1732.

Bhoi, A., Yadu, B., Chandra, J., & Keshavkant, S. (2022). Mutagenesis: A coherent technique to develop biotic stress resistant plants. *Plant Stress*, *3*, 100053.

Bhuyan, D. J., Vuong, Q. V., Bond, D. R., Chalmers, A. C., Bowyer, M. C., & Scarlett, C. J. (2018). Eucalyptus microcorys leaf extract derived HPLC-fraction reduces the viability of MIA PaCa-2 cells by inducing apoptosis and arresting cell cycle. *Biomedicine & Pharmacotherapy*, *105*, 449–460.

Blair, M. (2016). Diabetes mellitus review. *Urologic Nursing*, *36*(1), 27.

Bortesi, L., & Fischer, R. (2015). The CRISPR/Cas9 system for plant genome editing and beyond. *Biotechnology Advances*, *33*(1), 41–52.

Brennan, M., McLean, J. P., Altaner, C. M., Ralph, J., & Harris, P. J. (2012). Cellulose microfibril angles and cell-wall polymers in different wood types of Pinus radiata. *Cellulose*, *19*, 1385–1404.

Brondani, G. E., de Wit Ondas, H. W., Baccarin, F. J. B., Gonçalves, A. N., & de Almeida, M. (2012). Micropropagation of Eucalyptus benthamii to form a clonal micro-garden. *In Vitro Cellular & Developmental Biology-Plant*, *48*, 478–487.

Brondani, G. E., Dutra, L. F., Wendling, I., Grossi, F., Hansel, F. A., & Araujo, M. A. (2011). Micropropagation of an Eucalyptus hybrid (Eucalyptus benthamii x Eucalyptus dunnii). *Acta Scientiarum. Agronomy*, *33*, 655–663.

Brooks, C., Nekrasov, V., Lippman, Z. B., & Van Eck, J. (2014). Efficient gene editing in tomato in the first generation using the clustered regularly interspaced short palindromic repeats/CRISPR-associated9 system. *Plant Physiology*, *166*(3), 1292–1297.

Bunn, E. (2005). Development of in vitro methods for ex situ conservation of Eucalyptus impensa, an endangered mallee from southwest Western Australia. *Plant Cell, Tissue and Organ Culture*, *83*, 97–102.

Caceres, A. I., Liu, B., Jabba, S. V., Achanta, S., Morris, J. B., & Jordt, S. E. (2017). Transient receptor potential cation channel subfamily M member 8 channels mediate the anti-inflammatory effects of eucalyptol. *British Journal of Pharmacology*, *174*(9), 867–879.

Calviño-Cancela, M., & van Etten, E. J. (2018). Invasive potential of Eucalyptus globulus and Pinus radiata into native eucalypt forests in Western Australia. *Forest Ecology and Management*, *424*, 246–258.

Cao, P. B., Azar, S., SanClemente, H., Mounet, F., Dunand, C., Marque, G., Marque, C., & Teulières, C. (2015). Genome-wide analysis of the AP2/ERF family in Eucalyptus grandis: an intriguing over-representation of stress-responsive DREB1/CBF genes. *PLoS One*, *10*(4), e0121041.

Cao, P. B., Ployet, R., Nguyen, C., Dupas, A., Ladouce, N., Martinez, Y., Grima-Pettenati, J., Marque, C., Mounet, F., & Teulières, C. (2020). Wood architecture and composition are deeply remodeled in frost sensitive Eucalyptus overexpressing CBF/DREB1 transcription factors. *International Journal of Molecular Sciences*, *21*(8), 3019.

Carocha, V., Soler, M., Hefer, C., Cassan-Wang, H., Fevereiro, P., Myburg, A. A., Paiva, J. A., & Grima-Pettenati, J. (2015). Genome-wide analysis of the lignin toolbox of Eucalyptus grandis. *New Phytologist*, *206*(4), 1297–1313.

Cermelli, C., Fabio, A., Fabio, G., & Quaglio, P. (2008). Effect of eucalyptus essential oil on respiratory bacteria and viruses. *Current Microbiology*, 56, 89–92.

Chaín, J. M., Tubert, E., Graciano, C., Castagno, L. N., Recchi, M., Pieckenstain, F. L., Estrella, M. J., Gudesblat, G., Amodeo, G., & Baroli, I. (2020). Growth promotion and protection from drought in Eucalyptus grandis seedlings inoculated with beneficial bacteria embedded in a superabsorbent polymer. *Scientific Reports*, *10*(1), 18221.

Chakraborty, I., & Maity, P. (2020). COVID-19 outbreak: Migration, effects on society, global environment and prevention. *Science of the Total Environment, 728*, 138882.

Chalchat, J.-C., Muhayimana, A., Habimana, J., & Chabard, J. (1997). Aromatic plants of Rwanda. II. Chemical composition of essential oils of ten Eucalyptus species growing in Ruhande arboretum, Butare, Rwanda. *Journal of Essential Oil Research, 9*(2), 159–165.

Chandorkar, N., Tambe, S., Amin, P., & Madankar, C. (2021). A systematic and comprehensive review on current understanding of the pharmacological actions, molecular mechanisms, and clinical implications of the genus Eucalyptus. *Phytomedicine Plus, 1*(4), 100089.

Chang, S., Mahon, E. L., MacKay, H. A., Rottmann, W. H., Strauss, S. H., Pijut, P. M., Powell, W. A., Coffey, V., Lu, H., & Mansfield, S. D. (2018). Genetic engineering of trees: progress and new horizons. *In Vitro Cellular & Developmental Biology-Plant, 54*, 341–376.

Chauhan, R. D., Veale, A., Cathleen, M., Strauss, S. H., & Myburg, A. A. (2014). Genetic transformation of Eucalyptus—challenges and future prospects. In *Tree Biotechnology Ramawat, KG, Merillon, JM, Ahuja, MR*, (pp. 392–435). Boca Raton: CRC Press.

Chen, H., Li, J., Qiu, B., Zhao, Y., Liu, Z., Yang, J., & Kang, X. (2021). Long non-coding RNA and its regulatory network response to cold stress in Eucalyptus urophylla ST Blake. *Forests, 12*(7), 836.

Chen, Q., Guo, W., Feng, L., Ye, X., Xie, W., Huang, X., & Liu, J. (2015). Data for transcriptome and proteome analysis of Eucalyptus infected with Calonectria pseudoreteaudii. *Data in Brief, 3*, 24–28.

Chen, Z.-Z., Chang, S.-H., Ho, C.-K., Chen, Y.-C., Tsai, J.-B., & Chiang, V. L. (2001). Plant production of transgenic Eucalyptus camaldulensis carrying the Populus tremuloides cinnamate 4-hydroxylase gene. *Plant Biotechnology, 844*, 198.

Chriqui, D., Adam, S., Caissard, J.-C., Noin, M., & Azmi, A. (1991). Shoot regeneration and Agrobacterium-mediated transformation of Eucalyptus globulus and E. gunnii. Proceedings of I.U.F.R.O. Symposium Intensive Forestry: the Role of Eucalyptus., Durban, South Africa.

Christian, T., Jose, B., Claudia, P., Victor, H., Aurelio San, M., Manuel, S.-O., & Darcy, R. (2012). Induction of defensive responses in Eucalyptus globulus (Labill) plants, against Ctenarytaina eucalypti (Maskell) (Hemiptera: Psyllidae). *American Journal of Plant Sciences*, 2012.

Coppen, J. J. (2002). *Eucalyptus: the genus Eucalyptus*. CRC Press.

Corredoira, E., Vieitez, A. M., & Ballester, A. (2018). Eucalypts (Eucalyptus globulus Labill.). *Step Wise Protocols for Somatic Embryogenesis of Important Woody Plants: Volume I*, 269–282.

Dai, Y. (2020). Implementing the CRISPR-Cas9 technology in Eucalyptus hairy roots and functional characterization of auxin-dependent transcription factors involved in wood formation. Université Paul Sabatier-Toulouse III.

Dai, Y., Hu, G., Dupas, A., Medina, L., Blandels, N., San Clemente, H., Ladouce, N., Badawi, M., Hernandez-Raquet, G., & Mounet, F. (2020). Implementing the CRISPR/Cas9 technology in Eucalyptus hairy roots using wood-related genes. *International Journal of Molecular Sciences, 21*(10), 3408.

Davis, G. R. (2002). Cultivation and production of eucalypts in Australia: with special reference to the leaf oils. In *Eucalyptus* (pp. 197–215). CRC Press.

De Boer, A., & Volkov, V. (2003). Logistics of water and salt transport through the plant: structure and functioning of the xylem. *Plant, Cell & Environment, 26*(1), 87–101.

de la Torre, F., Rodríguez, R., Jorge, G., Villar, B., Álvarez-Otero, R., Grima-Pettenati, J., & Gallego, P. P. (2014). Genetic transformation of Eucalyptus globulus using the vascular-specific EgCCR as an alternative to the constitutive CaMV35S promoter. *Plant Cell, Tissue and Organ Culture (PCTOC), 117*, 77–84.

de Oliveira, L. S., Brondani, G. E., Molinari, L. V., Dias, R. Z., Teixeira, G. L., Gonçalves, A. N., & de Almeida, M. (2022). Optimal cytokinin/auxin balance for indirect shoot organogenesis of Eucalyptus cloeziana and production of ex vitro rooted micro-cuttings. *Journal of Forestry Research, 33*(5), 1573–1584.

De Rosa, I. M., Kenny, J. M., Maniruzzaman, M., Moniruzzaman, M., Monti, M., Puglia, D., Santulli, C., & Sarasini, F. (2011). Effect of chemical treatments on the mechanical and thermal behaviour of okra (Abelmoschus esculentus) fibres. *Composites Science and Technology, 71*(2), 246–254.

Demirci, Y., Zhang, B., & Unver, T. (2018). CRISPR/Cas9: an RNA-guided highly precise synthetic tool for plant genome editing. *Journal of Cellular Physiology, 233*(3), 1844–1859.

Dibax, R., Deschamps, C., Bespalhok Filho, J., Vieira, L., Molinari, H., De Campos, M., & Quoirin, M. (2010). Organogenesis and Agrobacterium tumefaciens-mediated transformation of Eucalyptus saligna with P5CS gene. *Biologia Plantarum, 54*, 6–12.

do Prado Frade, S. R., Souza, D. M. S. C., Fernandes, S. B., Avelar, M. L. M., Molinari, L. V., Gonçalves, D. S., Magalhães, T. A., & Brondani, G. E. (2023). Spectral quality influence on in vitro morphophysiological responses of Eucalyptus dunnii Maiden and Eucalyptus grandis W. Hill ex Maiden× E. urophylla ST Blake. *New Zealand Journal of Forestry Science*, *53*, 1–16.

Döll-Boscardin, P. M., Sartoratto, A., Sales Maia, B. H. L. d. N., Padilha de Paula, J., Nakashima, T., Farago, P. V., & Kanunfre, C. C. (2012). In vitro cytotoxic potential of essential oils of Eucalyptus benthamii and its related terpenes on tumor cell lines. *Evidence-Based Complementary and Alternative Medicine*, *2012*(1), 342652.

Du Toit, Y., Coles, D. W., Mewalal, R., Christie, N., & Naidoo, S. (2020). eCALIBRATOR: a comparative tool to identify key genes and pathways for eucalyptus defense against biotic stressors. *Frontiers in Microbiology*, *11*, 216.

Dudareva, N., & Pichersky, E. (2008). Metabolic engineering of plant volatiles. *Current Opinion in Biotechnology*, *19*(2), 181–189.

Ebadollahi, A. (2013). Essential oils isolated from Myrtaceae family as natural insecticides. *Annual Research & Review in Biology*, *3*(3), 148–175.

Elorriaga, E., Klocko, A. L., Ma, C., & Strauss, S. H. (2018). Variation in mutation spectra among CRISPR/Cas9 mutagenized poplars. *Frontiers in Plant Science*, 594.

Emamalipour, M., Seidi, K., Zununi Vahed, S., Jahanban-Esfahlan, A., Jaymand, M., Majdi, H., Amoozgar, Z., Chitkushev, L., Javaheri, T., & Jahanban-Esfahlan, R. (2020). Horizontal gene transfer: from evolutionary flexibility to disease progression. *Frontiers in Cell and Developmental Biology*, *8*, 229.

Esland, L., Larrea-Alvarez, M., & Purton, S. (2018). Selectable markers and reporter genes for engineering the chloroplast of Chlamydomonas reinhardtii. *Biology*, *7*(4), 46.

Eyles, A., Pinkard, E. A., O'GRADY, A. P., Worledge, D., & Warren, C. R. (2009). Role of corticular photosynthesis following defoliation in Eucalyptus globulus. *Plant, Cell & Environment*, *32*(8), 1004–1014.

Fan, D., Liu, T., Li, C., Jiao, B., Li, S., Hou, Y., & Luo, K. (2015). Efficient CRISPR/Cas9-mediated targeted mutagenesis in Populus in the first generation. *Scientific Reports*, *5*(1), 12217.

Fischer, J., & Dethlefsen, U. (2013). Efficacy of cineole in patients suffering from acute bronchitis: a placebo-controlled double-blind trial. *Cough*, *9*, 1–5.

Foresto, E., Gilardi, P., Ibarra, L. E., & Cogno, I. S. (2021). Light-activated green drugs: How we can use them in photodynamic therapy and mass-produce them with biotechnological tools. *Phytomedicine Plus*, *1*(3), 100044.

Frick, E. M., & Strader, L. C. (2018). Roles for IBA-derived auxin in plant development. *Journal of Experimental Botany*, *69*(2), 169–177.

Gago, D., Bernal, M. Á., Sánchez, C., Aldrey, A., Cuenca, B., Christie, C. B., & Vidal, N. (2022). Effect of sucrose on growth and stress status of Castanea sativa x C. crenata shoots cultured in liquid medium. *Plants*, *11*(7), 965.

Gago, D., Vilavert, S., Bernal, M. Á., Sánchez, C., Aldrey, A., & Vidal, N. (2021). The effect of sucrose supplementation on the micropropagation of Salix viminalis L. shoots in semisolid medium and temporary immersion bioreactors. *Forests*, *12*(10), 1408.

George, E. F., Hall, M. A., & Klerk, G.-J. D. (2008). The components of plant tissue culture media I: macro- and micro-nutrients. In *Plant Propagation by Tissue Culture: Volume 1. The Background* (pp. 65–113). Springer.

Ghanem, M. E., Hichri, I., Smigocki, A. C., Albacete, A., Fauconnier, M.-L., Diatloff, E., Martinez-Andujar, C., Lutts, S., Dodd, I. C., & Pérez-Alfocea, F. (2011). Root-targeted biotechnology to mediate hormonal signalling and improve crop stress tolerance. *Plant Cell Reports*, *30*, 807–823.

Gilani, S. A., Kikuchi, A., Yu, X., Ahmad, M. Z., Sugano, M., Fujii, Y., & Watanabe, K. N. (2017). Difference between non-transgenic and salt tolerant transgenic Eucalyptus camaldulensis for diversity and allelopathic effects of essential oils. *Pakistan Journal of Botany*, *49*(1), 345–351.

Gill, L. (1992). *Ethnomedical uses of plants in Nigeria*. Uniben press.

Girijashankar, V. (2011). Genetic transformation of Eucalyptus. *Physiology and Molecular Biology of Plants*, *17*, 9–23.

Goldbeck, J. C., do Nascimento, J. E., Jacob, R. G., Fiorentini, Â. M., & da Silva, W. P. (2014). Bioactivity of essential oils from Eucalyptus globulus and Eucalyptus urograndis against planktonic cells and biofilms of Streptococcus mutans. *Industrial Crops and Products*, *60*, 304–309.

Gomes, F., & Canhoto, J. M. (2003). Micropropagation of Eucalyptus nitens Maiden (shining gum). *In Vitro Cellular & Developmental Biology-Plant, 39*, 316–321.

González, E. R., de Andrade, A., Bertolo, A. L., Lacerda, G. C., Carneiro, R. T., Defávari, V. A. P., Labate, M. T. V., & Labate, C. A. (2002). Production of transgenic Eucalyptus grandis x E. urophylla using the sonication-assisted Agrobacterium transformation (SAAT) system. *Functional Plant Biology, 29*(1), 97–102.

Grattapaglia, D. (2008). Genomics of Eucalyptus, a global tree for energy, paper, and wood. In *Genomics of tropical crop plants* (pp. 259–298). Springer.

Grattapaglia, D., & Kirst, M. (2008). Eucalyptus applied genomics: from gene sequences to breeding tools. *New Phytologist, 179*(4), 911–929.

Guillén, A., Granados, S., Rivas, K. E., Estrada, O., Echeverri, L. F., & Balcázar, N. (2015). Antihyperglycemic activity of Eucalyptus tereticornis in insulin-resistant cells and a nutritional model of diabetic mice. *Advances in Pharmacological and Pharmaceutical Sciences, 2015*(1), 418673.

Gul, J., Munir, I., Gulzar, H., & Hassan, G. (2015). Agrobacterium-mediated transformation of Spinacia oleracea L. through a synthetic chitinase gene inducing resistance to fungal pathogens. *Pakistan Journal of Botany, 47*, 193–198.

Harcourt, R., Kyozuka, J., Floyd, R., Bateman, K., Tanaka, H., Decroocq, V., Llewellyn, D., Zhu, X., Peacock, W., & Dennis, E. (2000). Insect-and herbicide-resistant transgenic eucalypts. *Molecular Breeding, 6*, 307–315.

Hardel, D., & Laxmidhar Sahoo, L. S. (2011). A review on phytochemical and pharmacological of Eucalyptus globulus: a multipurpose tree.

Hassan, A., Balachandran, P., & Khamis, K. R. (2021). Early root development of Eucalyptus pellita F. Muell. seedlings from seed and stem cutting propagation methods at nursery stage. *International Journal of Forestry Research, 2021*, 1–10.

Hassanpour, S. H., & Dehghani, M. (2017). Review of cancer from perspective of molecular. *Journal of Cancer Research and Practice, 4*(4), 127–129.

Huang, C., Lin, W., Lai, C., Li, X., Jin, Y., & Yong, Q. (2019). Coupling the post-extraction process to remove residual lignin and alter the recalcitrant structures for improving the enzymatic digestibility of acid-pretreated bamboo residues. *Bioresource Technology, 285*, 121355.

Huang, Z., Xu, Q., Fang, X., & Wu, Z. (2022). Expression activity of artificial promoters for disease resistance in transgenic Eucalyptus urophylla. *Genes, 13*(10), 1813.

Husen, A., Bachheti, R. K., & Bachheti, A. (2021). *Non-timber forest products: Food, healthcare and industrial applications*. Springer.

Hussey, S. G., Grima-Pettenati, J., Myburg, A. A., Mizrachi, E., Brady, S. M., Yoshikuni, Y., & Deutsch, S. (2019). A standardized synthetic Eucalyptus transcription factor and promoter panel for re-engineering secondary cell wall regulation in biomass and bioenergy crops. *ACS Synthetic Biology, 8*(2), 463–465.

Hussey, S. G., Saïdi, M. N., Hefer, C. A., Myburg, A. A., & Grima-Pettenati, J. (2015). Structural, evolutionary and functional analysis of the NAC domain protein family in Eucalyptus. *New Phytologist, 206*(4), 1337–1350.

Ishfaq, P. M., Shukla, A., Beraiya, S., Tripathi, S., & Mishra, S. K. (2018). Biochemical and pharmacological applications of essential oils in human health especially in cancer prevention. *Anti-Cancer Agents in Medicinal Chemistry (Formerly Current Medicinal Chemistry-Anti-Cancer Agents), 18*(13), 1815–1827.

Ishige, N. (2004). Genetic improvement for environmental stress resistance in Eucalyptus. *Abstract of the International Plant and Animal Genome XII, 2004*.

Jarre, M., Petit-Boix, A., Priefer, C., Meyer, R., & Leipold, S. (2020). Transforming the bio-based sector towards a circular economy-What can we learn from wood cascading? *Forest Policy and Economics, 110*, 101872.

Joshi, I., Bisht, P., Sharma, V., & Uniyal, D. (2003). In vitro clonal propagation of mature Eucalyptus F1 hybrid (Eucalyptus tereticornis Sm. x E. grandis Hill ex. Maiden). *Silvae Genetica, 52*(3–4), 110–113.

Juergens, L. J., Worth, H., & Juergens, U. R. (2020). New perspectives for mucolytic, anti-inflammatory and adjunctive therapy with 1, 8-cineole in COPD and asthma: review on the new therapeutic approach. *Advances in Therapy, 37*(5), 1737–1753.

Juergens, U. R., Stöber, M., & Vetter, H. (1998). Inhibition of cytokine production and arachidonic acid metabolism by eucalyptol (1.8-cineole) in human blood monocytes in vitro. *European Journal of Medical Research*, *3*, 508–510.

Kaur, S. (2017). In vitro regeneration of shoots from nodal explants of Dendrobium chrysotoxum Lindl. *Journal of Horticultural Research*, *25*(1).

Kawaoka, A., Nanto, K., Ishii, K., & Ebinuma, H. (2006). Reduction of lignin content by suppression of expression of the LIM domain transcription factor in Eucalyptus camaldulensis. *Silvae Genetica*, *55*(6), 269–277.

Kawasu, T., Doi, K., Ohta, T., Shinohara, Y., Ito, K., & Shibata, M. (1990). Transformation of Eucalyptus saligna using electroporation. Abstract In 7th international congress on plant tissue and cell culture, Amsterdam. IAPTC.

Kendurkar, S. V., & Rangaswamy, M. (2018). Genetic transformation in Eucalyptus. *Biotechnologies of Crop Improvement, Volume 2: Transgenic Approaches*, 335–366.

Kennedy-Feitosa, E., Okuro, R. T., Ribeiro, V. P., Lanzetti, M., Barroso, M. V., Zin, W. A., Porto, L. C., Brito-Gitirana, L., & Valenca, S. S. (2016). Eucalyptol attenuates cigarette smoke-induced acute lung inflammation and oxidative stress in the mouse. *Pulmonary Pharmacology & Therapeutics*, *41*, 11–18.

Khan, N., Fahad, S., Faisal, S., Akbar, A., & Naushad, M. (2020). Socio-economic and medicinal review of Eucalyptus tree in the world. *Available at SSRN 3644215*.

Khazraei, H., Shamsdin, S. A., & Zamani, M. (2021). In Vitro cytotoxicity and apoptotic assay of Eucalyptus globulus essential oil in colon and liver cancer cell lines. *Journal of Gastrointestinal Cancer*, 1–7.

Klocko, A. L., Ma, C., Robertson, S., Esfandiari, E., Nilsson, O., & Strauss, S. H. (2016). FT overexpression induces precocious flowering and normal reproductive development in Eucalyptus. *Plant Biotechnology Journal*, *14*(2), 808–819.

Kolachevskaya, O. O., Sergeeva, L. I., Floková, K., Getman, I. A., Lomin, S. N., Alekseeva, V. V., Rukavtsova, E. B., Buryanov, Y. I., & Romanov, G. A. (2017). Auxin synthesis gene tms1 driven by tuber-specific promoter alters hormonal status of transgenic potato plants and their responses to exogenous phytohormones. *Plant Cell Reports*, *36*, 419–435.

Kondo, K. (2002). Analysis of the stress response genes in Eucalyptus and effect of introducing a stress tolerance-giving gene (rd29A-DREBIA) into Eucalyptus. *Plant and Cell Physiology*, *43*, S113.

Koonin, E. V., Makarova, K. S., & Zhang, F. (2017). Diversity, classification and evolution of CRISPR-Cas systems. *Current Opinion in Microbiology*, *37*, 67–78.

Krimi, Z., Raio, A., Petit, A., Nesme, X., & Dessaux, Y. (2006). Eucalyptus occidentalis plantlets are naturally infected by pathogenic Agrobacterium tumefaciens. *European Journal of Plant Pathology*, *116*, 237–246.

Lahr, F. A., Nogueira, M. C., Araujo, V. A. D., Vasconcelos, J. S., & Christoforo, A. L. (2018). Wood utilization of Eucalyptus grandis in structural elements: densities and mechanical properties. *Engenharia Agrícola*, *38*, 642–647.

Lapierre, C., Pollet, B., Petit-Conil, M., Toval, G., Romero, J., Pilate, G., Leplé, J.-C., Boerjan, W., Ferret, V., & De Nadai, V. (1999). Structural alterations of lignins in transgenic poplars with depressed cinnamyl alcohol dehydrogenase or caffeic acid O-methyltransferase activity have an opposite impact on the efficiency of industrial kraft pulping. *Plant Physiology*, *119*(1), 153–164.

Lauvergeat, V., Rech, P., Jauneau, A., Guez, C., Coutos-Thevenot, P., & Grima-Pettenati, J. (2002). The vascular expression pattern directed by the Eucalyptus gunnii cinnamyl alcohol dehydrogenase EgCAD2 promoter is conserved among woody and herbaceous plant species. *Plant Molecular Biology*, *50*, 497–509.

Lawal, T. O. (2014). Ulcer-healing promoting activities of methanol extracts of Eucalyptus camaldulensis Dehnh. and Eucalyptus torelliana F. Muell in rat. *Archives of Basic and Applied Medicine*, *2*(3).

Lebedev, V. G., Lebedeva, T. N., Chernodubov, A. I., & Shestibratov, K. A. (2020). Genomic selection for forest tree improvement: Methods, achievements and perspectives. *Forests*, *11*(11), 1190.

Leung, R. S., & Katial, R. (2008). The diagnosis and management of acute and chronic sinusitis. *Primary Care: Clinics in Office Practice*, *35*(1), 11–24.

Levy, I., Shani, Z., & Shoseyov, O. (2002). Modification of polysaccharides and plant cell wall by endo-1, 4-β-glucanase and cellulose-binding domains. *Biomolecular Engineering*, *19*(1), 17–30.

Liu, C. L. C., Kuchma, O., & Krutovsky, K. V. (2018). Mixed-species versus monocultures in plantation forestry: Development, benefits, ecosystem services and perspectives for the future. *Global Ecology and Conservation*, *15*, e00419.

Liu, W., & Stewart Jr, C. N. (2016). Plant synthetic promoters and transcription factors. *Current Opinion in Biotechnology*, *37*, 36–44.

Long, J. M., Conn, A. B., Batchelor, W. J., & Evans, R. (2000). Comparison of methods to measure fibril angle in wood fibres. *Appita Journal*, 206–209.

Lu, X., Tang, F., Wang, Y., Zhao, T., & Bian, R. (2004). Effect of Eucalyptus globulus oil on lipopolysaccharide-induced chronic bronchitis and mucin hypersecretion in rats. *Zhongguo Zhong yao za zhi= Zhongguo zhongyao zazhi= China Journal of Chinese Materia Medica*, *29*(2), 168–171.

Ma, X., Zhu, Q., Chen, Y., & Liu, Y.-G. (2016). CRISPR/Cas9 platforms for genome editing in plants: developments and applications. *Molecular Plant*, *9*(7), 961–974.

Machado, L. d. O. R., de Andrade, G. M., Barrueto Cid, L. P., Penchel, R. M., & Brasileiro, A. C. M. (1997). Agrobacterium strain specificity and shooty tumour formation in eucalypt (Eucalyptus grandis× E. urophylla). *Plant Cell Reports*, *16*(5), 299–303.

Maciel, M., Morais, S., Bevilaqua, C., Silva, R., Barros, R., Sousa, R., Sousa, L., Brito, E., & Souza-Neto, M. (2010). Chemical composition of Eucalyptus spp. essential oils and their insecticidal effects on Lutzomyia longipalpis. *Veterinary Parasitology*, *167*(1), 1–7.

MacMillan, C. P., Mansfield, S. D., Stachurski, Z. H., Evans, R., & Southerton, S. G. (2010). Fasciclin-like arabinogalactan proteins: specialization for stem biomechanics and cell wall architecture in Arabidopsis and Eucalyptus. *The Plant Journal*, *62*(4), 689–703.

Macrae, S., & van Staden, J. (1993). Agrobacterium rhizogenes-mediated transformation to improve rooting ability of eucalypts. *Tree Physiology*, *12*(4), 411–418.

Malfertheiner, P., Chan, F. K., & McColl, K. E. (2009). Peptic ulcer disease. *The Lancet*, *374*(9699), 1449–1461.

Manders, G., Dos Santos, A., d'Utra Vaz, F., Davey, M., & Power, J. (1992). Transient gene expression in electroporated protoplasts of Eucalyptus citriodora Hook. *Plant Cell, Tissue and Organ Culture*, *30*, 69–75.

Mangwanda, R., Myburg, A. A., & Naidoo, S. (2015). Transcriptome and hormone profiling reveals Eucalyptus grandis defence responses against Chrysoporthe austroafricana. *BMC Genomics*, *16*(1), 1–13.

Matusick, G., Ruthrof, K. X., Kala, J., Brouwers, N. C., Breshears, D. D., & Hardy, G. E. S. J. (2018). Chronic historical drought legacy exacerbates tree mortality and crown dieback during acute heatwave-compounded drought. *Environmental Research Letters*, *13*(9), 095002.

Mendonça, E. G., Stein, V. C., Balieiro, F. P., Lima, C. D. F., Santos, B. R., & Paiva, L. V. (2013). Genetic transformation of Eucalyptus camaldulensis by agrobalistic method. *Revista Árvore*, *37*, 419–429.

Merad, M., & Martin, J. C. (2020). Pathological inflammation in patients with COVID-19: a key role for monocytes and macrophages. *Nature Reviews Immunology*, *20*(6), 355–362.

Mishra, A. K., Sahu, N., Mishra, A., Ghosh, A. K., Jha, S., & Chattopadhyay, P. (2010). Phytochemical screening and antioxidant activity of essential oil of Eucalyptus leaf. *Pharmacognosy Journal*, *2*(16), 25–28.

Mizrachi, E., Verbeke, L., Christie, N., Fierro, A. C., Mansfield, S. D., Davis, M. F., Gjersing, E., Tuskan, G. A., Van Montagu, M., & Van de Peer, Y. (2017). Network-based integration of systems genetics data reveals pathways associated with lignocellulosic biomass accumulation and processing. *Proceedings of the National Academy of Sciences*, *114*(5), 1195–1200.

Morimoto, Y., Sato, H., Hiramatsu, C., & Seno, T. (2019). Material surface properties modulate vection strength. *Experimental Brain Research*, *237*, 2675–2690.

Mortensen, S., Bernal-Franco, D., Cole, L. F., Sathitloetsakun, S., Cram, E. J., & Lee-Parsons, C. W. (2019). EASI transformation: an efficient transient expression method for analyzing gene function in Catharanthus roseus seedlings. *Frontiers in Plant Science*, *10*, 755.

Mou, H., Wu, X., Huang, J., Liu, Y., & Fan, H. (2021). Eucalyptus lignin modification for dynamic adsorption with lignocellulose-degradation enzymes dependent on pH values. *Industrial Crops and Products*, *169*, 113650.

Mphahlele, M. M., Isik, F., Hodge, G. R., & Myburg, A. A. (2021). Genomic breeding for diameter growth and tolerance to Leptocybe gall wasp and Botryosphaeria/Teratosphaeria fungal disease complex in Eucalyptus grandis. *Frontiers in Plant Science*, *12*, 638969.

Mubarak, E. E., Ali, L. Z., Ahmed, I. F. A., Ahmed, A. B. A., & Taha, R. M. (2015). Essential oil compositions and cytotoxicity from various organs of Eucalyptus camaldulensis. *International Journal of Agriculture & Biology*, 320–326.

Mullins, K., Llewellyn, D., Hartney, V., Strauss, S., & Dennis, E. (1997). Regeneration and transformation of Eucalyptus camaldulensis. *Plant Cell Reports*, *16*, 787–791.

Myburg, A. A., Grattapaglia, D., Tuskan, G. A., Hellsten, U., Hayes, R. D., Grimwood, J., Jenkins, J., Lindquist, E., Tice, H., & Bauer, D. (2014). The genome of Eucalyptus grandis. *Nature*, *510*(7505), 356–362.

Nagle, M. F., Nahata, S. S., Zahl, B., Niño de Rivera, A., Tacker, X. V., Elorriaga, E., Ma, C., Goralogia, G. S., Klocko, A. L., & Gordon, M. (2023). Knockout of floral and meiosis genes using CRISPR/Cas9 produces male-sterility in Eucalyptus without impacts on vegetative growth. *Plant Direct*, *7*(7), e507.

Naidoo, S., Slippers, B., Plett, J. M., Coles, D., & Oates, C. N. (2019). The road to resistance in forest trees. *Frontiers in Plant Science*, *10*, 273.

Nakamura, T., Yoshida, N., Yamanoi, Y., Honryo, A., Tomita, H., Kuwabara, H., & Kojima, Y. (2020). Eucalyptus oil reduces allergic reactions and suppresses mast cell degranulation by downregulating IgE-FcεRI signalling. *Scientific Reports*, *10*(1), 20940.

Navarro, M., Ayax, C., Martinez, Y., Laur, J., El Kayal, W., Marque, C., & Teulieres, C. (2011). Two EguCBF1 genes overexpressed in Eucalyptus display a different impact on stress tolerance and plant development. *Plant Biotechnology Journal*, *9*(1), 50–63.

Ngomuo, M., Mneney, E., & Ndakidemi, P. A. (2014). The in vitro propagation techniques for producing banana using shoot tip cultures. *American Journal of Plant Sciences*, 2014.

Nourzadeh, M., Amini, A., Fakoor, F., Raoof, M., & Sharififar, F. (2017). Comparative antimicrobial efficacy of Eucalyptus galbie and Myrtus communis L. extracts, chlorhexidine and sodium hypochlorite against Enterococcus faecalis. *Iranian Endodontic Journal*, *12*(2), 205.

Nowak, K., Luniak, N., Meyer, S., Schulze, J., Mendel, R., & Hänsch, R. (2004). Fluorescent proteins in poplar: a useful tool to study promoter function and protein localization. *Plant Biology*, *7*(01), 65–73.

Ogunwande, I. A., Olawore, N. O., Adeleke, K. A., & Konig, W. A. (2003). Chemical composition of the essential oils from the leaves of three Eucalyptus species growing in Nigeria. *Journal of Essential Oil Research*, *15*(5), 297–301.

Ohara, K., Matsunaga, E., Nanto, K., Yamamoto, K., Sasaki, K., Ebinuma, H., & Yazaki, K. (2010). Monoterpene engineering in a woody plant Eucalyptus camaldulensis using a limonene synthase cDNA. *Plant Biotechnology Journal*, *8*(1), 28–37.

Oliveira, T. S. M. d. P. d. A. (2015). *Models to support eucalyptus plantations management under a changing environment* Repositório da Universidade de Lisboa]. ISA/UL.

Ouyang, L., He, W., Huang, Z., Zhao, L., Peng, S., Sha, Y., Zeng, F., & Lu, X. (2012). Introduction of the Rs-AFP2 gene into Eucalyptus urophylla for resistance to Phytophthora capsici. *Journal of Tropical Forest Science*, *24*, 198–208.

Ouyang, L., & Li, L. (2016). Effects of an inducible aiiA gene on disease resistance in Eucalyptus urophylla× Eucalyptus grandis. *Transgenic Research*, *25*, 441–452.

Panikar, S., Shoba, G., Arun, M., Sahayarayan, J. J., Nanthini, A. U. R., Chinnathambi, A., Alharbi, S. A., Nasif, O., & Kim, H.-J. (2021). Essential oils as an effective alternative for the treatment of COVID-19: Molecular interaction analysis of protease (Mpro) with pharmacokinetics and toxicological properties. *Journal of Infection and Public Health*, *14*(5), 601–610.

Pérez-Tornero, O., & Burgos, L. (2000). Different media requirements for micropropagation of apricot cultivars. *Plant Cell, Tissue and Organ Culture*, *63*, 133–141.

Phitsuwan, P., Sakka, K., & Ratanakhanokchai, K. (2013). Improvement of lignocellulosic biomass in planta: a review of feedstocks, biomass recalcitrance, and strategic manipulation of ideal plants designed for ethanol production and processability. *Biomass and Bioenergy*, *58*, 390–405.

Piao, Y., Lee, D., Park, S., Kim, H. G., & Jin, Y. (2022). Multi-hazard mapping of droughts and forest fires using a multi-layer hazards approach with machine learning algorithms. *Geomatics, Natural Hazards and Risk*, *13*(1), 2649–2673.

Pinheiro, R. E. E., Chaves, T. P., Melo, E. S., Ali, S., Ali, S. W., Umer, M., Gama, G. S. P., Lira, D. N., Souza, J. S. N., & Soares, M. J. d. S. (2020). Modulatory-antibiotic activity of the essential oil from Eucalyptus citriodora against MDR bacterial strains. *Cellular and Molecular Biology*, *66*(4), 60–64.

Plasencia, A., Soler, M., Dupas, A., Ladouce, N., Silva-Martins, G., Martinez, Y., Lapierre, C., Franche, C., Truchet, I., & Grima-Pettenati, J. (2016). Eucalyptus hairy roots, a fast, efficient and versatile tool to explore function and expression of genes involved in wood formation. *Plant Biotechnology Journal*, *14*(6), 1381–1393.

Popa, V. I., & Volf, I. (2018). *Biomass as renewable raw material to obtain bioproducts of high-tech value*. Elsevier.

Porto, M. S., Pinheiro, M. P. N., Batista, V. G. L., dos Santos, R. C., de Albuquerque Melo Filho, P., & de Lima, L. M. (2014). Plant promoters: an approach of structure and function. *Molecular Biotechnology*, *56*, 38–49.

Prakash, M., & Gurumurthi, K. (2010). Effects of type of explant and age, plant growth regulators and medium strength on somatic embryogenesis and plant regeneration in Eucalyptus camaldulensis. *Plant Cell, Tissue and Organ Culture (PCTOC)*, *100*, 13–20.

Pritzkow, C., Szota, C., Williamson, V. G., & Arndt, S. K. (2020). Phenotypic plasticity of drought tolerance traits in a widespread eucalypt (Eucalyptus obliqua). *Forests*, *11*(12), 1371.

Quezada, M., Giorello, F. M., Da Silva, C. C., Aguilar, I., & Balmelli, G. (2023). Single-step genome-wide association study for susceptibility to Teratosphaeria nubilosa and precocity of vegetative phase change in Eucalyptus globulus. *Frontiers in Plant Science*, *14*, 1124768.

Rajendran Ramjegathesh, R. R., & Jayaraj Jayaraman, J. J. (2015). Chitosan for plant disease management-prospects and problems. *Sustainable Crop Disease Management Using Natural Products*, 198–212.

Ramalho, M. A. P., Santos, H. G., & Souza, T. d. S. (2022). Eucalyptus breeding programs: a proposal for the use of inbred progênies. *CERNE*, *28*.

Rantzsch, U., Vacca, G., Dück, R., & Gillissen, A. (2009). Anti-inflammatory effects of Myrtol standardized and other essential oils on alveolar macrophages from patients with chronic obstructive pulmonary disease. *European Journal of Medical Research*, *14*, 1–5.

Rasooli, I., Shayegh, S., & Astaneh, S. D. A. (2009). The effect of Mentha spicata and Eucalyptus camaldulensis essential oils on dental biofilm. *International Journal of Dental Hygiene*, *7*(3), 196–203.

Resende, R. T., Resende, M. D. V., Silva, F. F., Azevedo, C. F., Takahashi, E. K., Silva-Junior, O. B., & Grattapaglia, D. (2017). Regional heritability mapping and genome-wide association identify loci for complex growth, wood and disease resistance traits in Eucalyptus. *New Phytologist*, *213*(3), 1287–1300.

Robson, M. C. (1997). Wound infection: a failure of wound healing caused by an imbalance of bacteria. *Surgical Clinics of North America*, *77*(3), 637–650.

Rocha-Sepúlveda, M. F., Williams, D., Vega, M., Harrison, P. A., Vaillancourt, R. E., & Potts, B. M. (2021). Genetic variation of microfibril angle and its relationship with solid wood and pulpwood traits in two progeny trials of Eucalyptus nitens in Tasmania. *Holzforschung*, *75*(8), 689–701.

Rochange, F., Serrano, L., Marque, C., Teulieres, C., & Boudet, A.-M. (1995). DNA delivery into Eucalyptus globulus zygotic embryos through biolistics: optimization of the biological and physical parameters of bombardment for two different particle guns. *Plant Cell Reports*, *14*, 674–678.

Rushton, P. J. (2016). What have we learned about synthetic promoter construction? *Plant Synthetic Promoters: Methods and Protocols*, 1–13.

Rushton, P. J., Reinstadler, A., Lipka, V., Lippok, B., & Somssich, I. E. (2002). Synthetic plant promoters containing defined regulatory elements provide novel insights into pathogen-and wound-induced signaling. *The Plant Cell*, *14*(4), 749–762.

Sakurai, K., Katoh, M., Someno, K., & FUJIMOTO, Y. (2001). Apoptosis and mitochondrial damage in INS-1 cells treated with alloxan. *Biological and Pharmaceutical Bulletin*, *24*(8), 876–882.

Sampaio Filho, I. d. J., Jardine, K. J., De Oliveira, R. C. A., Gimenez, B. O., Cobello, L. O., Piva, L. R. d. O., Candido, L. A., Higuchi, N., & Chambers, J. Q. (2018). Below versus above ground plant sources of abscisic acid (ABA) at the heart of tropical forest response to warming. *International Journal of Molecular Sciences*, *19*(7), 2023.

Sartoretto, L. M., Cid, L. P. B., & Brasileiro, A. C. M. (2002). Biolistic transformation of Eucalyptus grandis x E. urophylla callus. *Functional Plant Biology*, *29*(8), 917–924.

Saveyn, A., Steppe, K., Ubierna, N., & Dawson, T. E. (2010). Woody tissue photosynthesis and its contribution to trunk growth and bud development in young plants. *Plant, Cell & Environment*, *33*(11), 1949–1958.

Savidge, R. (2000). Biochemistry of seasonal cambial growth and wood formation-an overview of the challenges. *Cell and Molecular Biology of Wood Formation*, 1–30.

Schlabach, M. R., Hu, J. K., Li, M., & Elledge, S. J. (2010). Synthetic design of strong promoters. *Proceedings of the National Academy of Sciences*, *107*(6), 2538–2543.

Sebei, K., Sakouhi, F., Herchi, W., Khouja, M. L., & Boukhchina, S. (2015). Chemical composition and antibacterial activities of seven Eucalyptus species essential oils leaves. *Biological Research*, *48*, 1–5.

Serrano, L., Rochange, F., Semblat, J., Marque, C., Teulières, C., & Boudet, A. (1996). Genetic transformation of Eucalyptus globulus through biolistics: complementary development of procedures for organogenesis from zygotic embryos and stable transformation of corresponding proliferating tissue. *Journal of Experimental Botany*, *47*(2), 285–290.

Sewanu, S. O. (2012). *The chemical composition, antimicrobial and antioxidant properties of the essential oils of Tulbaghia violacea Harv and Eucalyptus grandis W. Hill ex Maiden*

Shani, Z., Dekel, M., Cohen, B., Barimboim, N., Kolosovski, N., Safranuvitch, A., Cohen, O., & Shoseyov, O. (2003). Cell wall modification for the enhancement of commercial eucalyptus species. *IUFRO tree biotechnology. Umea Plant Science Center, Umea*, S10–S26.

Shao, J., Yin, Z., Wang, Y., Yang, Y., Tang, Q., Zhang, M., Jiao, J., Liu, C., Yang, M., & Zhen, L. (2020). Effects of different doses of Eucalyptus oil from Eucalyptus globulus Labill on respiratory tract immunity and immune function in healthy Rats. *Frontiers in Pharmacology*, *11*, 1287.

Shao, Z. (2002). Studies on the introduction of the ceropin D gene into Eucalyptus urophylla to breed the resistant varieties to Pseudomonas solanacearum. *Scientia Silvae Sinica*, *38*, 92–97.

Sharma, V., Ankita, Karnwal, A., Sharma, S., Kamal, B., Jadon, V. S., Gupta, S., & Sivanasen, I. (2023). A comprehensive review uncovering the challenges and advancements in the in vitro propagation of Eucalyptus plantations. *Plants*, *12*(17), 3018.

Shen, G., Tan, S., Sun, X., Chen, Y., & Li, B. (2020). Experimental evidence for the importance of light on understory grass communities in a subtropical forest. *Frontiers in Plant Science*, *11*, 1051.

Shen, X.-J., Wen, J.-L., Mei, Q.-Q., Chen, X., Sun, D., Yuan, T.-Q., & Sun, R.-C. (2019). Facile fractionation of lignocelluloses by biomass-derived deep eutectic solvent (DES) pretreatment for cellulose enzymatic hydrolysis and lignin valorization. *Green Chemistry*, *21*(2), 275–283.

Shen, Y., Li, Y., Xu, D., Yang, C., Li, C., & Luo, K. (2018). Molecular cloning and characterization of a brassinosteriod biosynthesis-related gene PtoDWF4 from Populus tomentosa. *Tree Physiology*, *38*(9), 1424–1436.

Signat, B., Roques, C., Poulet, P., & Duffaut, D. (2011). Role of Fusobacterium nucleatum in periodontal health and disease. *Current Issues in Molecular Biology*, *13*(2), 25–36.

Singh, A., Avula, A., & Zahn, E. (2018). Acute Bronchitis [Internet]. Treasure Island (FL): StatPearls; Publishing; 2022 [Citado 22/10/2022].

Soler, M., Camargo, E. L. O., Carocha, V., Cassan-Wang, H., San Clemente, H., Savelli, B., Hefer, C. A., Paiva, J. A. P., Myburg, A. A., & Grima-Pettenati, J. (2015). The E ucalyptus grandis R 2 R 3-MYB transcription factor family: evidence for woody growth-related evolution and function. *New Phytologist*, *206*(4), 1364–1377.

Sonoda, T., Koita, H., Nakamoto-Ohta, S., Kondo, K., Suezaki, T., Kato, T., Nagai, K., Iida, N., Sato, S., & Umezawa, T. (2009). Increasing fiber length and growth in transgenic tobacco plants overexpressing a gene encoding the Eucalyptus camaldulensis HD-Zip class II transcription factor. *Plant Biotechnology*, *26*(1), 115–120.

Souza, D. M. S. C., Martins, A. R., Fernandes, S. B., Avelar, M. L. M., Molinari, L. V., Gonçalves, D. S., & Brondani, G. E. (2022). In vitro multiplication of Eucalyptus pilularis and Eucalyptus grandis x E. urophylla (Urograndis Eucalypt): Effect of light quality in temporary immersion bioreactor. *Mindanao Journal of Science and Technology*, *20*(1), 72–86.

Spokevicius, A. V., Van Beveren, K. S., & Bossinger, G. (2006). Agrobacterium-mediated transformation of dormant lateral buds in poplar trees reveals developmental patterns in secondary stem tissues. *Functional Plant Biology*, *33*(2), 133–139.

Stafford, J. L., Neumann, N. F., & Belosevic, M. (2002). Macrophage-mediated innate host defense against protozoan parasites. *Critical Reviews in Microbiology*, *28*(3), 187–248.

Strader, L. C., & Bartel, B. (2011). Transport and metabolism of the endogenous auxin precursor indole-3-butyric acid. *Molecular Plant*, *4*(3), 477–486.

Su, W., Xu, M., Radani, Y., & Yang, L. (2023). Technological Development and Application of Plant Genetic Transformation. *International Journal of Molecular Sciences*, *24*(13), 10646.

Subrahmaniam, H. J., Libourel, C., Journet, E. P., Morel, J. B., Muños, S., Niebel, A., Raffaele, S., & Roux, F. (2018). The genetics underlying natural variation of plant–plant interactions, a beloved but forgotten member of the family of biotic interactions. *The Plant Journal*, *93*(4), 747–770.

Sudhoff, H., Klenke, C., Greiner, J. F., Müller, J., Brotzmann, V., Ebmeyer, J., Kaltschmidt, B., & Kaltschmidt, C. (2015). 1, 8-Cineol reduces mucus-production in a novel human ex vivo model of late rhinosinusitis. *PLoS One*, *10*(7), e0133040.

Sugimoto, K., Nakagawa, K., Konoeda, S., Hayashi, S., Inui, H., & Nakano, Y. (2003). Composition for preventing fructose absorption. In: Google Patents.

Sundell, D., Mannapperuma, C., Netotea, S., Delhomme, N., Lin, Y.-C., Sjödin, A., Van de Peer, Y., Jansson, S., Hvidsten, T. R., & Street, N. R. (2015). The plant genome integrative explorer resource: PlantGenIE.org. *New Phytologist*, *208*(4), 1149–1156.

Suqin, C., Holtan, H. E., Repetti, P. P., & Reuber, T. L. (2011). Tissue-enhanced promoters. In: Google Patents.

Taheri, E., Ghorbani, S., Safi, M., Sani, N. S., Amoodizaj, F. F., Heidari, M., Chavoshi, R., Hajazimian, S., Isazadeh, A., & Heidari, M. (2020). Inhibition of colorectal cancer cell line CaCo-2 by essential oil of eucalyptus camaldulensis through induction of apoptosis. *Acta Medica Iranica*, 260–265.

Takasaki, M., Konoshima, T., Etoh, H., Singh, I. P., Tokuda, H., & Nishino, H. (2000). Cancer chemopreventive activity of euglobal-G1 from leaves of Eucalyptus grandis. *Cancer Letters*, *155*(1), 61–65.

Tannure Faria, J. C., Ribeiro-Kumara, C., da Rocha Costa, R. S., Nieri, E. M., de Carvalho, D., Brasil Pereira Pinto, J. E., de Sena Neto, A. R., & Brondani, G. E. (2022). Use of biodegradable polyester-based microvessels for micropropagation of mature Eucalyptus microcorys. *New Zealand Journal of Forestry Science, 52*, 736.

Taylor, L., Spokevicius, A., Mac Millan, C., Southerton, S., & Bossinger, G. (2007). Promoter studies during tension wood formation using induced somatic sector analysis (ISSA). *Poster presented at IUFRO Tree Biotechnology Congress*.

Teale, W. D., Paponov, I. A., & Palme, K. (2006). Auxin in action: signalling, transport and the control of plant growth and development. *Nature Reviews Molecular Cell Biology*, *7*(11), 847–859.

Teshome, D. T., Zharare, G. E., & Naidoo, S. (2020). The threat of the combined effect of biotic and abiotic stress factors in forestry under a changing climate. *Frontiers in Plant Science*, *11*, 1874.

Teulieres, C., Grima-Pettenati, J., Curie, C., Teissie, J., & Boudet, A. (1991). Transient foreign gene expression in polyethylene/glycol treated or electropulsated Eucalyptus gunnii protoplasts. *Plant Cell, Tissue and Organ Culture*, *25*, 125–132.

Thumma, B. R., Nolan, M. F., Evans, R., & Moran, G. F. (2005). Polymorphisms in cinnamoyl CoA reductase (CCR) are associated with variation in microfibril angle in Eucalyptus spp. *Genetics*, *171*(3), 1257–1265.

Tournier, V., Grat, S., Marque, C., El Kayal, W., Penchel, R., de Andrade, G., Boudet, A.-M., & Teulières, C. (2003). An efficient procedure to stably introduce genes into an economically important pulp tree (Eucalyptus grandis× Eucalyptus urophylla). *Transgenic Research*, *12*, 403–411.

Tran, N.-H. T., Oguchi, T., Matsunaga, E., Kawaoka, A., Watanabe, K. N., & Kikuchi, A. (2018). Environmental risk assessment of impacts of transgenic Eucalyptus camaldulensis events highly expressing bacterial Choline Oxidase A gene. *Plant Biotechnology*, *35*(4), 393–397.

Tran, N. H. T., Oguchi, T., Akatsuka, N., Matsunaga, E., Kawaoka, A., Yamada, A., Ozeki, Y., Watanabe, K. N., & Kikuchi, A. (2019). Development and evaluation of novel salt-tolerant Eucalyptus trees by molecular breeding using an RNA-Binding-Protein gene derived from common ice plant (Mesembryanthemum crystallinum L.). *Plant Biotechnology Journal*, *17*(4), 801–811.

Tsai, C.-J., & Xue, L.-J. (2015). CRISPRing into the woods. *GM Crops & Food*, *6*(4), 206–215.

Tu, Y., Rochfort, S., Liu, Z., Ran, Y., Griffith, M., Badenhorst, P., Louie, G. V., Bowman, M. E., Smith, K. F., & Noel, J. P. (2010). Functional analyses of caffeic acid O-methyltransferase and cinnamoyl-CoA-reductase genes from perennial ryegrass (Lolium perenne). *The Plant Cell*, *22*(10), 3357–3373.

Usachev, E. V., Pyankov, O. V., Usacheva, O. V., & Agranovski, I. E. (2013). Antiviral activity of tea tree and eucalyptus oil aerosol and vapour. *Journal of Aerosol Science*, *59*, 22–30.

Valério, L., Carter, D., Rodrigues, J. C., Tournier, V., Gominho, J., Marque, C., Boudet, A.-M., Maunders, M., Pereira, H., & Teulières, C. (2003). Down regulation of cinnamyl alcohol dehydrogenase, a lignification enzyme, in Eucalyptus camaldulensis. *Molecular Breeding*, *12*, 157–167.

Velini, E. D., Trindade, M. L. B., Alves, E., Catâneo, A. C., Marino, C. L., Maia, I. d. G., Mori, E. S., Furtado, E. L., Guerrini, I. A., & Wilcken, C. F. (2005). Eucalyptus ESTs corresponding to the protoporphyrinogen IX oxidase enzyme related to the synthesis of heme, chlorophyll, and to the action of herbicides. *Genetics and Molecular Biology*, *28*, 548–554.

Velmurugan, C., Geetha, C., Shajahan, S., Vijayakumar, S., & Kumar, P. L. (2014). Wound healing potential of leaves of Eucalyptus Citriodora lin rats. *World Journal of Pharmaceutical Sciences*, *2*(1), 1–136.

Voytas, D. F. (2013). Plant genome engineering with sequence-specific nucleases. *Annual Review of Plant Biology*, *64*, 327–350.

Wan, S., Li, C., Ma, X., & Luo, K. (2017). PtrMYB57 contributes to the negative regulation of anthocyanin and proanthocyanidin biosynthesis in poplar. *Plant Cell Reports*, *36*, 1263–1276.

Wang, J. P., Matthews, M. L., Williams, C. M., Shi, R., Yang, C., Tunlaya-Anukit, S., Chen, H.-C., Li, Q., Liu, J., & Lin, C.-Y. (2018). Improving wood properties for wood utilization through multi-omics integration in lignin biosynthesis. *Nature Communications*, *9*(1), 1579.

Wang, L., Ran, L., Hou, Y., Tian, Q., Li, C., Liu, R., Fan, D., & Luo, K. (2017). The transcription factor MYB115 contributes to the regulation of proanthocyanidin biosynthesis and enhances fungal resistance in poplar. *New Phytologist*, *215*(1), 351–367.

Wang, X., Chen, S., Zhang, H., Luo, P., Zhou, F., Zeng, B., Xu, J., & Fan, C. (2022). Agrobacterium-mediated genetic transformation of the most widely cultivated superior clone Eucalyptus urophylla× Eucalyptus grandis DH32–29 in Southern China. *Frontiers in Plant Science*, *13*, 1011245.

Wang, Z., Li, L., & Ouyang, L. (2021). Efficient genetic transformation method for Eucalyptus genome editing. *PLoS One*, *16*(5), e0252011.

Wegrzyn, J. L., Lee, J. M., Tearse, B. R., & Neale, D. B. (2008). TreeGenes: A forest tree genome database. *International Journal of Plant Genomics*, *2008, 412875*.

Wei, H., Geng, J., Shi, B., Liu, Z., Wang, Y.-H., Stevens, A. C., Sprout, S. L., Yao, M., Wang, H., & Hu, H. (2016). Cutting edge: Foxp1 controls naive CD8+ T cell quiescence by simultaneously repressing key pathways in cellular metabolism and cell cycle progression. *Journal of Immunology*, *196*(9), 3537–3541.

Weng, J. K., & Chapple, C. (2010). The origin and evolution of lignin biosynthesis. *New Phytologist*, *187*(2), 273–285.

Williams, J. E., & Woinarski, J. (1997). *Eucalypt ecology: individuals to ecosystems*. Cambridge University Press.

Wingfield, M. J. (2003). 2003 Daniel McAlpine memorial lecture increasing threat of diseases to exotic plantation forests in the Southern Hemisphere: lessons from Cryphonectria canker. *Australasian Plant Pathology*, *32*, 133–139.

Woodward, A. W., & Bartel, B. (2005). Auxin: regulation, action, and interaction. *Annals of Botany*, *95*(5), 707–735.

Wu, J., Jiang, Y., Liang, Y., Chen, L., Chen, W., & Cheng, B. (2019). Expression of the maize MYB transcription factor ZmMYB3R enhances drought and salt stress tolerance in transgenic plants. *Plant Physiology and Biochemistry*, *137*, 179–188.

Wu, W., Zhu, L., Wang, P., Liao, Y., Duan, L., Lin, K., Chen, X., Li, L., Xu, J., & Hu, H. (2023). Transcriptome-based construction of the Gibberellin metabolism and signaling pathways in Eucalyptus grandis× E. urophylla, and functional characterization of GA20ox and GA2ox in regulating plant development and abiotic stress adaptations. *International Journal of Molecular Sciences*, *24*(8), 7051.

Xu, C., Debnath, S., Syed, A., Elgorban, A. M., Bahkali, A. H., Eswaramaathy, R., Verma, M., Uddin Helal, M. M., & Jian, X. (2023). Action of the plant-based essential oil-derived compound Taxol for improvising drought tolerance in Eucalyptus by modulating the VIT1 channel protein: a cutting-edge computational approach. *Frontiers in Genetics*, *14*, 1165518.

Xu, C., Fu, X., Liu, R., Guo, L., Ran, L., Li, C., Tian, Q., Jiao, B., Wang, B., & Luo, K. (2017). PtoMYB170 positively regulates lignin deposition during wood formation in poplar and confers drought tolerance in transgenic Arabidopsis. *Tree Physiology*, *37*(12), 1713–1726.

Xu, R.-F., Li, H., Qin, R.-Y., Li, J., Qiu, C.-H., Yang, Y.-C., Ma, H., Li, L., Wei, P.-C., & Yang, J.-B. (2015). Generation of inheritable and "transgene clean" targeted genome-modified rice in later generations using the CRISPR/Cas9 system. *Scientific Reports*, *5*(1), 11491.

Yadav, N., & Chandra, H. (2017). Suppression of inflammatory and infection responses in lung macrophages by eucalyptus oil and its constituent 1, 8-cineole: Role of pattern recognition receptors TREM-1 and NLRP3, the MAP kinase regulator MKP-1, and NFκB. *PLoS One*, *12*(11), e0188232.

Yahiaoui, N., Marque, C., Corbière, H., & Boudet, A. M. (1998). Comparative efficiency of different constructs for down regulation of tobacco cinnamyl alcohol dehydrogenase. *Phytochemistry*, *49*(2), 295–306.

Yamada-Watanabe, K., Kawaoka, A., Matsunaga, K., Nanto, K., Sugita, K., Endo, S., Ebinuma, H., & Murata, N. (2003). Molecular breeding of Eucalyptus: analysis of salt stress tolerance in transgenic Eucalyptus camaldulensis that over-expressed choline oxidase gene (codA). *IUFRO tree biotechnology. Umea Plant Science Centre, Umea*, S7–S9.

Yang, L., Zhao, X., Ran, L., Li, C., Fan, D., & Luo, K. (2017). PtoMYB156 is involved in negative regulation of phenylpropanoid metabolism and secondary cell wall biosynthesis during wood formation in poplar. *Scientific Reports*, *7*(1), 41209.

Yu, H., Soler, M., Mila, I., San Clemente, H., Savelli, B., Dunand, C., Paiva, J. A., Myburg, A. A., Bouzayen, M., & Grima-Pettenati, J. (2014). Genome-wide characterization and expression profiling of the AUXIN RESPONSE FACTOR (ARF) gene family in Eucalyptus grandis. *PLoS One*, *9*(9), e108906.

Yu, H., Soler, M., San Clemente, H., Mila, I., Paiva, J. A., Myburg, A. A., Bouzayen, M., Grima-Pettenati, J., & Cassan-Wang, H. (2015). Comprehensive genome-wide analysis of the Aux/IAA gene family in Eucalyptus: evidence for the role of EgrIAA4 in wood formation. *Plant and Cell Physiology*, *56*(4), 700–714.

Yu, N., Sun, Y.-T., Su, X.-M., He, M., Dai, B., & Kang, J. (2018). Treatment with eucalyptol mitigates cigarette smoke-induced lung injury through suppressing ICAM-1 gene expression. *Bioscience Reports*, *38*(4), BSR20171636.

Yu, X., Kikuchi, A., Matsunaga, E., Morishita, Y., Nanto, K., Sakurai, N., Suzuki, H., Shibata, D., Shimada, T., & Watanabe, K. N. (2013). The choline oxidase gene codA confers salt tolerance to transgenic Eucalyptus globulus in a semi-confined condition. *Molecular Biotechnology*, *54*, 320–330.

Zhan, N., Huang, L., Wang, Z., Zhang, J., Xie, Y., Shang, X., Liu, G., & Wu, Z. (2022). Expression of genes encoding terpenoid biosynthesis enzymes during leaf development of Eucalyptus camaldulensis. *Biologia Plantarum*, *66*, 146–154.

Zhang, N., & Bevan, M. J. (2011). CD8+ T cells: foot soldiers of the immune system. *Immunity*, *35*(2), 161–168.

Zhang, Q., Cheetamun, R., Dhugga, K. S., Rafalski, J. A., Tingey, S. V., Shirley, N. J., Taylor, J., Hayes, K., Beatty, M., & Bacic, A. (2014). Spatial gradients in cell wall composition and transcriptional profiles along elongating maize internodes. *BMC Plant Biology*, *14*(1), 1–19.

Zhao, S., & Zhang, D. (2014). Supercritical CO2 extraction of Eucalyptus leaves oil and comparison with Soxhlet extraction and hydro-distillation methods. *Separation and Purification Technology*, *133*, 443–451.

Zhao, Y. (2014). Auxin biosynthesis. *The Arabidopsis Book/American Society of Plant Biologists*, *12*, 1–14.

Zhong, R., Burk, D. H., Morrison III, W. H., & Ye, Z.-H. (2002). A kinesin-like protein is essential for oriented deposition of cellulose microfibrils and cell wall strength. *The Plant Cell*, *14*(12), 3101–3117.

Zhou, T., Lin, Y., Lin, Y., Luo, J., & Ding, J. (2022). Regeneration and Agrobacterium-mediated genetic transformation of twelve Eucalyptus species. *Forestry Research*, *2*(1).

Zhou, X., Jacobs, T. B., Xue, L.-J., Harding, S. A., & Tsai, C.-J. (2015). Exploiting SNPs for biallelic CRISPR mutations in the outcrossing woody perennial Populus reveals 4-coumarate: CoA ligase specificity and redundancy. *New Phytologist*, *208*(2), 298–301.

Zwart, L., Berger, D. K., Moleleki, L. N., van der Merwe, N. A., Myburg, A. A., & Naidoo, S. (2017). Evidence for salicylic acid signalling and histological changes in the defence response of Eucalyptus grandis to Chrysoporthe austroafricana. *Scientific Reports*, *7*(1), 45402.

TABLE 4.2
Comprehensive overview of key aspects in the *Eucalyptus* gene transformation process

Aspect	Description	Key Factors/ Components	Methods/ Techniques Employed	Results/ Implications for *Eucalyptus* Transformation	Reference
Gene Transformation of Eucalyptus	A comprehensive overview of the multi-step gene transformation process in *Eucalyptus*, encompassing tissue culture, optimization protocols, and T-DNA insertion techniques.	Tissue culture media, growth regulators, nutrients, solidified agents, varied explants, and T-DNA insertion methods.	In vitro culture, protocol optimization, electroporation, polyethylene glycol methods.	Efficient tissue culture is pivotal for successful gene transformation, emphasizing the role of diverse explants.	(Abiri et al., 2020)
Factors Affecting Micropropagation	Exploration of variables influencing micropropagation efficiency, emphasizing the impact of growth regulators, culture medium composition, and light conditions.	Cytokinins, auxins, minerals, vitamins, carbon sources, pH, light intensity, photoperiod.	Semi-solid vs. liquid media, Murashige and Skoog, JADS, WPM, DKW media, light spectrum considerations.	Light intensity affects shoot initiation, with liquid media enhancing *Eucalyptus* shoot multiplication.	(Teale et al., 2006; Prakash et al., 2010; Brondani et al., 2011; Oliveira, 2015; Tannure Faria et al., 2022)
Promoters in *Eucalyptus* Engineering	In-depth examination of the pivotal role played by promoters in achieving optimal gene expression levels during *Eucalyptus* engineering.	Exogenous and endogenous promoters, constitutive, inducible, and tissue-specific promoters.	Use of CaMV 35S promoter, exploration of inducible and tissue-specific promoters.	Constitutive promoters may lead to adverse modifications; inducible promoters offer a promising solution.	(Zhao et al., 2014; Huang et al., 2019; Nagle et al., 2023)
Selection Markers and Reporter Genes	Insight into the utilization of markers and reporter genes for evaluating transformation success and efficiency in *Eucalyptus*.	Reporter genes (uidA/GUS, yfp, cfp, gfp), selection markers (NPT II, bar, kanamycin-resistant genes).	Use of uidA for histochemical staining, antibiotic/ herbicide resistance markers for selection.	uidA is commonly used with limitations; alternative markers such as yfp, cfp, and gfp have been explored.	(Zhou et al., 2022)
Spatial Organization of *Cis*-Elements	Discussion on the significance of spatial organization for promoter strength and activity in *Eucalyptus* gene transformation.	Promoters with multiple *cis*-elements, augmentation of *cis*-elements, importance of proper spacing.	Enhancement of promoter strength through *cis*-element augmentation, spatial separation impact.	Proper spacing of *cis*-elements is crucial for optimal synthetic promoter activity.	(Benfey et al., 1990; Rushton et al., 2002; Schlabach et al., 2010; Liu et al., 2016; Rushton, 2016; Huang et al., 2022)

TABLE 4.2 (Continued)
Comprehensive overview of key aspects in the Eucalyptus gene transformation process

Aspect	Description	Key Factors/ Components	Methods/ Techniques Employed	Results/ Implications for *Eucalyptus* Transformation	Reference
Synthetic Pathogen-Inducible Promoters (SPIP)	Exploration of synthetic promoters designed for disease resistance in *Eucalyptus*, focusing on their potential applications and benefits.	Combination of diverse dimers with CaMVmini35S promoter, creating Synthetic Pathogen-Inducible Promoters (SPIP).	Design and insertion of SPIPs (VGWSF, GWVSF, GWSFV) with virus-inducing ability.	SPIPs show potential for precise and timely expression of disease-resistant genes in transgenic Eucalyptus.	
Gene Transformation Methods in Eucalyptus	Overview of gene transformation strategies: in vivo mature stem, Agrobacterium-mediated, and biolistic transformations.	Strategies: in vivo mature stem, Agrobacterium-mediated, biolistic transformations.	Electroporation (1990), Agrobacterium-mediated (1992), biolistic methods (less preferred).	Biolistic methods less preferred due to low efficacy; Agrobacterium-mediated considered reliable with high efficacy.	(Mendonça et al., 2013; Zhao et al., 2014; Huang et al., 2019; Wang et al., 2022; Nagle et al., 2023; Su et al., 2023)
Hairy Root in Eucalyptus	Exploration of hairy root cultures as an alternative for the synthesis of various chemicals in *Eucalyptus*, including secondary metabolites.	Utilization of Agrobacterium rhizogenes, rapid growth, genetic stability, and synthesis of secondary metabolites.	Sonication-assisted Agrobacterium transformation (SAAT), Agrobacterium-mediated techniques for genetic modification.	Hairy roots present an appealing alternative for synthesizing secondary metabolites, contributing to pharmaceutical and nutraceutical fields.	(Plasencia et al., 2016)
Genome Editing in Eucalyptus	Overview of genome editing methods, particularly CRISPR/Cas9, for functional characterization of genes associated with wood production in *Eucalyptus*.	*Eucalyptus* genome analysis, CRISPR/Cas9 technology for targeted mutagenesis.	CRISPR gene editing, targeting R2R3-MYB, NAC, ARF, Aux/IAA genes.	CRISPR/Cas9 facilitates efficient and precise genome editing in *Eucalyptus*, addressing the limitations of stable transformation techniques.	(Myburg et al., 2014)

5 Selection and Conservation of Elite Clones of *Bacopa monnieri* Using Biotechnological Approaches

Sonalika Roy

5.1 INTRODUCTION

Bacopa monnieri, also known as Brahmi (Figure 5.1), is a plant that grows in wetlands and marshy regions of India, Australia, Europe, Africa, Asia, and America (Pandey, 2022). It has a history in Ayurvedic medicine for its properties, which include enhancing memory, reducing anxiety and boosting cognitive function. The active compounds found in Brahmi are bacosides that are known for their effects on the brain and have attracted attention in modern herbal medicine (Deo and Reddy, 2013). Elite clones of Brahmi refer to plant varieties with characteristics such as higher levels of beneficial compounds, strong resistance to pests and diseases and adaptability to different environments. These elite clones play a role in ensuring quality and effectiveness in herbal products derived from Brahmi (Mondal et al., 2023). Preserving and selecting these clones is essential for preserving the benefits and commercial success of products based out of Brahmi. Traditional methods of growing this plant face challenges due to variations amongst populations that can affect the levels of beneficial compounds present. Additionally, factors like soil quality, water availability and climate conditions can impact the growth and medicinal properties of the plant significantly (Silpa, 2018). Overcoming these obstacles requires strategies, for identifying and safeguarding elite clones effectively.

Biotechnological methods offer solutions to the issues linked with the growing of *B. monnieri*. These methods include a variety of techniques, like plant tissue culture, genetic alteration, cryopreservation and so on, all contributing to the selection, reproduction and preservation of elite clones. By utilizing these methods scientists can ensure the production of appropriate plant materials ultimately boosting the value of *B. monnieri*. Tissue culture stands as an aspect of plant biotechnology allowing for the multiplication of plants in a controlled environment. In the case of *B. monnieri* tissue culture techniques such as micropropagation can be employed to generate identical plants from a single high-quality clone that helps to address the drawbacks of traditional propagation techniques. Micropropagation not only speeds up the process of multiplying high quality clones but also helps in growing plants that are free, from diseases (Aasim et al., 2019). In the case of *B. monnieri*, Somatic Embryogenesis has also been effectively utilized to cultivate plants with high bacoside content. This technique involves triggering cells to transform into embryos, which can then develop into plants. This technique is particularly beneficial for propagating top quality clones and preserving plant resources (Khilwani et al., 2016).

Molecular Marker Assisted Selection is another technique for conserving *B. monnieri*. Molecular markers are DNA sequences linked to different traits in plants. These markers can be utilized to choose elite clones exhibiting desired features, such as high bacoside content. This approach

 DOI: 10.1201/9781003482659-5

FIGURE 5.1 *Bacopa monnieri* plant.

significantly speeds up the breeding process and improves the accuracy of selection. SSRs, also known as microsatellites are DNA sequences that exhibit significant variability among individuals and function as markers for research and identifying superior clones. Amplified Fragment Length Polymorphisms (AFLPs) are DNA markers produced through the amplification of restriction fragments. These exhibit a degree of variation and are valuable, for studying genetic diversity and mapping. In *B. monnieri* SSR and AFLP markers have been employed to evaluate diversity and select plants, with medicinal worthiness (Karthikeyan et al., 2011, Jha and Mughees, 2023).

Genetic transformation is the process of inserting genes into the plant to introduce new characteristics or enhance existing ones. In *B. monnieri* genetic transformation can be applied to boost mass production, enhance tolerance to stress and provide resistance against pests and diseases. The common techniques for transformation include Agrobacterium mediated transformation and biolistic (gene gun) transformation. Agrobacterium mediated transformation involves using the soil bacterium Agrobacterium tumefaciens to transfer genes into plant cells. This method has been successful in *B. monnieri* for incorporating genes that improve synthesis and enhance plant well-being. Biolistic Transformation, also known as particle bombardment uses high speed microprojectiles to deliver DNA into plant cells. This technique is particularly beneficial for plants that are resistant to Agrobacterium mediated transformation. In *B. monnieri* this approach has been explored for introducing genes that enhance content and stress tolerance (Güner, 2023).

Cryopreservation refers to the preservation of plant tissues, at temperatures (-196°C) using liquid nitrogen. Preserving clones of Bacopa, for long term sustainability is vital to safeguard genetic resources for future research and cultivation. Cryopreservation entails preparation of plant tissues, including dehydration and the application of cryoprotectants to prevent damage during freezing and thawing. Vitrification, a technique involving concentrated cryoprotectant solutions to inhibit ice crystal formation during freezing, has proven successful in preserving Brahmi tissues with survival and regeneration rates. Another method i.e., encapsulation dehydration encompasses encapsulating plant tissues in alginate beads followed by dehydration and freezing offering an approach for efficient preservation of elite *B. monnieri* clones (Gowthami et al., 2023).

Apart from the techniques mentioned above, there are other techniques that are involved in the selection and conservation of elite clones of *B. monnieri* that have been discussed in the upcoming chapters. Also, the fusion of methods presents a comprehensive strategy for selecting and conserving superior *B. monnieri* clones. By merging tissue culture techniques with markers, genetic transformation and cryopreservation, researchers can ensure production of high-quality plant material (Gosal and Kang, 2012; Sudheer et al., 2023). This holistic approach not only enhances the properties of *B. monnieri* but also promotes sustainable cultivation practices.

The careful selection and conservation of Bacopa clones is essential, for maximizing the medicinal benefits of this valuable herb. Utilizing methods such, as plant tissue culture, molecular marker assisted selection, genetic transformation and cryopreservation provides tools to address the limitations of traditional cultivation practices. These advanced techniques enable researchers and growers to maintain a supply of high-quality *B. monnieri* meeting the increasing demand for potent medicinal products. As studies in this area advance, incorporating innovations will be crucial for the management and protection of *B. monnieri* safeguarding its healing heritage, for future generations.

5.2 OVERVIEW OF *B. MONNIERI*

B. monnieri, known for its versatile range of health benefits and growing significance in scientific studies, is a plant with distinct features and uses. This overview encompasses its attributes like distribution in nature, chemical components, traditional and contemporary medicinal applications as well as the ongoing research dedicated to *B. monnieri.*

The plant's defining features include elongated leaves that grow opposite to each other along the stem. Typically measuring 1–2 cm in length the leaves have serrated edges. Bacopa has a growing structure that forms dense mats on the ground or in shallow water. Its succulent stems facilitate easy spread in environment. The flowers of *B. monnieri* are small and either white or pale blue, and these appear individually in the leaf axils on stalks. Each flower consists of five petals, five sepals and four stamens. The plant produces capsules containing numerous tiny seeds contributing to its ability for rapid growth through both vegetative and sexual reproduction. *B. monnieri* thrives in subtropical regions across the globe (Sudhakaran, 2020). It flourishes in moist habitats, like ponds, wetlands and riverbanks. The plant thrives in environments, with moisture and sunlight although it can also manage in shade. Bacopa originates from the subcontinent but has extended its presence to other regions like Asia, Africa, Australia and America. It is notably abundant in countries such as India, Nepal, Sri Lanka, China, Taiwan and parts of Southeast Asia. Its ability to adapt to surroundings contributes to its versatility across ecosystems (Thorat et al., 2018).

The beneficial qualities of *B. monnieri* are largely credited to its array of plant chemicals. The key active components are saponins specifically bacosides A and B. These substances play a role in enhancing functions and protecting the brain. Apart from bacosides, *B. monnieri* also contains compounds such as saponins, alkaloids, flavonoids and glycosides. Some desired compounds have been included in Table 5.1

For centuries *B. monnieri* has played an important role, in medicine. It falls under the category of "Medhya Rasayana," which consists of herbs known for rejuvenating the mind and nervous system. Ayurvedic experts recommend Bacopa to address cognitive issues like memory loss, anxiety and

TABLE 5.1
Important compounds present in *B. monnieri*

Alkaloids:	Brahmine and herpestine, which have been studied for their potential neurological benefits
Flavonoids:	Luteolin and apigenin, which exhibit antioxidant and anti-inflammatory properties
Glycosides:	Monnierin and plantainoside B, contributing to the overall medicinal profile of the plant

epilepsy. *B. monnieri* is also utilized in medicine to treat respiratory problems, digestive disorders and inflammatory conditions. It is believed to possess pain relieving and fever reducing properties, making it a versatile remedy for a variety of health concerns. Clinical trials have demonstrated that *B. monnieri* enhances memory, focus and overall cognitive function in individuals. The neuroprotective qualities of *B. monnieri* are linked to its ability to regulate neurotransmitter levels, diminish stress and boost plasticity. These attributes position it as an option for managing conditions such as Alzheimer's disease and Parkinsons disease. Studies suggest that Brahmi possesses anxiety reducing and mood uplifting properties due to its influence on neurotransmitter systems like serotonin and dopamine. These effects can be beneficial for individuals dealing with anxiety or depression by offering an alternative to modern medications. Bacopa demonstrates notable antioxidant properties, which play a role in shielding cells from harm. Its anti-inflammatory characteristics are also well documented making it beneficial for managing conditions and decreasing the likelihood of illnesses (Deo and Reddy, 2013).

Several clinical trials are currently in progress to delve deeper into the possibilities of this plant. These investigations seek to uncover its impact on health issues such as anxiety disorders and chronic inflammatory ailments. The results from these trials will offer evidence supporting its utilization in modern medicine. Concurrent research is also focusing on noting and separating phytochemicals found in Brahmi. Advanced analytical methods like high performance liquid chromatography (HPLC) and mass spectrometry are utilized to analyse the makeup of the plant. These studies enrich our comprehension of how its bioactive compounds work. Biotechnological techniques, including tissue culture, genetic modification and marker assisted selection are being utilized to improve the production of Bacopa and its active components (Dey et al., 2019). These methods aim to cultivate clones with increased yields of bacosides and enhanced resilience against pressures.

Given the rising interest in *B. monnieri* it is imperative to implement conservation measures to ensure its proper utilization. Researchers are exploring both on-site and off-site conservation strategies, like establishing germplasm banks and promoting harvesting practices.

These initiatives strive to safeguard the variety and environmental health of *B. monnieri* populations. *B. monnieri*, known for its usage in medicine and increased utilization in contemporary science represents a plant with significant healing capabilities. Its broad spectrum of advantages spanning from enhancement to anti-inflammatory properties highlights its worth as a versatile and potent botanical solution. By merging ancient wisdom, with biotechnological approaches, we can unlock the full potential of this extraordinary plant securing its benefits for future generations.

5.3 MORPHOLOGY OF *B. MONNIERI*

Understanding the characteristics of *B. monnieri* is essential in order to cultivate it for its beneficial properties. *B. monnieri* is a plant that forms carpets on the ground or in shallow water. It typically reaches a height of 10–25 cm and spreads horizontally with the help of its creeping stems. This kind of growth pattern makes it appropriate for areas where it can spread widely and help stabilize soil or sediment. The leaves of *B. monnieri* are distinctive and contribute significantly to its identification (Figure 5.2). It also produces small, solitary flowers that are significant for reproduction and the fruit and seed structures of *B. monnieri* are adapted for efficient dispersal (Jain et al., 2016; Choudhary et al., 2021).

The features of *B. monnieri* stem, leaf, flower and fruit are described in Tables 5.2, 5.3, 5.4 and 5.5.

The succulent leaves help in retaining water in areas prone to fluctuating water levels like wetlands. The opposite leaf arrangement along the stem maximizes exposure for photosynthesis especially in densely vegetated or shaded wetland areas. The small flowers serve well for both self-pollination and cross pollination by insects, ensuring diversity and successful reproduction. These morphological traits of *B. monnieri* also play a role in its applications. The lush leaves and stems

FIGURE 5.2 Spatulated green leaves with greenish-brown stem of Brahmi.

TABLE 5.2
Features of *B. monnieri* stem

Creeping and Prostrate:	The stems grow along the ground or float in water, rooting at the nodes to form new plants. This growth habit helps the plant spread efficiently and cover the substrate in its habitat
Succulent and Cylindrical:	The stems are smooth, cylindrical, and fleshy, capable of storing water and nutrients. This succulence is an adaptation to its wetland habitat, allowing the plant to endure periods of fluctuating water availability
Green in Color:	Young stems are green, contributing to photosynthesis. Older stems may turn brown and woody, especially at the base, providing structural support

TABLE 5.3
Features of *B. monnieri* leaves

Shape and Size:	The leaves are spatulate (spoon-shaped), measuring approximately 1–2 cm in length and 0.3–0.6 cm in width. They have a broad apex and a slightly serrated or entire margin
Arrangement:	Leaves are arranged oppositely along the stems, with each pair of leaves being perpendicular to the pair above and below. This arrangement maximizes light capture for photosynthesis
Sessile:	The leaves are sessile, meaning they lack petioles (leaf stalks) and attach directly to the stem. This characteristic helps in conserving space and resources in the plant's growth habit
Fleshy and Succulent:	Like the stems, the leaves are fleshy and capable of storing water. This succulence aids in water retention, crucial for survival in wet and fluctuating environments
Color:	The leaves are typically green, though they may exhibit a bluish tint under strong sunlight. This coloration is due to the presence of chlorophyll, essential for photosynthesis

TABLE 5.4
Features of *B. monnieri* flower

Structure:	Each flower is actinomorphic (radially symmetrical) and measures about 1–2 cm in diameter. The flowers are borne singly in the leaf axils on short pedicels (flower stalks)
Petals and Sepals:	The flowers have five petals, which are white or pale blue, and slightly rounded. There are five sepals that protect the developing flower bud
Reproductive Organs:	The flowers have five petals, which are white or pale blue, and slightly rounded. There are five sepals that protect the developing flower bud
Blooming Season:	*B. monnieri* flowers primarily during the warmer months, though it can bloom year-round in tropical and subtropical climates

TABLE 5.5
Features of *B. monnieri* fruit

Fruit:	The fruit is a small, oblong capsule that measures about 3–5 mm in length. The capsule splits open when mature to release the seeds
Seeds:	Each capsule contains numerous minute seeds, about 0.5–1 mm in diameter. The seeds are equipped with a mucilaginous coating that aids in water dispersion, ensuring that they can spread effectively in wetland environments

are rich in bacosides. Their juicy texture allows for harvesting without harming the plant much, promoting sustainable utilization. The root system of this plant is well suited for both aquatic and semi aquatic habitats. The plant develops roots at the points where its stems meet, aiding in propagation and the formation of new colonies. These roots, shallow and fibrous in nature, thrive in waterlogged soils. This root structure allows absorption of nutrients from the rich upper layers of wetland soils. The plants' ability to produce root at nodes, helps it anchor securely for its survival in the environment. The physical characteristics demonstrate adaptation to its surroundings, hence it can survive in both freshwater and salty conditions due to its adaptability (Sudhakaran, 2020). With its adaptability to environments, this plant is well suited for cultivation in different places ensuring a steady availability of medicinal materials.

The physical features and structure of *B. monnieri* showcases its ability to thrive in adverse environments. Its crawling stems, plump leaves, small flowers and effective root system collectively contribute to its resilience in harsh conditions. These physical attributes not only support its growth and propagation but also enhance its medicinal significance, establishing *B. monnieri* as a notable plant in both traditional and contemporary medicine practices. Recognizing these traits is essential, for identifying, growing and responsibly utilizing this plant.

5.4 PHYSIOLOGY OF *B. MONNIERI*

As discussed earlier, *B. monnieri,* a plant commonly used in medicine for centuries, is known for its ability to improve memory, reduce anxiety and alleviate depression. Recent scientific research has confirmed many of these benefits linking them to the chemical properties of the plant. To fully grasp the advantages of *B. monnieri*, it is crucial to understand its biology, such, as how it grows, its processes and how it interacts with its surroundings (Dadhich et al., 2022). The medicinal effects of this plant are linked to its chemical makeup, which includes compounds, like bacosides (saponins), alkaloids, flavonoids and sterols as discussed in Table 5.6.

TABLE 5.6
Compounds found in *B. monnieri*

Bacosides:	These are the primary active compounds in *B. monnieri*. Bacosides A and B are the most studied and are known to enhance synaptic activity and repair damaged neurons. They play a significant role in improving memory and cognitive function.
Alkaloids:	These compounds, such as brahmine and herpestine, contribute to the plant's neuropharmacological effects, including its anxiolytic and antidepressant properties
Flavonoids:	These antioxidants help mitigate oxidative stress, which is implicated in neurodegenerative diseases. Flavonoids also possess anti-inflammatory properties
Sterols:	Plant sterols in *B. monnieri*, such as beta-sitosterol, contribute to its anti-inflammatory and immune-modulating effects

This water plant undergoes photosynthesis using its leaves. It follows the C3 pathway commonly seen in plants found in temperate and aquatic areas. This process involves converting carbon dioxide into two-three carbon molecules of phosphoglycerate with the help of an enzyme called Rubisco. Brahmi has adapted its metabolism to survive in low oxygen conditions, often found in water-logged soils (Panda et al., 2021). The plant shows flexibility in its metabolic functions, allowing it to flourish under any circumstances. To withstand drought, salinity and oxidative stress, Brahmi has developed various mechanisms over time. As discussed in morphology of Brahmi, its fleshy leaves aid in retaining water and enhance resistance to drought (Dadhich et al., 2022). The plants' ability to thrive in environments is supported by specific ions and osmolytes that maintain cellular osmotic balance. High levels of antioxidants such as flavonoids and saponins play a role in combating stress by scavenging free radicals that could harm cellular structures. Moreover, the plant produces stress related proteins and enzymes that assist in repairing damaged proteins and maintaining stability during adversities (Qaiser et al., 2023).

Its growth and chemical makeup is significantly influenced by the plants' interactions with soil microorganisms. Helpful microbes in the soil such as mycorrhizal fungi and nitrogen fixing bacteria enhance absorption and overall plant health. These microbes stimulate the production of compounds like bacosides and flavonoids that contribute to the plants' beneficial properties like reducing anxiety and uplifting mood. The organic compounds released by the plant roots also provide nourishment for bacteria that promote plant growth and resilience to stress (Bharti et al., 2013; Pankaj et al., 2020). Recent studies have confirmed these uses while uncovering potential benefits of B. *monnieri*.

The plants' capability to thrive in environments along with its biochemical make-up, positions it as a valuable resource in both traditional healing practices and modern medicine. Exploring the processes that govern its growth and medicinal effects can lead to enhanced applications for promoting health and wellbeing. As the ongoing research dives into the workings of *B. monnieri* physiology, its reputation as a medicinal herb is likely to be further established.

5.5 DISTRIBUTION OF *B. MONNIERI*

Knowledge about the distribution of *B. monnieri* is crucial, for both research and its sustainable utilization. This section provides information about the spread, habitat preferences, environmental adaptability and factors influencing the expansion of Brahmi (Figure 5.3). Originating from the wetlands of India, it also thrives in parts of Asia like China, Nepal, Sri Lanka, Thailand, Malaysia, Indonesia and Vietnam. Its presence extends to continents such as North America, South America, Africa and Australia showcasing its ability to thrive in different climates. In Africa this herb can be found in countries like Nigeria, Sudan and Egypt. The plant has also taken root across North America (including regions in Central and South America). In the U.S. it can be found growing

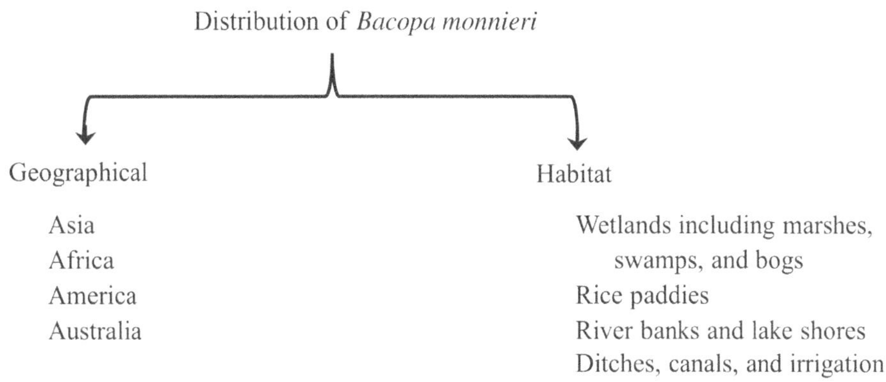

FIGURE 5.3 Distribution of *Bacopa monnieri.*

in states, like Florida, Louisiana and Texas. In South American countries, like Brazil and Mexico, *B. monnieri* can be found thriving. In Australia this plant is commonly seen in the eastern regions in Queensland and New South Wales. It also grows well in Pacific Islands that have subtropical climates (Saghatelyan, 2017; Halder et al., 2021; Pandey, 2022).

As *B. monnieri* prefers to inhabit in wet and marshy areas. It is commonly seen in rice fields, wetlands such as marshes, swamps and bogs and along riverbanks. It does best in places with moving water where the soil remains consistently damp (Joshi et al., 2019). One can often find Bacopa growing along riverbanks, lakeshores and, ponds where there is moisture and sunlight for it to grow well. Also, the plant can colonize in areas like ditches, canals and, irrigation channels due, to its ability to adapt both natural and human influenced environments (Swapna et al., 2011). The distribution of this plant is greatly influenced by temperature and rainfall patterns. It thrives in areas with enough rain to maintain the conditions it needs to grow. Changes in water availability caused by human related factors can impact how far the plant spreads (Smitha et al., 2021). Human activities like rice farming have helped Bacopa spread further. The plants presence in rice fields and irrigation systems show how humans have played a part in where it can grow.

The wide distribution of *B. monnieri* is attributed to its ability to tolerate a range of conditions like waterlogged, freshwater and, salty area conditions (Ribeiro et al., 2015). Also, the juicy leaves store moisture aiding to survival in adverse periods (Jain et al., 2016). This plant flourishes in subtropical climates, able to withstand temperatures between 15°C to 40°C with optimal growth occurring at 25°C to 35°C. Its photosynthetic efficiency helps it adjust to light levels although insufficient light might slow its growth and biomass production. Sandy, loamy or clayey soils with water retention capacity are ideal for its growth as it can handle a pH range from acidic to slightly alkaline (pH 5.5 to 7.5) though neutral to slightly acidic soils are best, for optimal development (Saran et al., 2022).

The wide distribution of Bacopa highlights its importance both ecologically and economically. In terms of ecology this plant supports the health of wetland ecosystems by offering shelter and food, for animals. The dense growth of *B. monnieri* plays a role, in stabilizing soil and preventing erosion. Economically, this plant is highly valued for its medicinal properties. It is processed into forms like extracts, powders and capsules which find applications in both traditional and modern medicine. The increasing demand for Bacopa in pharmaceutical and nutraceutical industries has led to its cultivation across many regions (Saloni et al., 2022).

Research on the genetic analysis of *Bacopa monnieri* for different populations have shown that there exists polymorphism and genetic differences in the plant due to geographical factors, climate and ecology. The genetic variability in Brahmi is evaluated using molecular markers like RAPD, ISSR, and AFLP. This diversity in the wild population is essential to improve the desirable traits in the plant breeding program including increased bacoside content, faster growth rate and stress tolerance

(Srivastava et al., 2016). While *B. monnieri* is widely distributed it is vital to focus on its conservation and sustainable utilization to ensure its availability in the long run. Factors like overharvesting and habitat destruction pose risks to its populations. To address these threats sustainable harvesting practices, habitat restoration efforts and cultivation programs are essential. Establishing guidelines for the harvesting of *B. monnieri* can help prevent exploitation by safeguarding natural habitats and promoting population regeneration. Also, restoring wetlands and other habitats can contribute to the conservation of this plant species. Actions such as improving water quality and reintroducing vegetation are pivotal for maintaining healthy *B. monnieri* populations. By encouraging the cultivation of this plant species, one can ensure its regular supply for medicinal purposes. Research on propagation techniques and pest control can further enhance productivity and product quality (Gudipati et al., 2012).

The widespread distribution of Bacopa showcases its adaptability and ecological versatility.

From the wetlands of India to the sprawling marshes of America, this plant flourishes in places that add to its widespread presence worldwide. It is crucial to grasp the factors that impact its spread for preserving it and using it sustainably. With the increasing interest in Brahmi, actions to safeguard its native habitats and endorse biotechnological approaches will guarantee its accessibility for generations to come.

5.6 MEDICINAL USES OF *BACOPA MONNIERI*

Bacopa monnieri, has been known in the field of Ayurvedic medicine for its diverse health benefits (Figure 5.4). One of the important benefits of *B. monnieri* is its capacity to enhance cognitive abilities. Traditionally it has been used to boost memory, focus and mental clarity and modern research supports these assertions. The plant contains bacoside compounds considered to aid in repairing damaged neurons and stimulating nerve growth which ultimately enhance function. Individuals who consume extracts of *B. monnieri* show enhancements in memory retention. It is also recognized for stress release and diminishes anxiety levels. Studies suggest that this herb can influence the function of neurotransmitters like serotonin and dopamine that are crucial, for mood regulation. *B. monnieri* also shows promise as an agent offering benefits for individuals with epilepsy and other seizure disorders. This herb can regulate neurotransmitter activity in the brain decreasing both the frequency and severity of seizures (Rai et al., 2017).

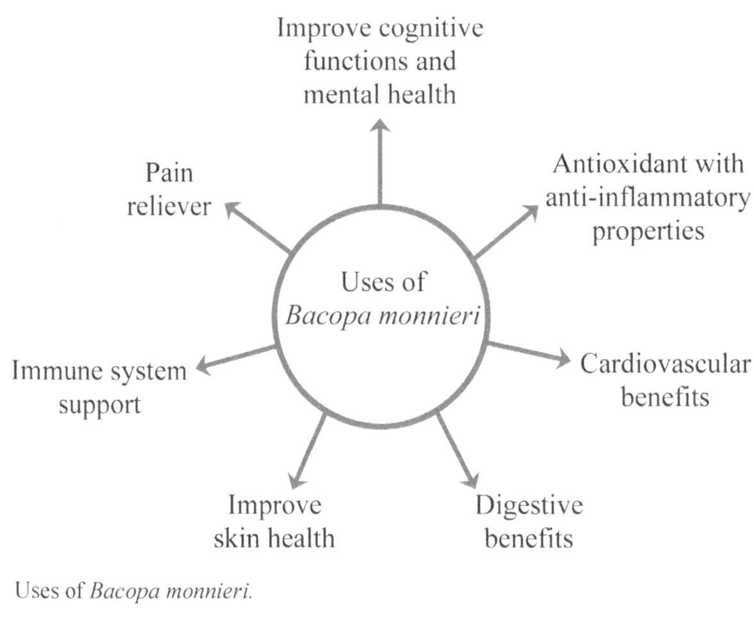

FIGURE 5.4 Uses of *Bacopa monnieri*.

B. monnieri has high levels of antioxidants that are vital for safeguarding the body against oxidative stress. The antioxidants present in *B. monnieri*, such as bacosides and flavonoids work to combat the damage caused by free radicals, promoting wellbeing and extending longevity. Moreover, apart from its antioxidant properties, *B. monnieri* also demonstrates inflammatory effects. By reducing inflammation levels, the plant may help lower the risk of developing conditions like heart disease, diabetes and cancer. Studies have found that it can help reduce inflammation and its related symptoms by inhibiting the production of inflammatory cytokines and has fever reducing properties. It also aids in regulating blood pressure by relaxing blood vessels leading to less strain on the heart. This relaxation effect is linked to nitric oxide in *B. monnieri*, which eases the muscles of blood vessels. It further helps in decreasing levels of LDL cholesterol and triglycerides while increasing HDL cholesterol. Also, the herb is recognized for its gastroprotective properties making it effective in treating ulcers and other gastrointestinal issues. It functions by boosting mucus production in the stomach shielding the stomach lining from the damaging effects of acid and proves beneficial for managing conditions, like irritable bowel syndrome. Moreover, this herb also supports the body in flushing out excess water and toxins for overall wellbeing. *B. monnieri* also aids the skin in wound healing and lowers infection risks. It is often utilized in treating skin conditions like eczema and psoriasis. *B. monnieri*s calming effects help soothe inflammation and discomfort while improving skin appearance by boosting collagen production and enhancing elasticity (Mukhopadhyay et al., 2012; Kamesh and Sumathi, 2012; Tamboli et al., 2022).

B. monnieri supports the immune system by enhancing the body's response and strengthens its ability to combat infections and diseases effectively. Studies have revealed that it has the ability to boost the production of cells, like macrophages and lymphocytes that are essential for the defence system, thus enhancing immunity it can help prevent illnesses and promote health.

Since ages this herb has been used as a remedy for pain relief, especially caused due to arthritis and neuralgia. The anti-inflammatory effects of Brahmi also contribute to its pain-relieving qualities by reducing inflammation and swelling in affected areas. Research indicates that *B. monnieri* can also inhibit enzymes involved, thus eliminating pain. This prevents unwanted side effects commonly associated with conventional pain medications (Valotto et al., 2024).

While *B. monnieri* is generally considered safe for individuals, it is important to seek advice from a medical expert before incorporating any new supplement into your routine, particularly if one has existing health conditions or taking other medications. As ongoing research sheds light on the advantages of *B. monnieri* it is gaining popularity as a remedy, in both traditional and contemporary medicine.

5.7 ECONOMIC IMPORTANCE OF *B. MONNIERI*

The rising popularity of *Bocopa monnieri* in global markets has led to its increased economic importance. This section explores the economic importance of this herb in examining its role in the pharmaceutical industry, agriculture, trade, and sustainable development.

The pharmaceutical industry greatly benefits from *B. monnieri*, which contains compounds, like bacosides, saponins and flavonoids known for their health benefits. These compounds are used to develop a range of products that aim to enhance cognitive function, reduce anxiety and provide neuroprotection as discussed in the previous sections. The demand for supplements has increased significantly amongst students, professionals and the elderly, leading to revenue for companies producing them. The expanding mental health market would create opportunities for manufacturers and suppliers of Brahmi based products which would further spark interest in developing medications and therapies using Brahmi further boosting its economic influence, in the pharmaceutical sector (Fatima et al., 2022).

The cultivation of *B. monnieri* carries implications for agriculture. Due to its ability to thrive in various environments, it can be cultivated in regions where other plants struggle to grow. This presents farmers with revenue source in areas with agricultural opportunities. As the demand for this herb rises there is a need for cultivation techniques. Increasing demand would spur advancements in practices and create employment opportunities, thus boosting economies. It can be grown organically, reducing reliance on chemical fertilizers and pesticides. This not only benefits the environment but also meets the rising demand for organic and sustainably produced goods thereby enhancing the economic value of *B. monnieri* cultivation (Panda et al., 2021).

The trade of Brahmi and its related products has significant impact on the economy. With the growing interest in health supplements *B. monnieri* has emerged as a commodity in the international market. Countries like India with a tradition of cultivating and using *B. monnieri* play a role as major exporters of this herb. The increasing worldwide demand for herbal supplements has created an export market contributing to foreign exchange earnings and supporting economic development. The wide variety of products derived from *B. monnieri* ranging from supplements to pharmaceuticals ensures that the market is not reliant on any one sector. The growth of online commerce has made it more convenient for consumers to access Brahmi products. Also, online retail platforms offer selection of Brahmi based supplements and health goods expanding their availability to markets. The financial advantages of cultivating and trading this herb also led to benefit in rural regions by creating employment opportunities and promoting financial stability (Patel et al., 2022). The economic significance of *B. monnieri* also focuses on sustainable development that promotes environmental, social and economic aspects. Cultivating *B. monnieri* can help support sustainability in ecosystems that play a key role in biodiversity and water purification (Sharma et al., 2012).

Investing in research and development (R&D) also highlights the plant's economic significance. Researchers aim to discover applications and improve cultivation methods promoting innovation and economic advancement. Research, on the properties of *B. monnieri* could lead to the creation of drugs and therapies expanding the market for Brahmi based products and attracting investments in biotechnology and pharmaceutical sectors. Also, enhancing cultivation techniques, pest control and sustainable farming practices, can boost the efficiency and profitability of the farming of these plants, benefiting farmers and enhancing agricultural sector stability. Introducing Brahmi based products like functional foods, beverages and cosmetics can tap into new market sectors stimulating economic growth (Brimson et al., 2021).

B. monnieri plays a role in industries such as pharmaceuticals, agriculture, trade and sustainable development due to its potential to improve cognitive function, support mental health and offer various health benefits. Trading the plant presents opportunities, for farmers, exporters and retailers, contributing to rural development and global commerce. As further research and advancements enhance, the economic influence of *B. monnieri* is set to expand even more, bringing advantages to economies globally.

5.8 BIOTECHNOLOGICAL APPROACHES FOR SELECTION AND CONSERVATION OF ELITE CLONES OF *B. MONNIERI*

Bacopa monnieri is an important medicinal herb. Its qualities, attribute to the presence of bacoside compounds that have captured the interest of both traditional and modern medical fields. With the increasing demand of high-quality *B. monnieri,* there is a pressing need to employ biotechnological methods for selecting and propagating elite clones of the plant. These techniques ensure better yield, quality and elevated levels of bioactive components (Rohini, 2020). This study explores strategies for identifying and multiplying superior *B. monnieri* clones emphasizing their benefits and potential uses.

The meticulous selection of Brahmi clones plays a vital role in maintaining the uniform quality and effectiveness of the end product. Elite clones are chosen based on useful characteristics. Brahmi clones with bacoside content are favoured for their heightened therapeutic properties. Also, clones

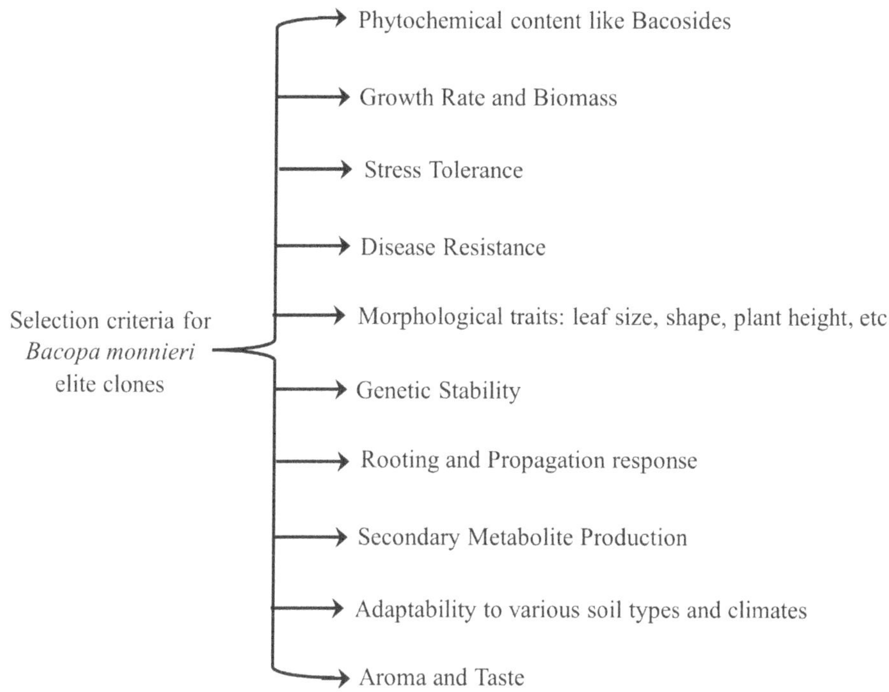

FIGURE 5.5 Criteria for selection for *Bacopa monnieri* elite clones.

TABLE 5.7
Limitations for traditional methods of selecting *B. monnieri*

Time-Consuming:	Multiple growing seasons are needed to assess and select superior plants.
Labor-Intensive:	Requires extensive manual labor for planting, maintenance, and evaluation.
Environmental Variability:	Phenotypic traits can be influenced by environmental factors, leading to inconsistent results.
Genetic Variability:	Natural populations show considerable genetic diversity, affecting the uniformity of medicinal properties.

that demonstrate resilience to challenges like drought, salinity and temperature variations are more suitable for large scale cultivation. Clones possessing resistance to diseases, reduce the reliance on chemical treatments and rapid growing clones that yield biomass prove advantageous for farming fostering agricultural practices (Karthikeyan et al., 2011; Bansal et al., 2014). Figure 5.5 depicts selection criteria for *Bacopa monnieri* elite clones.

Traditional methods of selecting *B. monnieri* involve phenotypic evaluation and field trials. While these methods have been foundational, they come with significant limitations as shown in Table 5.7. These limitations necessitate the adoption of biotechnological approaches that provide precision, efficiency, and consistency in selecting elite clones (Inthima and Supaibulwatana, 2024). Biotechnological approaches such as Tissue Culture and Micropropagation, Protoplast fusion, Cryopreservation, Synthetic seed technology, Genetic transformation, Molecular Marker Assisted Selection, Metabolic engineering, Omics approaches and Epigenetic modifications have been discussed in this section.

5.8.1 Tissue Culture and Micropropagation

Micropropagation is a biotechnological method that presents solution for mass production and conservation of superior clones of *B. monnieri*. This technique involves growing plants in a controlled environment using plant parts like nodal segments, shoot tips or leaf pieces. Micropropagation ensures the generation of clones with genetic characteristics similar to the parent plant, ensuring consistency in desired trait expression. Specific procedures have been developed along with optimizing the use of plant growth regulators such, as cytokinins and auxins. The steps used in micropropagation and the techniques involved are shown in Tables 5.8 and 5.9. By using explants of elite clones, micropropagation keeps the genetic traits intact, preserving the desirable traits in the propagated plants. The controlled indoor conditions and the sterilization processes minimize the chances of pathogen contamination leading to disease resistant plants. Also, unlike traditional methods, micropropagation is not influenced by climatic changes, allowing for continuous production and availability of

TABLE 5.8
Steps in micropropagation

1.	Selection of Explants:	The procedure starts with the collection of right explants from the chosen elite clones. Nodes as well as shoot tips are preferred explants for *B. monnieri*.
2.	Explant Sterilization:	It is important to get rid of any microbial substances from explants and hence the surface is disinfected. This step is important to reduce the incidences of infections during tissue culture. Some of the frequently used sterilizing agents are ethanol, and HgCl.
3.	Initiation of Culture:	An explant collected aseptically is transferred and placed in a suitable culture medium which has basic nutrient and supplements that aid growth of tissue. In regard to the growth regulators like cytokinins and auxin the role is crucial to stimulate the shoot formation and rooting process.
4.	Shoot Multiplication:	The initiated cultures are then moved to a multiplication medium that enhances formation of multiple shoots from one bud. Subculturing is done from time to time to freshen up the cultures and at the same time increase the shoot multiplication.
5.	Rooting:	These are then trans-planted to a rooting medium for the purpose of root initiation. Some of the auxins which are widely used for rooting include indole-3-butyric acid (IBA) and naphthaleneacetic acid (NAA) for *B. monnieri*.
6.	Acclimatization:	The plantlets develop sufficient root systems and gradually harden off in ex vitro environment. The process involves taking the plantlets to a hardening bench or a green house where they are exposed to natural weathers prior to being put in the field.

TABLE 5.9
Techniques in micropropagation

Nodal culture:	In this method, the axillary shoot buds are taken during active or resting period and are transferred to artificial nutrient medium and turned into whole plant.
Callus Culture:	Initiating callus culture from various explants can be a preliminary step in plant tissue culture. Callus induction requires optimizing the concentration of plant growth regulators (PGRs) like auxins and cytokinins. Callus culture serves as a foundation for further regeneration of plants and secondary metabolite production.
Somatic Embryogenesis:	Somatic embryogenesis involves the development of embryos from somatic cells. This technique is advantageous for large-scale propagation and genetic transformation studies. It provides a uniform platform for producing genetically identical plants with desired trait

planting material throughout the year. This technique offers a way to preserve *B. monnieri* elite clones, for future cultivation and research purposes (Binita et al., 2005; Rahe et al., 2020).

Although there are benefits to micropropagation, there are also some challenges to consider. It is essential to ensure that aseptic conditions are consistently upheld during the process to avoid any contamination that could hinder the growth of cultures. Despite the goal of producing plants that are genetically identical, variations known as soma clonal mutations may arise during the tissue culture process. The establishment and upkeep of tissue culture facilities demand resources and specialized knowledge which could pose limitations, for certain growers (Abdalla et al., 2022).

Micropropagation is commonly used in the production of Bacopa to ensure a steady supply of high-quality planting material, for large scale farming. By using micropropagation elite clones of Bacopa, the plant can be preserved *in vitro* maintaining their characteristics and desirable traits for future breeding efforts and scientific research. The pharmaceutical sector benefits from micropropagation as it provides plant material with a bacoside content guaranteeing the effectiveness and standardization of products derived from *B. monnieri*. Micropropagation supports Bacopa research by offering a source of uniform plant material, for experimental investigations and breeding programs focused on enhancing varieties (Jaspreet et al., 2013).

The future of growing *B. monnieri* through micropropagation shows promise, due to improvements in tissue culture methods. Micropropagation techniques can boost efficiency and scalability making it possible for mass production of the plant (Pant, 2016). Micropropagation can be combined with other biotechnological methods like cryopreservation to offer a way to preserve elite clones for the future ensuring their genetic makeup remains intact and with epigenetic research to help in better understanding of how traits are expressed and maintained.

The capacity of micro-propagation to produce identical disease-free plants throughout the year makes it a crucial approach in meeting the rising demand for this medicinal plant. By addressing the limitations of macro-propagation techniques, micro-propagation secures the growth and preservation of *B. monnieri* safeguarding its healing properties for future generations. With advancements in micropropagation methods we can further improve the conservation of *B. monnieri* clones contributing to the overall advancement and utilization of this beneficial medicinal herb.

5.8.2 Protoplast Fusion

Protoplasts are plant cells without cell walls. To fuse protoplasts, researchers combine two or more of these naked cells from different plants which creates a hybrid cell that has genetic material from different parents thus mixing good traits. It helps to create new plant types with better features. For protoplast fusion, picking appropriate *B. monnieri* clones with good qualities, require key steps. It may be that one parent might have lots of bacosides, while another grows fast or handles stress well. Researchers take protoplasts from these plants and join them to make hybrid protoplasts with mixed genes. The process to fuse protoplasts has some important steps. Firstly, enzymes like cellulase and pectinase break down cell walls in plant parts such as leaves, stems, or roots. This releases protoplasts. Next, methods like spinning and filtering, clean these protoplasts. After that, protoplasts from different parent plants mix in a liquid. This liquid has chemicals like polyethylene glycol (PEG) or calcium ions to make the protoplasts join after which hybrid protoplasts form with mixed genetic material. These fused protoplasts then grow in a medium with growth chemicals. This helps cells divide and form callus. The callus that is formed differentiates into shoots and roots developing into whole plants. The newly formed plants are checked for desired good traits like high bacoside levels with the help of HPLC, better growth, and more resistance to stress for *B. monnieri*. Plants that show these good traits become elite clones (Dey et al., 2019; Chamak, 2022). These top clones can multiply more through tissue culture to keep a steady supply of high-quality *B. monnieri* plants. Once elite clones have been selected, it is important to preserve their superior quality for future use.

Researchers have managed to create bacoside-enriched *B. monnieri* clones using protoplast fusion (Aasim et al., 2019). Protoplasts can be separated from two plants, one having high bacoside content and the other having better growth and fused and cultured in a regeneration medium where these can develop into whole plants. The regenerated plants' bacoside content can be screened using HPLC, high-bacoside clones can be identified, and can be selected for further propagation. These elite clones can have bacosides at much higher levels than the parent plants. Similarly, the technique of protoplast fusion can also be utilized to increase the stress tolerance of *B. monnieri*. Protoplasts are separated from two parent plants, one of which having high stress tolerance and the other one having high bacoside content. The two protoplasts after fusion can be cultured in regeneration medium under stress conditions to select out the hybrids that are stress-tolerant. The regenerated plants can then be checked for both stress tolerance and bacoside content. These elite clones can then be later propagated through tissue culture for further conservation and commercial use. Thus, protoplast fusion is an effective method of improving the medicinal properties of the plants and has been utilized in studies (Soman et al., 2019; Samridha and Chandra, 2024).

The fusing of protoplasts for selection and conservation of elite clones of *B. monnieri* is a great innovation. Protocols for protoplast isolation, fusion, and regeneration should be further developed to boost the effectiveness and efficiency of the protoplast fusion process. This practice could be done by using various enzymes, fusion agents, and culture conditions. The researchers can make hybrid plants having better characteristics like greater bacoside content, good growth rate and higher stress tolerance by merging the genetic material of various parent plants. Also, tissues of protoplasts or elite clones can be cryopreserved to ensure long-term preservation without genetic or phenotypic changes. This approach provides a reliable means of preserving high-quality genetic material for future use. As research in this field progresses, protoplast fusion will play an increasingly important role in ensuring the availability and sustainability of this valuable medicinal herb, contributing to its continued therapeutic potential and commercial success.

5.8.3 Cryopreservation

Cryopreservation is a method used to preserve cells and tissues by cooling them to temperatures below zero. It is a technique for preserving clones by storing biological material at extremely low temperatures in liquid nitrogen (-196°C). This process stops all metabolic and biochemical activities, allowing the material to be stored for long periods without any physical changes. The key steps in cryopreservation include pretreatment, cryoprotection, cooling, storage and thawing (Figure 5.6). Before cryopreservation, plant tissues like shoot tips, meristems or somatic embryos undergo pretreatment to increase their ability to withstand freezing. This can involve using agents such as sucrose or mannitol to remove water from the tissues and prepare them for the process. Cryoprotectants are substances that shield tissue from freezing damage. Common cryoprotectants include dimethyl sulfoxide (DMSO) glycerol, ethylene glycol and sucrose. These substances enter the cells, preventing ice crystal formation that could harm the cells. Concentration of cryoprotectants to be used is crucial as these need to offer sufficient protection to the tissues, without being harmful (Gowthami et al., 2023).

Next comes the cooling step. This step includes cooling the plant tissues to a temperature of -196°C. This can be done using two methods, i.e. slow freezing and vitrification. In the slow freezing approach, tissues are gradually cooled at a controlled pace (0.5–1°C, per minute) to a subzero temperature before being immersed in liquid nitrogen. Slow freezing allows for water removal thus reducing the chances of ice crystal formation within the cells. On the other hand, vitrification entails cooling tissues in the presence of highly concentrated cryoprotectants, resulting in a glass like state formation without ice crystals. This is beneficial as it minimizes cell damage and is suitable for tissues. Once cooled to -196°C, the tissues are moved to nitrogen storage containers. At this temperature all metabolic activities come to an end and the tissues can be preserved indefinitely without

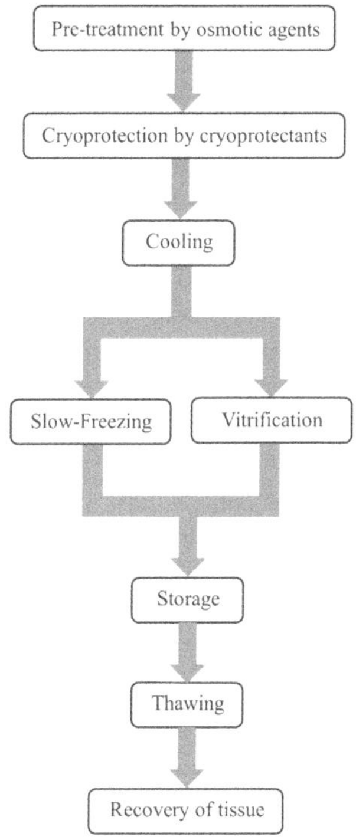

FIGURE 5.6 Steps in cryopreservation.

any physiological alterations. Finally, to recover cryopreserved tissues rapid warming is necessary to prevent ice crystal formation during thawing. Tissues are swiftly transitioned from nitrogen to a water bath (37–40°C), for quick thawing. After thawing the tissues are gently nurtured in a growth medium to encourage regeneration. The effectiveness of the revival process relies on the quality of the cells and the refinement of the conditions for growth (Sharma et al., 2009).

Cryopreservation allows plant material to be stored indefinitely without worrying about changes or variations. Storing plant tissues in nitrogen takes up less space as compared to maintaining large living collections. Also, the tissues are kept safe from pests and diseases, ensuring the genetic material remains intact. Once thawed successfully, cryopreserved tissues can be quickly multiplied using micropropagation techniques ensuring a supply of high-quality plant material. Apart from the advantages of the technique, there are some limitations. Developing cryopreservation methods for Brahmi is crucial because different genetic variations may react differently to cryoprotectants and freezing conditions. Conducting experiments is necessary to identify the pretreatment cryoprotection and cooling techniques for the particular tissue. It's important to be mindful of the toxicity of cryoprotectants at certain concentrations and strike a balance between protection and harm. Moreover, achieving recovery rates after thawing poses a challenge that involves rehydration and optimizing of growing conditions to support the growth of healthy plants. The use of antioxidants and growth regulators, in the recovery process can boost survival rates and promote regeneration (Sharma et al., 2020).

The shoot tips of *B. monnieri* were effectively preserved using vitrification method. Prior to this, the shoot tips underwent pre-treatment with sucrose solution followed by exposure to a vitrification solution containing glycerol and DMSO. After treatment the shoot tips were quickly cooled in nitrogen. Upon thawing, the shoot tips showed good survival and regeneration proving the success of vitrification for *B. monnieri* preservation. Additionally for embryos, pre-treatment with sucrose and encapsulation in alginate beads containing cryoprotectants was found effective. The encapsulated embryos were stored in nitrogen. Post thaw recovery resulted in plantlet regeneration without any genetic changes depicting the potential of this technique, for preserving elite clones (Gowthami et al., 2024).

Thus, cryopreservation of *B. monnieri* is a boon for the conservation and future use of this important medicinal plant. A two-fold protocol integrating cryopreservation with micropropagation to maintain an uninterrupted source of quality *B. monnieri* plantlets can be developed. Synthetic seed technology can also be used along with cryopreservation to boost the conservation protocol. Thus, it can be concluded that cryopreservation is beneficial for the conservation of this valuable medicinal herb by preserving its genetic material and maintaining the genetic integrity of the material.

5.8.4 Synthetic Seed Technology

Synthetic seed technology is a process of encapsulation of somatic embryos, shoot buds or other plant tissues in a protective coat which looks like seed. It is a suitable method of storage, transportation, and germination of plant material under certain environmental conditions. The major constituents of synthetic seeds are the plant tissue or propagule and the encapsulation matrix which is normally hydrogels, such as sodium alginate. Synthetic seeds can be applied for the purpose of easy transportation and preservation of the clones of the desired plants (Sharma et al., 2019).

The selected somatic embryos or shoot buds are encapsulated in a hydrogel matrix. The encapsulation solution is usually prepared with sodium alginate, calcium chloride and other agents to improve the properties of the gel and the supply of nutrients. Sodium alginate is widely used in encapsulation because of its biocompatibility and gelation ability. The encapsulated beads (synthetic seeds) are sintered to strengthen their mechanical properties and enhance the beads' resistance to breakage. This requires putting the beads into a calcium chloride solution that interacts with the alginate to form a rigid gel network. The hardened synthetic seeds are partially desiccated to decrease their moisture content to increase the shelf life and storage stability. Moreover, it is possible to preserve them at low temperatures (Sharma et al., 2012). The success of synthetic seeds is determined by the extent to which they can germinate and regenerate into healthy plants. For this the synthetic seeds are sown on to a germination medium that consists of suitable nutrient and growth regulating substances. The encapsulation matrix starts to break down and the somatic embryos or shoot buds grow into seedlings. After germination of the synthetic seeds and the development of seedlings, the seeds are taken to a greenhouse for hardening. This step is very important to enable the plants to grow and survive under natural conditions. The plants are observed for their growth and development after they have been regenerated and desirable plants are cloned by plant tissue culture technique in order to have a steady stream of elite clones (Pramanik et al., 2021).

Synthetic seeds can be stored for long periods of time under some form of controlled environment. The low metabolic rate of the material enclosed in a capsule provides for the long-term storage of plant material without altering the genotype or phenotype. Synthetic seeds are convenient to handle and transport; enabling the exchange of genetic materials between research institutes, conservation agencies and commercial growers. The process of encapsulation also offers protection from diseases. Synthetic seeds are cheaper than plants propagated through the traditional method and occupy less space, labour, and resources (Hegazi, 2016).

Somatic embryos of *B. monnieri* have been raised and developed into synthetic seeds. These embryos were then coated in a calcium chloride solution to immobilise these in a sodium alginate matrix. The synthetic seeds demonstrated good viability and germination percentages, and the regenerative plants were proved to possess comparable bacoside content and growth profile like the original plants. In another study, synthetic seeds of *B. monnieri* were kept at low temperature and they were germinated after six months. The seeds were later sown under a regulated environment and it was found that the synthetic seeds were capable of germinating and developing healthy plants with high bacoside content, and thus ensuring the efficiency of synthetic seed technology for long term conservation (Khilwani et al., 2016; Ali et al., 2021).

Optimizing the process of encapsulation, hardening, and desiccation for improved synthetic seeds germination is essential. Molecular characterization of plants regenerated from synthetic seeds should be done to verify the genetic stability and authenticity to the source clones. Integrating synthetic seed technology with other advanced techniques like genetic transformation can also be done for the improvement of bacosides and other beneficial phytochemicals in *B. monnieri*. With the advancement of research in this field, synthetic seed technology will assume a significant position in the availability and sustainability of this useful medicinal herb for its therapeutic value and marketability.

5.8.5 Molecular Marker-Assisted Selection

Molecular Marker Assisted Selection (MAS) has become a tool, for choosing and preserving quality *B. monnieri* clones enabling precise genetic enhancements and the retention of favourable traits. MAS utilizes the distinct DNA sequences of markers that are linked to desirable traits of the plant. These markers help pinpoint and monitor diversity among populations making it easier to pick plants with desired features without depending on observable traits that may be affected by environmental conditions (Pant et al., 2024). Several types of molecular markers are utilized in MAS as shown in Table 5.10 (Kumar et al., 2009).

Though this tool is effective, there are obstacles to overcome. The complexity of the genome can be challenging to establish associations between markers and traits. Moreover, the expenses involved in developing and using markers may pose a barrier for some breeding projects. To address these challenges future studies should concentrate on creating cost genotyping techniques and refining associations between markers and traits to enhance the accuracy of MAS. Integrating MAS with methods like genetic transformation could further improve the selection and preservation of superior clones. Recent advancements in markers and genomic selection offer promising opportunities to enhance the efficiency and precision of Bacopa breeding programs leading to the creation of improved cultivars, with enhanced therapeutic benefits (Holland, 2004).

Studying the diversity of *B. monnieri* is important, for choosing and safeguarding clones. Researchers have employed RAPD and SSR markers to analyze the variability among types of *B. monnieri* (Choudhury et al., 2024). For instance, RAPD markers were utilized to assess the

TABLE 5.10
Types of molecular markers

Random Amplified Polymorphic DNA (RAPD):	These markers are short sequences of DNA that are subjected to PCR amplification in order to show variations. These are used for assessing the genetic diversity and identification of unique genetic profiles amongst elite clones.
Simple Sequence Repeats (SSRs):	These represent a type of tandemly repeated highly polymorphic DNA sequences. These give details about the variation existing between different genotypes and provide tools for mapping QTLs associated with important characters.
Amplified Fragment Length Polymorphism (AFLP):	AFLP markers detect polymorphisms in DNA restriction fragments. These are highly sensitive and can produce a large number of markers and are used for constructing high-density genetic maps.
Single Nucleotide Polymorphisms (SNPs):	SNPs are single base pair variations in the genome. These are abundant, widely distributed and ideal for fine-scale genetic mapping and genome-wide association studies (GWAS).

diversity of Bacopa from diverse regions identifying specific genetic characteristics that could be prioritized for conservation efforts and breeding programs. Through Marker Assisted Selection (MAS) scientists can pinpoint clones with traits like high bacoside content, resistance to drought and immunity to diseases. By associating markers with these traits researchers can effectively evaluate populations of *B. monnieri* to identify individuals with the desired genetic attributes. A study conducted by Bansal et al. (2014) utilized SSR markers to detect Quantitative Trait Loci (QTLs) linked to production of suitable chemicals. By selecting plants with alleles at these loci they successfully developed clones with bacoside content, crucial for the medicinal properties of *B. monnieri* (Choudhury et al., 2024).

Marker assisted backcrossing (MABC) is a method used to transfer characteristics from a donor parent into the recipient through multiple backcrosses while utilizing molecular markers for trait selection in each generation (Asati et al., 2022). In *B. monnieri,* MABC can be deployed to introduce traits such as resistance against pests and increased metabolite production. Utilization of MABC to enhance drought tolerance in bacoside producing clones by combining traits from different parent lines can be achieved. As cryopreservation and slow growth storage are commonly employed techniques for *in vitro* conservation, MAS can be utilized to evaluate the integrity of preserved clones ensuring that they maintain their superior characteristics over time. Application of markers to monitor the stability of *B. monnieri* clones preserved through cryopreservation have been showcased, validating that the conserved material stayed true to its original form. Genomic selection (GS) is a MAS approach that predicts individual breeding value of a material by associating their traits with their high-density genetic marker scores. GS can capture the effects of genes simultaneously offering a comprehensive method for selecting superior clones and recent progress has facilitated the integration of GS, in breeding practices.

5.8.6 Genetic Transformation

Genetic modification methods allow for the addition or alteration of genes, improving desirable traits and enhancing and preserving top clones while maintaining accuracy. The process involves inserting genes into *B. monnieri* genome to introduce qualities or enhance existing ones playing a crucial role in developing superior plants. The process of genetic transformation is explained as in the Table 5.11 (Jackson and Linskens, 2003).

Agrobacterium tumefaciens is known for its ability to transfer a segment of its DNA called T DNA, into the makeup of plants. Scientists insert the desired gene into the T DNA section of the Agrobacterium plasmid, which is then utilized to infect plant tissues. Through Agrobacterium mediated transformation scientists insert copies of the synthase gene in *B. monnieri* to increase squalene production and subsequently enhance bacoside synthesis. This method of transformation using Agrobacterium is widely effective for dicotyledonous plants (Mahender et al., 2012). Biolistics (Gene Gun) is another method which entails covering particles, like gold or tungsten

TABLE 5.11
Process of Genetic Transformation

1	**Gene Identification and Isolation**	Identifying and isolating the genes responsible for desirable traits, such as high bacoside production or stress resistance
2	**Gene Insertion**	Inserting the isolated genes into the plant genome using methods such as Agrobacterium-mediated transformation or biolistics (gene gun)
3	**Selection and Regeneration**	Selecting transformed cells that have successfully incorporated the foreign gene and regenerating them into whole plants
4	**Molecular Analysis**	Confirming the presence and expression of the introduced gene through molecular techniques such as PCR, Southern blotting, and RT-PCR

with DNA and propelling them into plant cells using a gene gun. These DNA coated particles can breach the cell wall and introduce DNA into the plant's genetic makeup. Biolistics proves beneficial, for plants that exhibit vulnerability to Agrobacterium infection (Rustgi and Luo, 2020). Lastly, CRISPR/Cas9 Genome Editing is a method that enables modifications to the plants genetic makeup by generating specific double strand breaks in the DNA, which are subsequently fixed by the cells inherent repair processes. This technique allows for the removal or insertion of genes with remarkable accuracy and effectiveness. Furthermore, the technology has been used to deactivate genes that hinder bacoside production allowing improvement of bacoside biosynthesis pathways while preserving other metabolic processes as discussed in the previous sections (Shukla et al., 2022).

Genetic modification allows for incorporating characteristics like high bacoside levels, resilience to stress and immunity to diseases, all crucial for producing top class *B. monnieri*. This method enables manipulation of the plant's genetic makeup in a way that the desired traits manifest accurately in the modified plants. It speeds up the creation of clones as compared to traditional breeding techniques, which are slow and less reliable. Modified plants can be preserved using methods such as tissue culture and cryopreservation safeguarding their purity and favourable attributes. While genetic transformation offers benefits, there are some hurdles to overcome. It involves the use of expertise and equipment that might not be easily accessible. Also, transgenic plants face regulations and concerns about approval, which could impede their widespread use. Techniques like CRISPR/Cas9 may introduce off-target mutations, which need to be carefully monitored and minimized (Su et al., 2023).

By cultivating modified Brahmi plants with resistance to stress and diseases we can reduce the reliance on chemicals promoting ecofriendly farming practices. This is especially crucial for large scale farming. The genetic transformation enables scientific exploration by allowing the examination of gene functions and regulations. Such research could unveil genes linked to characteristics and contribute to the development of superior plant varieties. Through modification and subsequent conservation techniques elite clones of Bacopa can be safeguarded for the future, safeguarding the species genetic diversity and therapeutic properties. The future of genetic transformation in Brahmi shows a lot of potential with ongoing progress in biotechnology. Improvements in editing methods like CRISPR/Cas9 will make modifications more precise and efficient, reducing side effects and enhancing transformation effectiveness. Progress in metabolic engineering will help optimize metabolic pathways to increase the production of bacosides and other beneficial secondary metabolites. Through genomics, transcriptomics, proteomics and metabolomics we can gain an understanding of the mechanisms behind trait expression guiding the development of superior clones. By combining genetic transformation methods with other biotechnological methods, the future for *B. monnieri* cultivation appears optimistic contributing to advancements in modern medicine and preservation efforts.

5.8.7 Metabolic Engineering

Metabolic engineering entails improving the pathways that create metabolites like bacosides. This technique involves twisting metabolic pathways to boost the production of desired compounds. Also manipulating metabolic pathways in an organism increases the production of metabolites. In *B. monnieri* the focus is on ramping up the biosynthesis of bacosides which are the active compounds known for brain boosting. Techniques such as gene editing, pathway optimization and elicitation precursor feeding play role in metabolic engineering endeavors (Figure 5.7) (Hughes and Shanks., 2002; Niazian and Sabbatini., 2021).

In metabolic engineering a crucial approach is boosting the activity of genes that code for enzymes in the pathway. Researchers have marked enzymes like synthase and cycloartenol synthase

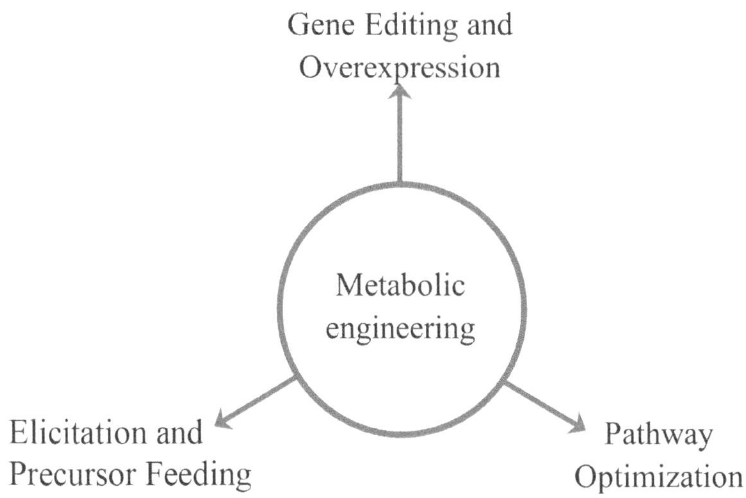

FIGURE 5.7 Metabolic engineering methods.

that are essential for generating bacosides. Enhancing synthase in Brahmi can significantly raise the bacoside levels (Murthy, 2022).

Pathway optimization focuses on adjusting the flow of metabolites through the pathway by regulating enzyme expression levels to ensure efficient conversion of precursors and intermediates leading to the final products. By employing system biology techniques, researchers have developed metabolic models of Brahmi to identify control points in the bacoside biosynthesis process. By tuning enzyme expression at these control points, overall yield of bacosides has successfully been boosted (Sharma et al., 2017; Leonard et al., 2018).

Elicitation involves using living agents to stimulate the production of secondary metabolites. Substances like methyl jasmonate, salicylic acid and yeast extract have been applied to increase the production of bacosides. These substances imitate conditions prompting the plants' defence mechanisms and resulting in an increased production of metabolites. Treating Brahmi suspension cultures with methyl jasmonate led to an increase in the expression of genes involved in the biosynthesis of bacosides thereby elevating bacoside levels. Another method employed in metabolic engineering is precursor feeding, where specific precursors necessary for synthesis are externally provided to the culture medium (Hegazi et al., 2017; Jeyasri et al., 2023).

The use of genomic tools like transcriptomics, proteomics and metabolomics in conjunction with metabolic engineering gives a complete view of molecular mechanisms underlying bacoside biosynthesis (Halder et al., 2021). Thus, by this approach researchers can find new target genes and regulatory elements to be manipulated to increase bacoside yield. Recent advances in next generation sequencing (NGS) and bioinformatics have allowed high resolution genetic and transcriptomic data generation for *B. monnieri*. These datasets are invaluable as far as gene expression patterns, metabolic pathway interactions as well as possible regulatory networks in relation to secondary metabolite production are concerned (Tungphatthong et al., 2018) .

Although the metabolic engineering of *B. monnieri* has resulted in considerable breakthroughs, some challenges still remain. The bacoside biosynthetic pathway is complicated which necessitates careful fine-tuning and validation of engineered traits. Future research needs to be about more efficient gene-editing tools, new elicitors as well as improving the scalability of metabolic engineering techniques. Moreover, combining omics technologies with traditional breeding and conservation approaches will enhance overall efficiency and precision in *B. monnieri* improvement programs (Leonard et al., 2018).

Metabolic engineering offers a solution, for selecting and preserving the strains of *B. monnieri* to enhance its bacoside production and support sustainable cultivation of this valuable medicinal plant. Improvements have been achieved by modifying metabolic pathways, such as boosting enzyme expression or optimizing growth conditions resulting in quality and higher yields, from *B. monnieri*. As tools and biotechnological techniques progress, incorporating biotechnological methods will further drive the field forward ensuring the ongoing availability and therapeutic benefits of *B. monnieri*.

5.8.8 Omics Approaches

Techniques of genomics, transcriptomics, proteomics and metabolomics are the broad-spectrum approaches, which are exquisite for the study and control of the genetic and biochemical network for synthesis of molecules (Figure 5.8) (Husen and Ahmad, 2023). This section focuses on the application of omics technologies to preserve the best clones of *B. monnieri*. Genomics is the analysis of an organism's genome. It involves identification of suitable genes that are responsible for desirable traits such as high bacoside yield, disease resistance and stress tolerance (Sidharthan and Baranwal, 2021). The complete genome sequencing of *B. monnieri* generates information regarding the plant genome and the published genomic data by the researchers contain genes related to bacosides and biosynthesis of secondary metabolite (Jeena et al., 2017). This genomic information forms the basis for preparation of a gene list on which genetic improvement can be focused on. Genetic markers, such as SSRs (Simple Sequence Repeats) and SNPs (Single Nucleotide Polymorphisms) are used to construct high-density genetic maps in medicinal plants (Ting et al., 2014). These markers help in identifying Quantitative Trait Loci (QTLs) associated with important traits. For instance, QTL mapping studies have marked regions of the genome that

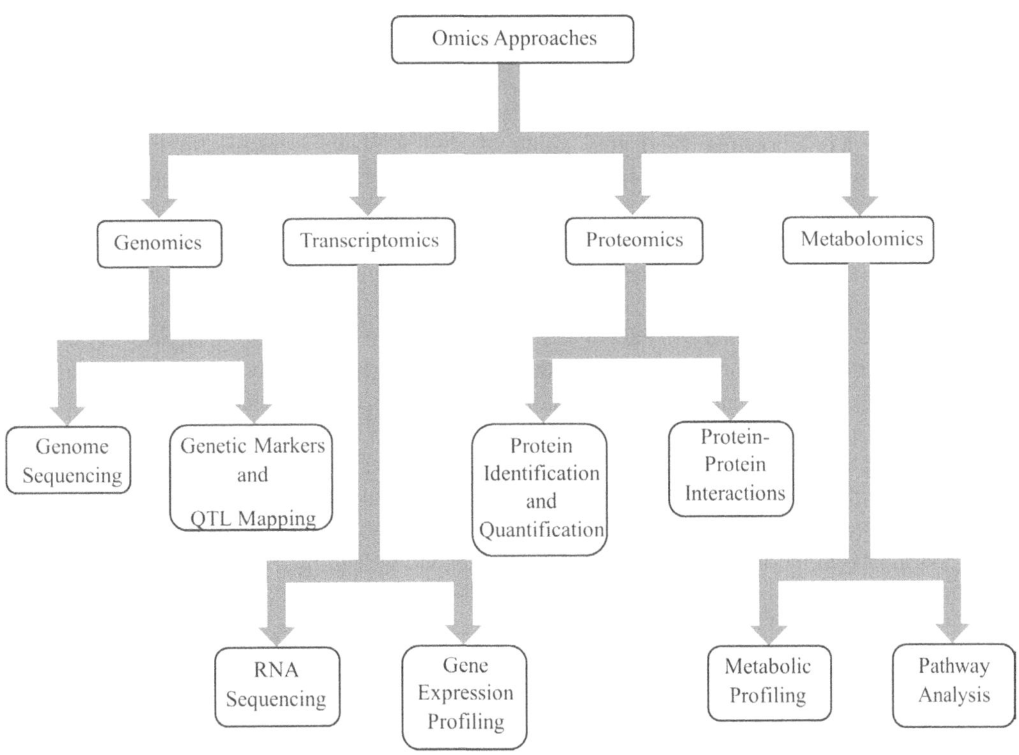

FIGURE 5.8 Omics approaches.

are linked to bacoside production, enabling the selection of elite clones with enhanced bacoside content (Barut et al., 2022).

Transcriptomics delves into examining the range of RNA transcripts generated by the genome. This method aids in grasping the patterns of gene expression and marking the networks implicated in metabolic pathways. By utilizing RNA sequencing, researchers gain insights into the gene expression profiles of *B. monnieri* across all environmental conditions and stages of development. The process involves comparing the transcriptomes of bacoside producing clones to study genes that are expressed and play roles in biosynthesis. To validate the RNA sequencing results and measure the expression levels of genes, gene expression profiling techniques like qRT-PCR (Real-Time Quantitative Reverse Transcription PCR) are employed. Notably, genes responsible for enzymes such as cycloartenol synthase exhibit heightened expression levels in elite clones of the plant (Jeena et al., 2017; Kumar et al., 2024).

Proteomics addresses the large-scale analysis of proteins, their structures and functions. This approach provides details regarding the Post Translational Modifications and protein constitution in cellular metabolic processes. Through this, proteins associated in the synthesis of bacosides for *B. monnieri* have been identified. When the proteome of different clones is compared, researchers are able to find out which enzymes and regulatory proteins are upregulated in the high bacoside producing clone. Knowledge of how proteins interact with other proteins is important in the determination of the regulatory mechanisms of metabolic pathways. A study not only showed the utilization of multiple defence signalling pathways in crypttransgenic plants of *B. monnieri* but also further showed first evidence of the involvement of JA-ET signalling in cryptogein-induced elicitation (Paul et al., 2017).

Metabolomics is the study of the metabolites of a system, and it is the measurement of the metabolome. This method helps in measuring biochemical activity and the metabolic state (Parate et al., 2024). GC-MS and LC-MS have been used to carry out phytochemical analysis which helps in determining the amount of bacosides and other secondary metabolites found in *B. monnieri* and identify metabolic pathways involved in high bacoside content by comparing metabolic contents of both elite and non-elite clones (Thakur et al., 2024). Pathway analysis combines metabolomic data with genomic, transcriptomic, and proteomic data for building models of metabolic pathways. These help to define potential regulatory points as well as plausible targets for manipulation of metabolism. The mevalonate pathway and methylerythritol phosphate pathway are important for bacoside synthesis in *B. monnieri* according to the studies on pathways (Ahuja et al., 2024).

Omics approaches aid in verifying the consistency of preserved clones by comparing their transcriptomic and proteomic profiles against those of the original elite clones. Regular evaluation of stability through markers and omics data ensures that conserved clones maintain their exceptional traits. There are still a number of challenges in selection and conservation of superior *B. monnieri* clones despite the significant advances made through omics techniques. Examples here include cost implications on high-throughput omics technologies, the complex bacoside biosynthetic pathway, and influences of environmental factors. Further studies should be conducted to establish affordable and easily adaptable omics technologies for comprehensive metabolic pathways analysis as well as novel bioinformatics tools development for effective data analysis. The integration of omics data with traditional conservation methods ensures the genetic stability and long-term preservation of elite clones, paving the way for sustainable cultivation and utilization of *B. monnieri* in traditional and modern medicine.

5.8.9 Epigenetic Modifications

Epigenetics can be defined as the changes that occur in an organism's phenotype without changes in its genotype. These changes are mainly regulated by the factors which include DNA methylation, histone modification, and non-coding RNAs. Epigenetic changes are involved in the mechanism governing gene activities and may be affected by external factors. In plants, epigenetics plays a critical role in the developmental process, stress response mechanism and synthesis of secondary metabolites (Purusothaman et al., 2021; Abdulraheem et al., 2024). DNA methylation is a process that incorporates a methyl group to DNA, particularly in the cytosine base at CpG site. This alteration can cause gene suppression and is associated with the management of several processes in plants. Using *B. monnieri* it has been possible to link the DNA methylation profiles to the expression of genes that incorporate bacoside biosynthesis and stress. Hence, based on the genes that are specific to high bacoside content, researchers can choose the elite clones more precisely (Roy et al., 2024).

Histones are proteins on which DNA wraps, to make up nucleosomes. Modifications of histones that occur after histone synthesis include methylation, acetylation, phosphorylation and ubiquitination which can affect chromatin compactness and gene activity. Hence histones that control the activation and repression of genes in the cases of secondary metabolites are like the trailed markers (Rajan et al., 2015). MicroRNAs (miRNAs) and long non-coding RNAs (lncRNAs) are involved in regulation at the post-transcriptional level by interacting with chromatin or binding to messenger RNAs (mRNAs). Determining specific non-coding RNAs (ncRNA) that map genes involved in the synthesis of bacoside will showcase unique characters in elite clones of *B. monnieri* (Preethi et al., 2012; Janakiram et al., 2019). The epigenetic modifications include techniques like chemical epigenetic modifiers, RNA interference, CRISPR-dcas9, and markers like DNA methylation, histone modifications, non-coding RNAs.

To improve characteristics in *B. monnieri*, epigenetic modification strategies can be used. A number of chemical reagents are used for epigenetic modification that includes 5-azacytidine which targets DNA methylation and trichostatin A which targets histone deacetylase. These chemicals can be used in *B. monnieri* cultures to influence the gene profiles in order to alter the synthesis of bacosides. According to the kind of modifier that is used, the desired Brahmi clones can be obtained (Preethi et al., 2016). RNA interference (RNAi) is a technique where the functioning of specific genes is suppressed by using small interfering RNAs (siRNAs). Thus, in *B. monnieri*, RNAi can be employed effectively where undesirable genes are inhibited. Through silencing of these genes, bacosides synthesis can be increased (Das et al., 2016). Epigenome editing can be made possible using the modified CRISPR-dCas9 by linking it to epigenetic enzymes namely DNA methyltransferases or histone acetyltransferases (Enríquez, 2016). This can be easier to target and alter specific locations on the genome with higher efficiency, thus, activating or repressing the genes related to the bacoside biosynthetic pathway in case of Brahmi.

While genetic changes can create variability and potential side effects, epigenetic marks can be passed on from generation to generation without changing the DNA code. This stability helps maintain the useful characteristics of clones that have been inherited. Investigations have pointed out certain detection of DNA methylation profile in the plants that yield high bacoside content in *B. monnieri*. In plants, methylation-sensitive amplified polymorphism (MSAP) analysis allowed to find out the methylation status and establish the connection between methylation levels and biosynthetic genes expression (Guo et al., 2014). In Brahmi this can lead to superior cloning that can enable the selection of clones belonging to the elite group with superior medicinal value. Also, histone modification pattern of plants under stress was investigated by chromatin immunoprecipitation sequencing. It was further observed that the certain histone marks were involved in the activation of stress responsive genes and secondary metabolic pathways (Statham et al., 2012). In this way, when the clones with beneficial histone modification are chosen, the stress tolerance and medicinal properties of *B. monnieri* can be improved.

Epigenetic modifications for selection and conservation of clones of *B. monnieri* appear to have wonderful prospective. Screening process can be adopted to help in the identification of specific epigenetic marks that are linked with the desirable characteristics in *B. monnieri*. Next generation sequencing technologies or bioinformatics can be incorporated to the epigenomic data. Epigenomic changes can be applied in conventional breeding strategies in order to embark on the generation of elite clones. Hence, when other biotechnological approaches are complemented with the process of epigenetic reprogramming, the scientists make new varieties of the *B. monnieri* plant stronger in medicinal terms. Information on how temperature, light and nutrient availability affect the epigenetic marks, may help in the management of the cultivation practices. This is why epigenetic adaptations are one of the most promising and multifunctional tools in the selection and preservation of elite clones of *B. monnieri*. The fact that epigenetic marks are stable and heritable means that these traits would remain consistent, leading to the plant being beneficial for long-term usage making epigenetic changes important for the propagation of *B. monnieri*.

5.9 CONCLUSION

Selection and conservation of selected clones of *Bacopa monnieri* using modern biotechnological methodologies is a remarkable innovation in the field of plant biotechnology. This is very important not only for the conservation of this useful plant but also for the improvement of its medicinal properties. The plants' cognition stimulating properties, wide use in medicine and its endangered status due to overexploitation, makes the plant important for conservation through biotechnological tools. Effective methods like the use of micropropagation, synthetic seed technology, and cryopreservation provide a reliable approach with regard to preserving elite clones and having access to the plant in the long run, for various research projects and for the production of medicines. The use of molecular markers and genetic engineering escalates the selection and conservation procedures to another level. The molecular markers-based identification and selection of superior clones, assessment of genetic diversity, use of methods like CRISPR/Cas9 and Agrobacterium mediated modification can be further used to modify the medicinal value, stress tolerance and physical growth of the plant. Preservation of superior genotypes of *Bacopa monnieri* through biotechnological approach will be beneficial and will respond to global medicinal demand for human health, while also contributing to conservation of ethno-botanical knowledge and genetic resources.

REFERENCES

Aasim, M., Sameeullah, M., Karataş, M., Bakirci, S., Bakhsh, A. and Akhtar, M.S., 2019. An insight into biotechnological approaches used for the improvement of secondary metabolites from the medicinal aquatic plant, water hyssop (*Bacopa monnieri* L.). In M. S. Akhtar and M. K. Swamy (Eds.), *Natural bio-active compounds: Volume 3: Biotechnology, bioengineering, and molecular approaches* (pp. 123–152). Springer.

Abdalla, N., El-Ramady, H., Seliem, M.K., El-Mahrouk, M.E., Taha, N., Bayoumi, Y., Shalaby, T.A., and Dobránszki, J., 2022. An academic and technical overview on plant micropropagation challenges. *Horticulturae*, *8*(8), p. 677.

Abdulraheem, M.I., Xiong, Y., Moshood, A.Y., Cadenas-Pliego, G., Zhang, H., and Hu, J., 2024. Mechanisms of plant epigenetic regulation in response to plant stress: Recent discoveries and implications. *Plants*, *13*(2), p. 163.

Ahuja, A., Sharma, M., and Mallubhotla, S., 2024. Bacosides neurotropic molecules production by tissue cultures of *Bacopa monnieri* (L.) Wettst. In P. B. Kavi Kishor, P. Suprasanna and A. Romano (Eds.), *Biotechnological Production of bioactive phytochemicals of medicinal value* (pp. 353–373). Elsevier.

Ali, D., Alarifi, S., and Pandian, A., 2021. Somatic embryogenesis and in vitro plant regeneration of *Bacopa monnieri* (Linn.) Wettst., a potential medicinal water hyssop plant. *Saudi Journal of Biological Sciences*, *28*(1), pp. 353–359.

Asati, R., Tripathi, M.K., Tiwari, S., Yadav, R.K., and Tripathi, N., 2022. Molecular breeding and drought tolerance in chickpea. *Life*, *12*(11), p. 1846.

Bansal, M., Kumar, A., and Sudhakara Reddy, M., 2014. Diversity among wild accessions of *Bacopa monnieri* (L.) Wettst. and their morphogenetic potential. *Acta Physiologiae Plantarum*, *36*, pp. 1177–1186.

Barut, M., Nadeem, M.A., Akgür, Ö., Tansi, L.S., Aasim, M., Altaf, M.T., and Baloch, F.S., 2022. Medicinal and aromatic plants in the omics era: Application of plant breeding andbiotechnology for plant secondary metabolite production. *Turkish Journal of Agriculture and Forestry*, *46*(2), pp. 182–203.

Bharti, N., Yadav, D., Barnawal, D., Maji, D., and Kalra, A., 2013. Exiguobacterium oxidotolerans, a halotolerant plant growth promoting rhizobacteria, improves yield and content of secondary metabolites in *Bacopa monnieri* (L.) Pennell under primary and secondary salt stress. *World Journal of Microbiology and Biotechnology*, *29*, pp. 379–387.

Binita, B.C., Ashok, M.D., and Yogesh, T.J., 2005. *Bacopa monnieri* (L.) Pennell: A rapid, efficient and cost effective micropropagation. *Plant Tissue Culture and Biotechnology*, *15*(2), pp. 167–175.

Brimson, J.M., Brimson, S., Prasanth, M.I., Thitilertdecha, P., Malar, D.S., and Tencomnao, T., 2021. The effectiveness of *Bacopa monnieri* (Linn.) Wettst. as a nootropic, neuroprotective, or antidepressant supplement: Analysis of the available clinical data. *Scientific Reports*, *11*(1), p. 596.

Chamak, F.H., 2022. *Micropropagation of medicinal plant in Bangladesh-a short review* (Doctoral dissertation, Brac University).

Choudhary, S., Kumari, I., Thakur, S., Kaurav, H. and Chaudhary, G., 2021. Brahmi (*Bacopa monnieri*)–A potential ayurvedic cognitive enhancer and neuroprotective herb. *International Journal of Ayurveda and Pharma Research*, *9*(5), pp. 41–49.

Choudhury, D.R., Sharma, L., Suma, A., Singh, G.P. and Singh, R., 2024. Exploring the genetic diversity and population structure of *Bacopa monnieri* (L.) using random amplified (RAPD and ISSR) and gene-targeted (SCoT and CBDP) markers. *Genetic Resources and Crop Evolution*, pp. 1–21.

Dadhich, A., Dhiman, M., Sharma, L., Kumar, R., Jain, R. and Sharma, M.M., 2022. Unveiling the interaction of divergent abiotic stresses and their consequences in terms of bacosides in *Bacopa monnieri* (L.) Wettst. *Journal of Applied Research on Medicinal and Aromatic Plants*, *31*, p. 100423.

Das, D.N., Naik, P.P., Nayak, A., Panda, P.K., Mukhopadhyay, S., Sinha, N. and Bhutia, S.K., 2016. *Bacopa monnieri*-induced Protective Autophagy Inhibits Benzo [a] pyrene-mediated Apoptosis. *Phytotherapy Research*, *30*(11), pp. 1794–1801.

Deo, Y.K. and Reddy, K.R.C., 2013. Critical review on pharmacological properties of Brahmi. *International Journal of Ayurvedic Medicine*, *4*(2), pp. 92–99.

Dey, A., Hazra, A.K., Nongdam, P., Nandy, S., Tikendra, L., Mukherjee, A., Banerjee, S., Mukherjee, S. and Pandey, D.K., 2019. Enhanced bacoside content in polyamine treated in-vitro raised *Bacopa monnieri* (L.) Wettst. *South African Journal of Botany*, *123*, pp. 259–269.

Eeckhaut, T., Lakshmanan, P.S., Deryckere, D., Van Bockstaele, E. and Van Enríquez, P., 2016. Focus: Epigenetics: CRISPR-mediated epigenome editing. *Yale Journal of Biology and Medicine*, *89*(4), p. 471.

Fatima, U., Roy, S., Ahmad, S., Ali, S., Elkady, W.M., Khan, I., Alsaffar, R.M., Adnan, M., Islam, A. and Hassan, M.I., 2022. Pharmacological attributes of *Bacopa monnieri* extract: Current updates and clinical manifestation. *Frontiers in Nutrition*, *9*, p. 972379.

Gosal, S.S. and Kang, M.S., 2012. Plant tissue culture and genetic transformation for crop improvement. In N. Tuteja, S. Singh Gill, A. F. Tiburcio and R. Tuteja (Eds.), *Improving Crop Resistance to Abiotic Stress*, pp. 357–397.

Gowthami, R., Rajashekaran, P.E., Martinez-Montero, M.E., Chander, S., Singh, A.M., Puneeth, G.M., Shankar, M. and Deepak, D.A., 2024. In vitro techniques for safeguarding medicinal plants with antiallergic properties. In S. Gantait, J. Majumder and A. B. Sharangi (Eds.), *Biotechnology of medicinal plants with antiallergy properties: Research trends and prospects* (pp. 635–672). Springer Nature Singapore.

Gowthami, R., Sharma, N., Chandra, R., Kurian, J.S., Malhotra, E.V. and Agrawal, A., 2023. Development of an improved and simple shoot tip cryoconservation protocol for cryobanking of *Bacopa monnieri* (L.) Wettst. germplasm. *In Vitro Cellular & Developmental Biology-Plant*, *59*(6), pp. 744–756.

Gudipati, T., Srivastava, P., Goswami, H. and Bhadauria, R., 2012. *In Vitro* Conservation of *Bacopa monnieri* Germplasm: A Critically Endangered Medicinal Plant of Madhya Pradesh. *Journal of Indian Botanical Society*, *91*(1–3), pp. 83–91.

Güner, B.Ç., 2023. Genome engineering in medicinal plants for improved. In S. K. Upadhyay (Ed.), *Applications of Genome Engineering in Plants*, p. 260.

Guo Ning, G.N., Xi XiaoGuang, X.X., Zhang XiaoXu, Z.X., Fan HongHong, F.H., Cai YongPing, C.Y., Lin Yi, L.Y. and Sun Xu, S.X., 2014. Methylation sensitive amplified polymorphism (MSAP) analysis of DNA epigenetic changes of cotton (Gossypium hirsutum L.) cotyledons under ethephon treatment. *Journal of Agricultural Biotechnology*, *22*(9): 1131–1140.

Halder, M., Majumder, A., Ray, S. and Jha, S., 2021. Medicinal plant research at crossroads: Biotechnological approaches for conservation, production and stability in tissue cultures and regenerated plants. In H. M. Ekiert, K. G. Ramawat and J. Arora (Eds.), *Medicinal Plants: Domestication, Biotechnology and Regional Importance*, pp. 459–544.

Halder, S., Anand, U., Nandy, S., Oleksak, P., Qusti, S., Alshammari, E.M., Batiha, G.E.S., Koshy, E.P. and Dey, A., 2021. Herbal drugs and natural bioactive products as potential therapeutics: A review on pro-cognitives and brain boosters perspectives. *Saudi Pharmaceutical Journal*, *29*(8), pp. 879–907.

Hegazi, G., Taha, H., Sharaf, A.M.M. and Elaish, S.R., 2017. Enhancing in vitro production of Bacoside a from *Bacopa monnieri* using precursor and elicitors feeding. *Journal of Basic and Applied Scientific Research*, *7*(9), pp. 27–35.

Hegazi, G.A.E.M., 2016. In vitro preservation of *Bacopa monnieri* (L.) Pennell as a rare medicinal plant in Egypt. *Journal of Basic and Applied Scientific Research*, *6*(12), pp. 35–43.

Holland, J.B., 2004, September. Implementation of molecular markers for quantitative traits in breeding programs—challenges and opportunities. In *Proceedings of the 4th international crop science congress* (Vol. 26, pp. 1–13). Regional Institute.

Hughes, E.H. and Shanks, J.V., 2002. Metabolic engineering of plants for alkaloid production. *Metabolic Engineering*, *4*(1), pp. 41–48.

Husen, A. and Ahmad, A. eds., 2023. *Genomics, transcriptomics, proteomics and metabolomics of crop plants*. Elsevier.

Inthima, P. and Supaibulwatana, K., 2024. Comparative growth and bacoside production in diploid and tetraploid *Bacopa monnieri* (L.) Wettst. cultivated indoors via hydroponic and soil culture systems. *Horticulturae*, *10*(6), p. 574.

Jackson, J.F. and Linskens, H.F. eds., 2003. *Genetic transformation of plants* (Vol. 23). Springer Science & Business Media.

Jain, P.K., Das, D., Jain, P. and Jain, P., 2016. Pharmacognostic and pharmacological aspect of *Bacopa monnieri*: A review. *Rays (Fig. 2)*, *8*, p. 9.

Janakiram, T., Sunitha, P. and Aparna, V., 2019. Biotechnological vistas in medicinal plants-A way forward. *Medicinal Plants-International Journal of Phytomedicines and Related Industries*, *11*(3), pp. 207–217.

Jaspreet Kaur, J.K., Kalpana Nautiyal, K.N. and Manu Pant, M.P., 2013. In vitro propagation of Bacopa monnieri (L.) Wettst-A medicinally priced herb. International Journal of Current Microbiology and Applied Sciences, 2(8), pp. 131–138.

Jeena, G.S., Fatima, S., Tripathi, P., Upadhyay, S. and Shukla, R.K., 2017. Comparative transcriptome analysis of shoot and root tissue of *Bacopa monnieri* identifies potential genes related to triterpenoid saponin biosynthesis. *BMC Genomics*, *18*, pp. 1–15.

Jeyasri, R., Muthuramalingam, P., Karthick, K., Shin, H., Choi, S.H. and Ramesh, M., 2023. Methyl jasmonate and salicylic acid as powerful elicitors for enhancing the production of secondary metabolites in medicinal plants: An updated review. *Plant Cell, Tissue and Organ Culture (PCTOC)*, *153*(3), pp. 447–458.

Jha, S.R. and Mughees, M., 2023. Tools for safety evaluation and conservation of medicinal plants. In A. Ahmad and A. Asif (Eds.), *Omics studies of medicinal plants* (pp. 263–278). CRC Press.

Joshi, N., Pandey, S.T., Singh, V.P., Kumar, A. and Gautam, P., 2019. Weed management in direct-seeded rice (*Oryza sativa*)+ brahmi (*Bacopa monnieri*) intercropping system. *Indian Journal of Agronomy*, *64*(1), pp. 23–27.

Kamesh, V. and Sumathi, T., 2012. Antihypercholesterolemic effect of Bacopa monniera linn. on high cholesterol diet induced hypercholesterolemia in rats. *Asian Pacific Journal of Tropical Medicine*, *5*(12), pp. 949–955.

Karthikeyan, A., Madhanraj, A., Pandian, S.K. and Ramesh, M., 2011. Genetic variation among highly endangered *Bacopa monnieri* (L.) Pennell from Southern India as detected using RAPD analysis. *Genetic Resources and Crop Evolution*, *58*, pp. 769–782.

Khilwani, B., Kaur, A., Ranjan, R. and Kumar, A., 2016. Direct somatic embryogenesis and encapsulation of somatic embryos for in vitro conservation of *Bacopa monnieri* (L.) Wettst. *Plant Cell, Tissue and Organ Culture (PCTOC)*, *127*, pp. 433–442.

Kumar, P., Gupta, V.K., Misra, A.K., Modi, D.R. and Pandey, B.K., 2009. Potential of molecular markers in plant biotechnology. *Plant Omics*, *2*(4), pp. 141–162.

Kumar, S., Singh, N., Lahane, V., Tripathi, V., Yadav, A.K. and Shukla, R.K., 2024. Methyl jasmonate inducible UGT79A18 is a novel glycosyltransferase involved in the bacoside biosynthetic pathway in *Bacopa monnieri*. *Physiologia Plantarum*, *176*(2), p. e14260.

Leonard, J., Seth, B., Sahu, B.B., Singh, V.R. and Patra, N., 2018. Statistical optimization for enhanced bacoside A production in plant cell cultures of *Bacopa monnieri*. *Plant Cell, Tissue and Organ Culture (PCTOC)*, *133*, pp. 203–214.

Mahender, A., Mallesham, B., Srinivas, K., Kranthi Kumar, G., Venugopal Rao, K., Rajesh, Y., Zhang, P. and Sadanandam, A., 2012. A rapid and efficient method for in vitro shoot organogenesis and production of transgenic Bacopa monnieri L. mediated by Agrobacterium tumefaciens. *In vitro Cellular & Developmental Biology-Plant*, *48*, pp. 153–159.

Mondal, S., Bhar, K., Mondal, P., Panigrahi, N., Sahoo, S.K., Swetha, P., Chakraborty, S., Teja, N.Y. and Parveen, N., 2023. In quest of the mysterious holistic Vedic herb Bacopa monnieri (L.) Pennell. *Pharmacognosy Research*, *15*(3), pp. 410–454.

Mukhopadhyay, M.K., Banerjee, P. and Nath, D., 2012. Phytochemicals–biomolecules for prevention and treatment of human diseases–A review. *IJSER*, *3*(7), pp. 1–32.

Murthy, H.N., 2022. Biotechnological production of bacosides from cell and organ cultures of *Bacopa monnieri*. *Applied Microbiology and Biotechnology*, *106*(5), pp. 1799–1811.

Niazian, M. and Sabbatini, P., 2021. Traditional in vitro strategies for sustainable production of bioactive compounds and manipulation of metabolomic profile in medicinal, aromatic and ornamental plants. *Planta*, *254*, pp. 1–22.

Panda, D., Barik, J.R., Barik, J., Behera, P.K. and Dash, D., 2021. Suitability of Brahmi (*Bacopa monnieri* L.) cultivation on fly ash-amended soil for better growth and oil content. *International Journal of Phytoremediation*, *23*(1), pp. 72–79.

Pandey, A.K., 2022. *Bacopa monnieri* (Linn.) Pennell-A possible plant for impossible diseases (A review). *Advances in Pharmacology and Pharmacy*, *10*(1), pp. 35–53.

Pankaj, U., Singh, D.N., Mishra, P., Pooja, G.A.U.R., Babu, C.V., Shanker, K. and Verma, R.K., 2020. Autochthonous halotolerant plant growth-promoting rhizobacteria promote bacoside A yield of *Bacopa monnieri* (L.) Nash and phytoextraction of salt-affected soil. *Pedosphere*, *30*(5), pp. 671–683.

Pant, M., and Mehta, R., 2016. The economics of plant tissue culture: An Indian perspective. *Asian Journal of Pharmaceutical and Clinical Research*, *9*, pp. 36–37.

Pant, M., Pant, K., Chauhan, A., Negi, A.S., Nainwal, P., Srivastava, A. and Kumar, V., 2024. Molecular marker-assisted genetic diversity analysis in soybean cultivars from Himalayan region of Uttarakhand, India. *Ecological Genetics and Genomics*, *31*, p. 100237.

Parate, S.S., Upadhyay, S.S., Karthikkeyan, G., Pervaje, R., Abhinand, C.S., Modi, P.K. and Prasad, T.S.K., 2024. Comparative metabolomics and network pharmacology analysis reveal shared neuroprotective mechanisms of *Bacopa monnieri* (L.) Wettst and Centella asiatica (L.) Urb. *Molecular Neurobiology*, *61*, pp. 1–23.

Patel, A., Jaiswal, N., Srivastava, P.K. and Patra, D.D., 2022. Enhancing secondary metabolite production and antioxidants in *Bacopa monnieri* grown on tannery sludge contaminated soil. *Industrial Crops and Products*, *187*, p. 115365.

Paul, P., Nandagopal, K., Agarwala, N., Das, S. and Jha, S., 2017. A proteomic approach to evaluate the effects of endogenous expression of cryptogein gene in crypt-transgenic plants of *Bacopa monnieri*. *Journal of Applied Biotechnology & Bioengineering*, *4*(3), pp. 588–603.

Pramanik, B., Sarkar, S., Bhattacharyya, S. and Gantait, S., 2021. meta-Topolin-induced enhanced biomass production via direct and indirect regeneration, synthetic seed production, and genetic fidelity assessment of *Bacopa monnieri* (L.) Pennell, a memory-booster plant. *Acta Physiologiae Plantarum*, *43*, pp. 1–14.

Preethi, J., Singh, H.K. and Rajan, K.E., 2016. Possible involvement of standardized Bacopa monniera extract (CDRI-08) in epigenetic regulation of reelin and brain-derived neurotrophic factor to enhance memory. *Frontiers in Pharmacology*, *7*, p. 166.

Preethi, J., Singh, H.K., Charles, P.D. and Emmanuvel Rajan, K., 2012. Participation of microRNA 124-CREB pathway: A parallel memory enhancing mechanism of standardised extract of Bacopa monniera (BESEB CDRI-08). *Neurochemical Research*, *37*, pp. 2167–2177.

Purusothaman, D., Chalichem, N.S.S., Bethapudi, B., Murugan, S. and Mundkinajeddu, D., 2021. *Bacopa monnieri* for cognitive health—A review of molecular mechanisms of action. *Nutraceuticals in Brain Health and Beyond*, pp. 15–30.

Qaiser, H., Abdullah, R., Iqtedar, M., Kaleem, A. and Sajer, B.H., 2023. Brahmi. In M. Zia-Ul-Haq, A. A. AL-Huqail, M. Riaz, U. F. Gohar (Eds.), *Essentials of medicinal and aromatic crops* (pp. 789–809). Springer International Publishing.

Rahe, M.A., Mollika, S.R., Khan, M.S., Banu, T.A., Al Amin, G.M., Habib, M.A., Akter, S., Islam, M. and Sharmin, R.A., 2020. In vitro micropropagation of *Bacopa monnieri* (L.) Penn.-an important medicinal plant. *Plant Tissue Culture and Biotechnology*, *30*(1), pp. 57–63.

Rai, K., Gupta, N., Dharamdasani, L., Nair, P. and Bodhankar, P., 2017. *Bacopa monnieri*: A wonder drug changing fortune of people. *International Journal of Applied Sciences and Biotechnology*, *5*(2), pp. 127–132.

Rajan, K.E., Preethi, J. and Singh, H.K., 2015. Molecular and functional characterization of Bacopa monniera: A retrospective review. *Evidence-Based Complementary and Alternative Medicine*, *2015*(1), p. 945217.

Ribeiro, J.P.N., Tiberio, F.C. and de Oliveira, A.A., 2015. The effect of competition on *Bacopa monnieri* zonation in an temporarily open/closed tropical estuary. *Estuarine, Coastal and Shelf Science*, *163*, pp. 231–234.

Rohini, M.R., 2020. Biotechnological interventions for conservation and multiplication of threatened medicinal plants. In P.E. Rajasekharan and S. H. Wani (Eds.), *Conservation and Utilization of Threatened Medicinal Plants*, pp. 135–158. Springer.

Roy, S., Shanmugam, G., Rakshit, S., Pradeep, R., George, M. and Sarkar, K., 2024. Exploring the immunomodulatory potential of Brahmi (*Bacopa monnieri*) in the treatment of invasive ductal carcinoma. *Medical Oncology*, *41*(5), p. 115.

Rustgi, S. and Luo, H., 2020. *Biolistic DNA delivery in plants*. Springer.

Saghatelyan, A.A., 2017. Phytogeographical relationships and analysis of the flora of South Texas plains, USA. *Journal of the Botanical Research Institute of Texas*, *11*, pp. 527–561.

Saloni, S.M., Rai, D.C., Panda, P. and Kumar, S., 2022. A comprehensive review on *Bacopa monnieri* (L.) Pennell (Brahmi): Utilization as a functional food ingredient and health-promoting attributes. *Annals of Phytomedicine*, *11*(1), pp. 142–150.

Samridha, V. and Chandra, S., 2024. In-vitro propagation to conserve medicinally important plants: Insight, procedures, and opportunities. In L. Kumar, N. Bharadvaja, R. Singh and R. Anand (Eds.), *Medicinal and aromatic plants: Current research status, value-addition to their waste, and agro-industrial potential* (Vol. I, pp. 13–25). Springer Nature Switzerland.

Saran, P.L., Damor, H.I., Lodaya, D.H., Suthar, M.K., Kalariya, K.A. and Roy, S., 2022. Identification of potential accessions of *Bacopa monnieri* L. for herbage yield and bacosides A content. *Industrial Crops and Products*, *176*, p. 114348.

Sharma, A., Verma, N., Verma, P., Verma, R.K., Mathur, A. and Mathur, A.K., 2017. Optimization of a *Bacopa monnieri*-based genetic transformation model for testing the expression efficiency of pathway gene constructs of medicinal crops. *In vitro Cellular & Developmental Biology-Plant*, *53*, pp. 22–32.

Sharma, N., Gowthami, R. and Pandey, R., 2019. Synthetic seeds: A valuable adjunct for conservation of medicinal plants. In P.E. Rajasekharan and S. H. Wani (Eds.), *Synthetic seeds: Germplasm regeneration, preservation and prospects* (pp. 181–216). Springer.

Sharma, N., Pandey, R. and Gowthami, R., 2020. In vitro conservation and cryopreservation of threatened medicinal plants of India. In P.E. Rajasekharan and S. H. Wani (Eds.), *Conservation and Utilization of Threatened Medicinal Plants*, pp. 181–228.

Sharma, N., Satsangi, R. and Pandey, R., 2009, April. Cryopreservation of shoot tips of *Bacopa monnieri* (L.) Wettst. by vitrification technique. In *International Symposium on Cryopreservation in Horticultural Species 908* (pp. 283–288). ISHS Acta Horticulturae 908.

Sharma, N., Satsangi, R., Pandey, R., Singh, R., Kaushik, N. and Tyagi, R.K., 2012. In vitro conservation of *Bacopa monnieri* (L.) using mineral oil. *Plant Cell, Tissue and Organ Culture (PCTOC)*, *111*, pp. 291–301.

Shukla, J., Pant, B., Tufchi, N., Paul, G.C., Pant, K., Pant, M. and Sihna, S., 2022, December. CRISPR Cas9 genome editing approach for gRNA of the genes associated with Schizophrenia: A computational approach. In *2022 2nd International Conference on Innovative Sustainable Computational Technologies (CISCT)* (pp. 1–4). IEEE.

Sidharthan, V.K. and Baranwal, V.K., 2021. Mining of the water hyssop (*Bacopa monnieri*) transcriptome reveals genome sequences of two putative novel rhabdoviruses and a solendovirus. *Archives of Virology*, *166*, pp. 1985–1990.

Silpa, S., 2018. *Master of science (Horticulture)*. Influence of drying methods and packaging materials on shelf life and quality of Brahmi (*Bacopa monnieri* l.). University of Horticultural Sciences Bagalkot, Karnataka, India – 587104.

Smitha, G.R., Kalaivanan, D. and Sujatha, S., 2021. Influence of shade and organic nutrition on growth, yield and quality of memory enhancing herb, Brahmi (*Bacopa monnieri* L.). *Medicinal Plants-International Journal of Phytomedicines and Related Industries*, *13*(1), pp. 72–90.

Soman, J., Hema, J. and Subramanian, S., 2019. Plant tissue culture and DNA delivery methods. *Advances in Plant Transgenics: Methods and Applications* (pp. 3–22).

Srivastava, A., Garg, G., Sharma, P., Shah, N., Sharma, S. and Shrivastava, N., 2016. Genetic diversity in chemically diverse accessions of *Bacopa monnieri*. *JPC–Journal of Planar Chromatography–Modern TLC*, *29*, pp. 203–208.

Statham, A.L., Robinson, M.D., Song, J.Z., Coolen, M.W., Stirzaker, C. and Clark, S.J., 2012. Bisulfite sequencing of chromatin immunoprecipitated DNA (BisChIP-seq) directly informs methylation status of histone-modified DNA. *Genome Research*, *22*(6), pp. 1120–1127.

Su, W., Xu, M., Radani, Y. and Yang, L., 2023. Technological development and application of plant genetic transformation. *International Journal of Molecular Sciences*, *24*(13), p. 10646.

Sudhakaran, M.V., 2020. Botanical Pharmacognosy of *Bacopa monnieri* (Linn.) Pennell. *Pharmacognosy Journal*, *12*(6), pp. 1559–1572.

Sudheer, W.N., Thiruvengadam, M. and Nagella, P., 2023. A comprehensive review on tissue culture studies and secondary metabolite production in *Bacopa monnieri* L. Pennell: A nootropic plant. *Critical Reviews in Biotechnology*, *43*(6), pp. 956–970.

Swapna, M.M., Prakashkumar, R., Anoop, K.P., Manju, C.N. and Rajith, N.P., 2011. A review on the medicinal and edible aspects of aquatic and wetland plants of India. *Journal of Medicinal Plants Research*, *5*(33), pp. 7163–7176.

Tamboli, F.A., Rangari, V.D., Tarlekar, S.D., Jadhav, R.D., Alaskar, K.M., Desai, V. and Kanthe, R.U., 2022. Brahmi (*Bacopa monnieri*): An ayurvedic herb in the management of various disease. *Journal of Postharvest Technology*, *10*(4), pp. 59–74.

Thakur, M., Yadav, A., Dubey, K.K., Dakal, T.C. and Yadav, V., 2024. Antimicrobial activity against antibiotic-resistant pathogens and antioxidant activity and LCMS/MS phytochemical content analysis of selected medicinal plants. *Journal of Pure & Applied Microbiology*, *18*(1), pp. 722–738.

Thorat, B.S., Bagkar, T.A. and Patil, R.R., 2018. Brahmi-The memory booster medicinal herb. *Journal of Medicinal Plants Studies*, *6*(1), pp. 185–187.

Ting, N.C., Jansen, J., Mayes, S., Massawe, F., Sambanthamurthi, R., Ooi, L.C.L., Chin, C.W., Arulandoo, X., Seng, T.Y., Alwee, S.S.R.S. and Ithnin, M., 2014. High density SNP and SSR-based genetic maps of two independent oil palm hybrids. *BMC Genomics*, *15*, pp. 1–11.

Tungphatthong, C., Somnuek, J., Phadungcharoen, T., Ingkaninan, K., Denduangboripant, J. and Sukrong, S., 2018. DNA barcoding of species of Bacopa coupled with high-resolution melting analysis. *Genome*, *61*(12), pp. 867–877.

Valotto Neto, L.J., Reverete de Araujo, M., Moretti Junior, R.C., Mendes Machado, N., Joshi, R.K., dos Santos Buglio, D., Barbalho Lamas, C., Direito, R., Fornari Laurindo, L., Tanaka, M. and Barbalho, S.M., 2024. Investigating the neuroprotective and cognitive-enhancing effects of *Bacopa monnieri*: A systematic review focused on inflammation, oxidative stress, mitochondrial dysfunction, and apoptosis. *Antioxidants*, *13*(4), p. 393.

6 Cryopreservation for Germplasm Conservation of Commercially Important Medicinal Plants

Yash Bisht, Nandini Sharma, Pallavi Dheer, Shreya Mundepi, Shubham Mangla, Somya Vaishnav, Aditi Sharma and Manu Pant*

*Corresponding Author

6.1 INTRODUCTION

Ever since ancient times, humans have counted on plants and nature for food along with their multiple medicinal benefits. Medicinal plants have long been used globally for their healing properties as they play an important role in modern pharmacology by producing bioactive compounds like phenols, alkaloids, terpenes etc., thus acting against several pathogens. For instance, Morphine, which belongs to a class of medicines called narcotic analgesics, is a powerful pain reliever derived from latex of *Papaver somniferum* (Opium poppies) (Davies and Hollman, 2002). Some glycosides like Anthraquinone are found in senna and aloe vera act as laxatives which stimulate bowel movement (Epifano et al., 2015).

The use of such plant-based treatments in India has been reported in Rigveda since (4500–1500 BC), which has listed around 67 plants with therapeutic values. Ancient civilizations like the Egyptians, used medicinal plants for mummification to preserve dead bodies using cedar oil, palm wine and myrrh for the liquification, cleansing and antimicrobial properties, respectively (Abdel-Maksoud and El-Amin, 2011). According to WHO, 80 per cent of the population world-wide depends on plant-based medicines for various treatment strategies. Also, WHO has reported around 21,000 plant species to be potential candidates for medicinal use (WHO, 2013). India is a rich source of medicinal plants with about 8,000 different forest species (Rajasekharan and Wani, 2020). The increasing demand for these medicinal plants leads to their irrational cutting causing deforestation and depletion of medicinal plants diversity. People living in rural areas and nearby places mostly depend on their surrounding vegetation for food, wood and medicines, which leads to loss and degradation of vegetation followed by the environment and ecology (Sarma, 2010).

From the ancient times where vitiligo used to be treated with bishop's weed, treatment of burns with Aloe vera (natural), improvement of immunity with Echinacea, treatment of gastrointestinal discomfort with Ginger (Bhowmik et al., 2010), use of aspirin derived from willow tree (*Salix* spp.) and now to the treatment of leukemia with vinblastine and vincristine, plants have always served as a key player in the treatment of many moderate to severe diseases (Dhyani et al., 2022). Although there are various plant species present on the Earth, few of them have been screened to show biologically beneficial activities. There is a need to discover the use of new species, while also learning to conserve the genetic traits of these valuable species.

DOI: 10.1201/9781003482659-6

6.2 GERMPLASM CONSERVATION

Germplasm conservation is used to preserve plant diversity by ensuring sustainable utilization of genetic resources. To encourage continued research and therapeutic usage, this approach promotes food security, biodiversity preservation, and conservation of medicinal plants (Priyanka et al., 2021).

There are various methods for germplasm conservation including in situ conservation in natural protected areas, on farm conservation of medicinal plants, sacred forest areas and ex situ conservation, via seed banks and botanical gardens. Seed banks act as potential candidates to store the diverse germplasm of many medicinal plants by collecting the seeds, followed by drying and maintaining a suitable temperature. It is a cost-effective method for the seeds which maintain their viability and to preserve the genetic diversity of wild medicinal plants. For example, *Acmella oppositifolia, Curcuma domesticca, Crocus sativus, Origanum vulgare,* etc. However, if the species is incapable of producing seeds and seeds are non-viable or recalcitrant then it becomes difficult to maintain and conserve plants by this method. For instance, Neem seeds cannot be stored easily in seed banks as they lose viability very quickly. Also, the major challenge faced by this method is the reintroduction of plant species back into the wild and to actively assist the restoration of wild populations (Murthy and Saxena, 1998; Botanical Survey of India).

Biotechnological methods are increasingly essential for conserving plant genetic resources. *In vitro* conservation, which involves the storage of plant tissues in controlled environmental conditions and is ideal for long-term storage, is particularly effective for vegetatively propagated and recalcitrant seed species (Engelmann et al., 2008). *In vitro* culture method involves the growth of plants on nutrient media under sterile conditions and controlled temperature and light. Micropropagation is a technique in which a large number of identical plants are produced by inoculating the explants (small tissue samples) on a nutrient rich medium under aseptic conditions. This technique is advantageous to obtain mass production of any superior variety of plants (Thorpe, 2012). For example, *Withania somnifera* (Ashwaganda) was successfully propagated using nodal explants (Udayakumar et al., 2014). The study by Máthé et al. (2015) demonstrated a potential for large scale production of medicinal plants. Other species like *Curcuma longa* (turmeric) (Sunitibala, 2001) and *Centella aiatica* (Das et al., 2008) also showed the successful application of micropropagation techniques in medicinal plants. Although this method has shown successful results in many plants, it is quite challenging to perform in less-equipped laboratories. Additionally, it may result in somaclonal variations, which can cause the plants to lose their original traits. Furthermore, these plants cannot be stored for extended periods, which limits their utility or applicability.

6.3 CRYOPRESERVATION

Cryopreservation is the method in which the plant germplasm is stored at extremely low temperatures (around -196°C) in liquid nitrogen (Danso and Ford-Lloyd, 2003). This method provides long-term storage by maintaining the viability for an extended period of time and minimizing the risk of contamination (Benelli et al., 2013; Bettoni et al., 2016). Further, it has been observed from various studies that no metabolic activities and cell division occur at this temperature which helps in maintaining the plant cells without any change in their metabolism. This emphasizes the need for further research to optimize cryopreservation protocols (Shatnawi, 2011). One of the major roles of this method is to reduce the threats posed by habitat destruction, overexploitation, and climate change to medicinal plants through endgame protection of the species genetic resources using this technique.

Cryopreservation has been proven to be the most competent method due to its long-term preservation, unlike the *in vitro* culture methods which are expensive and much more sensitive to contamination. For a successful cryopreservation, it is important to prevent crystallization of cells which can be achieved by using cryoprotectants like Dimethyl sulfoxide (DMSO), Ethylene Glycol (EG), glycerol, and sucrose. There are some protocols which use vitrification solutions to convert cellular water from liquid to glassy solid. This is similar to cryopreservation but involves more concentrated cryoprotectants at ultra-low temperatures (Sakai and Englemann, 2007).

Rapid thawing is done by rapidly warming the sample to minimize the damage from ice crystals. Cryoprotectants are removed and gradually acclimatized to normal conditions for obtaining their culture. Cryopreservation can be done in various plant materials like callus, somatic embryos, and seeds. Cryopreservation techniques such as encapsulation-dehydration, vitrification and droplet-vitrification have made the conservation of medicinal plants possible for next generations (Bing and Da-Wen, 2002).

6.4 CRYOPRESERVATION METHODS

6.4.1 Vitrification

It is a technique in which the explant is treated with a highly concentrated cryoprotectant solution, which dehydrates the cytosol prior to freezing. The penetrating agents present in the solution elevates the solute concentration by entering the cell gradually, while the non-penetrating agent osmotically withdraws water (Figure 6.1). This enhances the cytoplasmic viscosity, forming an amorphous glassy state and hence preventing ice formation in the cell (Teixeira et al., 2014). Vitrification simplifies and enhances cryopreservation by reducing risk of ice damage along with eliminating the need for optimal cooling and warming rates. However, it does increase the risk of toxicity due to the addition and removal of cryoprotectants. During vitrification, Reed et al. (2005) observed a significant increase in cellular viscosity, allowing various plant species to withstand osmotic and evaporative dehydration stresses.

Dimethyl sulfoxide (DMSO) is a bipolar, penetrating, water-soluble, aprotic solvent which is an important component of plant vitrification solution (PVS). It also prevents the excessive

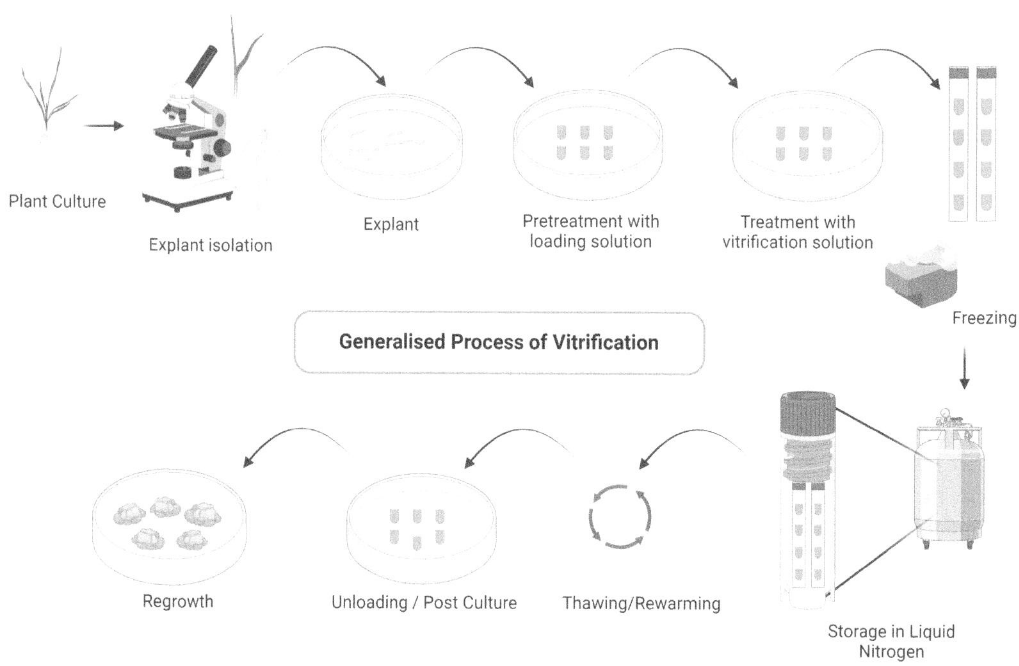

FIGURE 6.1 Diagrammatic illustration showing the process of vitrification.

accumulation of electrolytes along with other substances during cooling. In addition, it quickly penetrates into the cell membrane due to its low molecular weight, along with stabilizing the membrane. Ethylene glycol is another extensively used cryoprotectant due to its high coefficient of cell permeability. After sample rewarming, it also prevents tissue superhydration, which is the uncontrolled cell expansion upon inoculation into a medium with low osmoticum concentration. Polyethylene glycol (PEG) is a non-penetrating cryoprotectant that prevents injury caused by cold due to PVS, along with preventing ice crystals (Pomeisl et al., 2020).

In early studies of vitrification, 60% PVS2 solution was used for dehydrating nucellar somatic cells from *Citrus sinensis* Osb. var. brasiliensis tanaka, at 25°C for 5 min followed by chilling PVS2 at 0°C for 3 minutes, with a growth rate of about 80%. The duration of exposure of the explant varies according to the explant type, size, and susceptibility. Additionally, studies have shown a reduction in toxicity during osmodehydration at 0°C instead of 25°C. After dehydration, the cooling was done by rapid immersion in liquid nitrogen (LN) for the vitrification of internal solutes. Cryopreserved samples were then thawed at 40°C for 1 to 3 minutes. Finally, sucrose and sorbitol were used to remove the vitrification solution from cryopreserved explants. Therefore, vitrification has been proved to be a reliable technique for the long-term germplasm storage of numerous medicinal species (Sakai and Engelmann, 2007).

6.4.2 Encapsulation-Dehydration Based Methods

This technique is known for its simplicity and minimal manipulation requirement. Explants are first encapsulated and then treated with high concentration of sucrose (which is a readily available and an economical cryoprotectant) solution prior to partial dehydration before freezing (Gonzalez-Arnao and Engelmann, 2006; Martinez et al.,1999; Maneerattanarungroj et al., 2007). Standardized protocols have to be developed in order to maximize the regeneration rate of the cryopreserved plants. Furthermore, the optimal duration of dehydration needs to be determined for long-term preservation of plants.

Encapsulation-Dehydration (Ed): Pioneered by Fabre and Dereuddre, this method is based on synthetic seed technology (Sakai and Engelmann, 2007). The process involves encapsulation of explants in alginate beads followed by gradual increase in sucrose concentration (0.7–1.5 M) around the beads to elevate the internal solute concentration of the explants. Dehydration is done physically via controlled methods such as sterile airflow or silica gel until the moisture content reaches around 20–25% then, plunging the beads into LN for storage followed by the thawing of the beads in a water bath (Benito et al., 2004; Reed et al.; 2004, Englemann et al., 2008). Along with major advantages of showing no effects on plant growth during addition or elimination of toxic cryoprotectants and preventing the problem of explant sensitivity to the PVS2 vitrification solution, there are certain limitations of being labour intensive and time-consuming (Englemann et al., 2008).

Encapsulation-vitrification (EV): This combines techniques of encapsulation-dehydration and vitrification. Similar to Encapsulation-dehydration, explants are first encapsulated followed by dehydration using a vitrification solution instead of physical methods like airflow or silica gel. This significantly reduces the dehydration time compared to ED. The choice between ED and EV depends on the specific plant species being preserved. ED offers the benefit of avoiding potentially harmful solutions, but it requires more time and effort. While, EV offers faster processing but might introduce some level of toxicity depending on the type of vitrification solution used (Benito et al., 2004).

6.4.3 Droplet Vitrification Cryopreservation

Droplet-vitrification is a method of cryopreservation which involves suspending a single plant bud or tissue sample (explant) in a dripping solution on a thin aluminium (used due to its heat conductivity) strip. The strip is then rapidly immersed in liquid nitrogen at -196°C. Placing the explant solution directly on the aluminium strip facilitates rapid and even temperature changes during both cooling and warming. This ensures a consistent freezing rate throughout the plant cells within the explant. The speed and cooling rates, reaching around 130°C per minute, initiate a process called intracellular vitrification. This essentially turns the cell contents into a droplet state, preventing ice crystal formation. Additionally, rapid reheating minimizes the risk of ice crystals reforming within the cells, ultimately leading to better survival rates for the frozen plant material. The success of droplet-vitrification has been demonstrated in preserving shoot tips from various flowering and woody plants. These include rose bushes (*Rosa × hibrida L.*), taro (*Colocasia esculenta var. esculenta*), apple trees (*Malus domestica Borkh*), and even several types of lilies (*Lilium x lancifolium, Lilium × longiflorum, Lilium x siberia*) (Santos et al., 2021).

The droplet-vitrification method incorporated several sequential procedures. Initially, excised shoot tips (ST) underwent pre-cultivation before cryopreservation at -196°C. Further, they are exposed to loading solutions consisting of basic Murashige-Skoog (MS) medium along with 2.0 M glycerol and 0.4 M sucrose. Plant vitrification solutions were made containing mix of cryoprotectants: PVS2 (30% glycerol, 15% EG, 15% DMSO, and 0.4M sucrose); PVS3 (50% each of glycerol and sucrose); and PSV (35% glycerol, 20% EG and 0.6M sucrose-diluted in MS medium). PVS2 is the most commonly employed less-toxic vitrification solution (Sakai et al., 1990, 1991). All the mediums were sterilised using 0.22 μm, 150 ml analytical filter unit. For the optimal conditions for vitrification, two different temperatures 25 ± 2°C and 0°C and exposure for 20, 40 and 60 minutes were used (Coutinho Silva et al., 2013).

6.4.4 Cryo-Plate Techniques

The latest technique, cryo-plate, is an improvement of the droplet vitrification method developed in Japan. It is suitable for use with sizable explants vulnerable to harm from physical forces and the toxicity of cryoprotectants. Aluminium plates with high rates of warming (3000–4500°C/min) and cooling (4000–5000°C/min) are known as cryo-plates. Using this method, explants are put in the aluminium cryo-plate wells and given tiny drops of calcium alginate to stick to the well's bottom, followed by dehydration. For dehydration, either V cryo-plate or D cryo-plate protocol are used, which generally differ in the dehydration protocol. The ease of handling, better stability and explant tolerance to sudden change in temperature are some of the advantages of cryo-plate method. Furthermore, it minimizes the risk of explant loss and mechanical damage and has high regeneration rates after cryopreservation (Niino et al., 2014, Matsumoto et al., 2015, Rafique et al., 2016).

6.4.4.1 V-Cryo-Plate

This technique was developed by Yamamoto et al. (2011), by combining droplet vitrification with encapsulation-vitrification, where encapsulation is a vital step. This dual approach helps to protect delicate cells and tissues from damaging effects of freezing, thereby maintaining viability after storage in ultra-low temperatures. Explants are first encapsulated and then treated with loading solution, PVS dehydration on cooling plates, flash freezing in LN, and lastly thawing at room temperature.

6.4.4.2 D-Cryo-Plate

Developed by Niino et al. (2013), this technique is based on physical dehydration. It can be obtained with drying beads, silica gel, or using the flow of air current in a laminar air-flow chamber before submersion in LN. Although this approach is time-consuming, but offers the advantage of using less chemicals.

6.5 VACUUM-INFILTRATION-VITRIFICATION

Nadarajan and Pritchard (2014) developed vacuum-infiltration-vitrification. This method enhances the penetration of cryoprotective agents by using vacuum during explant incubation in CPA, thereby reducing incubation time. Furthermore, it reduces the temperature dependency for cryoprotectant application and limits the problem of specimen heterogeneity.

6.6 CRYO-MESH

Devised in 2017 by Funnekotter et al., cryo-mesh approach was used for cryopreservation of shoot tips of kangaroo paws (*Anigozanthos viridis*). The fundamental distinction between the VC method and the regular cryo-mesh procedure is the use of wire stainless steel mesh strips, or cryo-mesh, in place of aluminium cryo-plates. They are easily available, and biomaterials can be rapidly cooled and re-warmed by using the cryo-mesh technique. Changes to the steel mesh's composition, size, wire diameter, and porosity can all impact the mesh's specific heat capacity, which in turn affects the rate of cooling and rewarming. While using the alginate method, it spreads in the mesh apertures, attaching the shoot tips to the mesh. As the material in the cryo-mesh is exposed to loading and vitrification solutions from all sides, its construction permits quick penetration as well as a great range of material sizes and shapes.

In their experiment, Funnekotter et al. (2017) applied a 2% high-viscosity sodium alginate solution to pre-cultured shoot tips before they were placed in the cryo-mesh. Following a gentle transfer into LS of each cryo-mesh containing ten coated shoot tips, 100 mM $CaCl_2$ was added and allowed to settle at room temperature for 20 minutes to set the alginate and attach the shoot tips to the cryo-mesh. The PVS2 was transferred to a cryo-mesh, then left on ice for various durations. The cryovials were filled with LN and kept for at least an hour. After removing the cryo-mesh, it was rewarmed for 20 minutes in an unloading solution. The shoot tips were then placed on recovery media.

6.7 APPLICATIONS OF CRYOPRESERVATION FOR MEDICINAL PLANTS

The process of cryopreservation is applied for conserving several medicinal plants (Figure 6.2). In a study on *Anemarrhena asphodeloides* (Hong and Yin, 2012), the embryogenic calli were encapsulated using calcium alginate beads in a loading solution consisting of 2 M glycerol and 0.4 M sucrose for one hour. After dehydration, the encapsulated calli were immersed in LN for prolonged storage, and 80 % of the frozen calli successfully survived even after one year of storage. Embryogenic calli of Castor Aralia (*Kalopanax septemlobus*) were tested against six different vitrification solutions in which 3–80% combination of glycerol, DMSO, EG, and sucrose was found to be most effective for cryopreservation (Shin et al., 2012). Sharaf ct al. (2012) showed that the shoot tips of *Artemisia herba-alba* could be sodium-alginate beads. The most successful outcome (68% survival and 12% regrowth) involved dehydrating encapsulated and vitrified shoot tips using 100% PVS2 for 30 minutes before cryopreservation.The nodal explants of *Perilla frutescens* could be encapsulated using Na alginate beads and successfully cryopreserved using V-Cryo-plate under LN (Matsumoto et al., 2014). In this study, the growth recovery rate was found to be approximately 80%. Similarly, seeds of *Arundina graminifolia* were encapsulated using Na-alginate beads and cryopreserved utilizing D cryo-plate technique utilizing LN. This approach produced the highest regeneration (82%), followed by encapsulation dehydration (74%) (Thammasiri et al., 2018).

In *Cannabis sativa*, the shoot tips have been cryopreserved in liquid nitrogen (LN) after pretreatment on agar plates containing 5% dimethyl sulfoxide (DMSO) for 48 hours. The technique has proven to be very useful in preservation of a wider range of *Cannabis* cultivars, requiring minimal maintenance and eliminating the risk of pests or diseases (Uchendu et al., 2019). Shoot tips

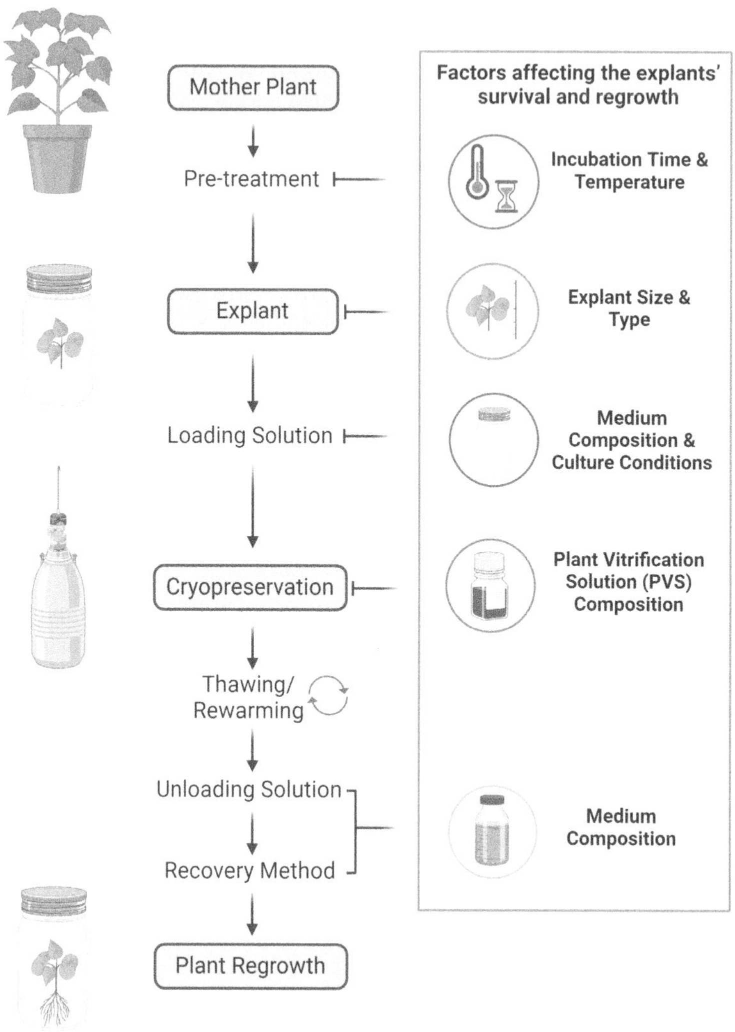

FIGURE 6.2 Explant survival and regrowth during cryopreservation.

have also been preferred explant for other medicinal plant species. In a study on cryopreservation of *Gentiana kurroo* (Sharma et al., 2021), 8-week-old shoot tips of three accessions were precultured for two days in the dark on 5% DMSO containing MS medium. Cryopreservation was achieved using vitrification (35% regeneration rate) and droplet-vitrification (60% regeneration rate). The genetic stability was confirmed using 40 ISSRs and 30 EST-SSRs. The study further confirmed the use of droplet-vitrification as a suitable approach for long-term germplasm conservation of *G. kurroo.* In *Dioscorea deltoidei*, 4-week-old shoot meristems of 15 different accessions were cryopreserved using vitrification and droplet-vitrification methods. Shoot regrowth post-liquid nitrogen (LN) treatment was statistically comparable between vitrification (30–54%) and droplet-vitrification (21–51%) techniques in 66.7% of the accessions, whereas vitrification significantly gave better results for 33.3% accessions. The study showed that for long-term cryo-banking of *D. deltoidei*, vitrification was found more effective than droplet-vitrification due to its enhanced regrowth and ease of handling (Sharma et al., 2022).

6.8 CONCLUSION

The demand for medicinal plants is increasing over time along with improving significance of preservation technology in the present time. Hence, there is a need to have simple, stable, and cost-effective methods for long-term storage of germplasm and to address the applications of these methods in various medicinally important plant species. Cryopreservation is potentially the safest method for sustainable use of these plants by storing them at a very-low temperature for a long period of time and hence reviving the germplasm back to its viable stage. This method can further be explored in terms of various other medicinal plants and its advancements to be studied at a global level.

REFERENCES

Abdel-Maksoud, G., and El-Amin, A. R. 2011. A review on the materials used during the mummification processes in ancient egypt. *Mediterranean Archaeology and Archaeometry, 11*(2), 129–150.

Benelli, C., De Carlo, A., and Engelmann, F. 2013. Recent advances in the cryopreservation of shoot-derived germplasm of economically important fruit trees of Actinidia, Diospyros, Malus, Olea, Prunus, Pyrus and Vitis. *Biotechnology Advances, 31*(2), 175–185.

Benito, M. G., Clavero-Ramírez, I., and López-Aranda, J. M. 2004. The use of cryopreservation for germplasm conservation of vegetatively propagated crops. *Spanish Journal of Agricultural Research, 2*(3), 341–351.

Bettoni, J. C., Costa, M. D., Gardin, J. P. P., Kretzschmar, A. A., and Pathirana, R. 2016. Cryotherapy: A new technique to obtain grapevine plants free of viruses. *Revista Brasileira de Fruticultura, 38*, e–833.

Bhowmik, D., Tripathi, K. K., Chandira, M. R., and Kumar, K. P. 2010. Zingiber officinale the herbal and traditional medicine and its therapeutically importance. *Research Journal of Pharmacognosy and Phytochemistry, 2*(2), 102–110.

Bing, L., and Da-Wen, S. 2002. Novel methods for rapid freezing and thawing of foods–A review. *Journal of Food Engineering, 54*(3), 175–182.

Botanical Survey of India. *Ex-Situ conservation.* (n.d.). Available at: https://bsi.gov.in/page/en/ex-situ-conse rvation

Chamberlain, D. F. 1982. *Revision of rhododendron 2, Notes from the Royal Botanical Garden, Edinburgh* (Vol. 39, no. 2), 209–486.

Coutinho Silva, L., Paiva, R., Swennen, R., Andrè, E., and Panis, B. 2013. Shoot-tip cryopreservation by droplet vitrification of Byrsonima intermedia A. Juss.: A woody tropical and medicinal plant species from Brazilian cerrado. *CryoLetters, 34*(4), 338–348.

Danso, K. E., and Ford-Lloyd, B. V. 2003. Encapsulation of nodal cuttings and shoot tips for storage and exchange of cassava germplasm. *Plant Cell Reports, 21*, 718–725.

Das, R., Hasan, M. F., Hossain, M. S., and Rahman, M. 2008. Micropropagation of Centella asiatica L. an important medicinal herb. *Progressive Agriculture, 19*(2), 51–56.

Davies, M. K., and Hollman, A. 2002. The opium poppy, morphine, and verapamil. *Heart, 88*(1), 3–3.

Dhyani, P., Quispe, C., Sharma, E., Bahukhandi, A., Sati, P., Attri, D. C., ... and Cho, W. C. 2022. Anticancer potential of alkaloids: A key emphasis to colchicine, vinblastine, vincristine, vindesine, vinorelbine and vincamine. *Cancer Cell International, 22*(1), 206.

Engelmann, F. 2011. Use of biotechnologies for the conservation of plant biodiversity. In *Vitro Cellular and Developmental Biology-Plant, 47*, 5–16.

Engelmann, F., Arnao, M. T. G., Wu, Y., and Escobar, R. 2008. Development of encapsulation dehydration. *Plant Cryopreservation: A Practical Guide*, 59–75.

Epifano, F., Fiorito, S., Locatelli, M., Taddeo, V. A., and Genovese, S. 2015. Screening for novel plant sources of prenyloxyanthraquinones: Senna alexandrina Mill. and Aloe vera (L.) Burm. F. *Natural Product Research, 29*(2), 180–184.

Funnekotter, B., Bunn, E., and Mancera, R. L. 2017. Cryo-mesh: A simple alternative cryopreservation protocol. *CryoLetters, 38*(2), 155–159.

Gonzalez-Arnao, M. T., and Engelmann, F. 2006. Cryopreservation of plant germplasm using the encapsulation-dehydration technique: Review and case study on sugarcane. *CryoLetters, 27*(3), 155–168.

Hong SenRong, H. S., and Yin MingHua, Y. M. 2012. High-efficiency encapsulation-vitrification protocols for cryopreservation of embryogenic calli of the oriental medicinal plant Anemarrhena asphodeloides Bunge. CryoLetters, 33(3), 190–200.

Maneerattanarungroj, P., Bunnag, S., and Monthatong, M. 2007. In vitro conservation of Cleisostoma areitinum (Rchb. f.) Garay, rare Thai orchid species by an encapsulation-dehydration method. *Asian Journal of Plant Sciences*, *6*(8), 1235–1240.

Martinez, D., Tamés, R. S., and Angeles Revilla, M. 1999. Cryopreservation of in vitro-grown shoot-tips of hop (Humulus lupulus L.) using encapsulation/dehydration. *Plant Cell Reports*, *19*, 59–63.

Máthé, A., Hassan, F., and Abdul Kader, A. 2015. In vitro micropropagation of medicinal and aromatic plants. In: Máthé, Á. (Eds.), *Medicinal and Aromatic Plants of the World: Scientific, Production, Commercial and Utilization Aspects*, 1, 305–336.

Matsumoto, T., Yamamoto, S. I., Fukui, K., Rafique, T., Engelmann, F., and Niino, T. 2015. Cryopreservation of persimmon shoot tips from dormant buds using the D cryo-plate technique. *The Horticulture Journal*, *84*(2), 106–110.

Matsumoto, T., Yoshimatsu, K., Kawahara, N., Yamamoto, S., and Niino, T. 2014. Development of in vitro propagation by node culture and cryopreservation by V-Cryo-plate method for Perilla frutescens. *Advances in Horticulture Science*, *28*, 79–83.

Murthy, B. N. S., and Saxena, P. K. 1998. Somatic embryogenesis and plant regeneration of neem (Azadirachta indica A. Juss.). *Plant Cell Reports*, *17*, 469–475.

Nadarajan, J., and Pritchard, H. W. 2014. Biophysical characteristics of successful oilseed embryo cryoprotection and cryopreservation using vacuum infiltration vitrification: An innovation in plant cell preservation. *PLoS One*, *9*(5), e96169.

Niino, T., Watanabe, K., Nohara, N., Rafique, T., Yamamoto, S. I., Fukui, K., ... and Engelmann, F. 2014. Cryopreservation of mat rush lateral buds by air dehydration using aluminum cryo-plate. *Plant Biotechnology*, *31*(3), 281–287.

Niino, T., Yamamoto, S. I., Fukui, K., Martínez, C. R. C., Arizaga, M. V., Matsumoto, T., and Engelmann, F. 2013. Dehydration improves cryopreservation of mat rush (Juncus decipiens Nakai) basal stem buds on cryo-plates. *CryoLetters*, *34*(6), 549–560.

Pomeisl, K., Richter, J., Golan, M., and Kratochvílová, I. 2020. Simple syntheses of new pegylated trehalose derivatives as a chemical tool for potential evaluation of cryoprotectant effects on cell membrane. *Molecules*, *25*(3), 497.

Priyanka, V., Kumar, R., Dhaliwal, I., and Kaushik, P. 2021. Germplasm conservation: Instrumental in agricultural biodiversity—A review. *Sustainability*, *13*(12), 6743.

Rafique, T., Yamamoto, S. I., Fukui, K., Tanaka, D., Arizaga, M. V., Abbas, M., ... and Niino, T. 2016. Cryopreservation of shoot-tips from different sugarcane varieties using D cryo-plate technique. *Pakistan Journal of Agricultural Sciences*, *53*(1), 151–158.

Rajasekharan, P. E., and Wani, S. H. 2020. *Distribution, diversity, conservation and utilization of threatened medicinal plants* (pp. 3–30). Springer International Publishing.

Reed, B. M., Kovalchuk, I., Kushnarenko, S., Meier-Dinkel, A., Schoenweiss, K., Pluta, S., ... and Benson, E. E. 2004. Evaluation of critical points in technology transfer of cryopreservation protocols to international plant conservation laboratories. *CryoLetters*, *25*(5), 341–352.

Reed, B. M., Schumacher, L., Dumet, D., and Benson, E. E. 2005. Evaluation of a modified encapsulation-dehydration procedure incorporating sucrose pretreatments for the cryopreservation of Ribes germplasm. *In Vitro Cellular and Developmental Biology-Plant*, *41*, 431–436.

Sakai, A., and Engelmann, F. 2007. Vitrification, encapsulation-vitrification and droplet-vitrification: A review. *CryoLetters*, *28*(3), 151–172.

Sakai, A., Kobayashi, S., and Oiyama, I. 1990. Cryopreservation of nucellar cells of navel orange (Citrus sinensis Osb. var. brasiliensis Tanaka) by vitrification. *Plant Cell Reports*, *9*(1), 30–33.

Sakai, A., Kobayashi, S., and Oiyama, I. 1991. Survival by vitrification of nucellar cells of navel orange (Citrus sinensis var. brasiliensis Tanaka) cooled to-196 C. *Journal of Plant Physiology*, *137*(4), 465–470.

Santos, I. R. I., Martins, I. S., Salomão, A. N., and Oliveira, D. V. D. 2021. Comparative study of three Vitrification solutions for an effective droplet-vitrification cryopreservation procedure for Pfaffia glomerata (Spreng.) Pedersen. *European Journal of Medicinal Plants*, *32*(7), 30–38.

Sarma, S. 2010. *Meghalaya the land and forest a remote sensing based study*. Geophil Publishing House, Guwahati.

Sharaf, S. A., Shibli, R. A., Kasrawi, M. A., and Baghdadi, S. H. 2012. Cryopreservation of wild Shih (Artemisia herba-alba Asso.) shoot-tips by encapsulation-dehydration and encapsulation-vitrification. *Plant Cell, Tissue and Organ Culture (PCTOC)*, *108*, 437–444.

Sharma, N., Gowthami, R., Devi, S. V., Malhotra, E. V., Pandey, R., and Agrawal, A. 2021. Cryopreservation of shoot tips of Gentiana kurroo Royle–A critically endangered medicinal plant of India. *Plant Cell, Tissue and Organ Culture (PCTOC)*, *144*(1), 67–72.

Sharma, N., Malhotra, E. V., Chandra, R., Gowthami, R., Sultan, S. M., Bansal, S., ... and Agrawal, A. 2022. Cryopreservation and genetic stability assessment of regenerants of the critically endangered medicinal plant Dioscorea deltoidea Wall. ex Griseb. for cryobanking of germplasm. *In Vitro Cellular and Developmental Biology-Plant*, *58*(4), 521–529.

Shatnawi, M. A. 2011. Cryopreservation of Capparis spinosa shoot tips via vitrification, encapsulation dehydration and encapsulation vitrification. *World Applied Sciences Journal*, *15*(3), 318–325.

Shin, D. J., Kong, H., Popova, E. V., Moon, H. K., Park, S. Y., Park, S. U., ... and Kim, H. H. 2012. Cryopreservation of Kalopanax septemlobus embryogenic callus using vitrification and droplet-vitrification. *CryoLetters*, *33*(5), 402–410.

Sunitibala, H., Damayanti, M., and Sharma, G. J. 2001. In vitro propagation and rhizome formation in Curcuma longa Linn. *Cytobios*, *105*(409), 71–82.

Teixeira, A. S., Faltus, M., Zámečník, J., González-Benito, M. E., and Molina-García, A. D. 2014. Glass transition and heat capacity behaviors of plant vitrification solutions. *Thermochimica Acta*, *593*, 43–49.

Thammasiri, K., Prasongsom, S., Kongsawadworakul, P., Chuenboonngarm, N., Jenjittikul, T., Soonthornchainaksaeng, P., ... and Muangkroot, A. 2018. Cryopreservation of Arundina graminifolia (D. Don) Hochr. seeds using D cryo-plate method. In *III International Symposium on Plant Cryopreservation 1234* (pp. 301–308).

Thorpe T. 2012. History of plant tissue culture. *Methods in Molecular Biology (Clifton, N.J.)*, *877*, 9–27. https://doi.org/10.1007/978-1-61779-818-4_2

Uchendu, E., Lata, H., Chandra, S., Khan, I. A., and ElSohly, M. A. 2019. Cryopreservation of shoot tips of elite cultivars of Cannabis sativa L. by droplet vitrification. *Medical Cannabis and Cannabinoids*, *2*(1), 29–34.

Udayakumar, R., Kasthurirengan, S., Mariashibu, T. S., Sahaya Rayan, J. J., Ganapathi, A., Kim, S. C., ... and Choi, C. W. 2014. Agrobacterium-mediated genetic transformation of *Withania somnifera* using nodal explants. *Acta Physiologiae Plantarum*, *36*, 1969–1980.

World Health Organization. 2013. *WHO traditional medicine strategy: 2014–2023*. World Health Organization.

Yamamoto, S. I., Rafique, T., Priyantha, W. S., Fukui, K., Matsumoto, T., and Niino, T. 2011. Development of a cryopreservation procedure using aluminium cryo-plates. *CryoLetters*, *32*(3), 256–265.

7 Hairy Root Culture

An Effective Method for Production of Plant Bioactive Compounds

Bhaswatimayee Mahakur, Aryaputri Bedaprabha, Durga P. Barik, Sugam Gupta, Rahul Kumar, Debasis Mitra and Devvret Verma**

*Corresponding authors

7.1 INTRODUCTION

The word "hairy root" first came into use by Stewart et al. (1900) to describe diseased fruit crops. Hildebrandt (1934) was the next to coin the phrase "hairy root disease" to describe the unique symptoms of this condition, which include the appearance of a small, thin mass of roots that look like hairs as an effect of microbial infection (Mehrotra et al., 2015). *Agrobacterium rhizogenes* infection caused a wound to form, which results in the development of hairy root disease. The root-inducing (Ri) plasmid of *A. rhizogenes* carries root loci (rol) genes, which are injected into the genome of the host plant, which showed the proliferation of malignant roots and root hair (Ono and Tian, 2011). Häkkinen and Oksman-Caldentey (2018), identified that hairy root disease mainly affects dicotyledonous plants and causes substantial losses in vineyards, vegetable nurseries, and orchards.

7.1.1 Significance and Biosynthetic Origin of Plant Bioactive Compounds

Plant-derived secondary metabolites are substances produced by plants that do not directly contribute to their growth, development, or reproduction, but instead play essential roles in plant defense, signalling, and interaction with their environment. These compounds have significant commercial value due to their medicinal, aromatic, and flavouring properties. Examples of commercially important plant-derived secondary metabolites include alkaloids, terpenoids, phenolics, glycosides, and flavonoids, along with their biosynthetic origins. The synthesis of these secondary metabolites involves intricate pathways involving numerous enzymes and intermediates (Figure 7.1). According to Ramakrishna et al. (2020), a functional diet is made up of substances that have physiological and biological effects, and offer benefits to health beyond simply nutrition. "Bioactive compounds" is the broad term for the ingredients in these foods that give them their nutritional value. Bioactive compounds are obtained from plants, animals, or other sources that are "generally regarded as safe (GRAS)" and that can interrelate with one or multiple components of living tissue. Bioactive compounds have a broad choice of potential therapeutic effects (Banwo et al., 2021). The biological and useful activities of bioactive compounds, such as their antiviral, anti-diabetic, anti-inflammatory, anticancer, and antitumor properties, are highly fascinating. These activities protect the human from excessive levels of free radicals and reactive oxygen species, which can simply

DOI: 10.1201/9781003482659-7

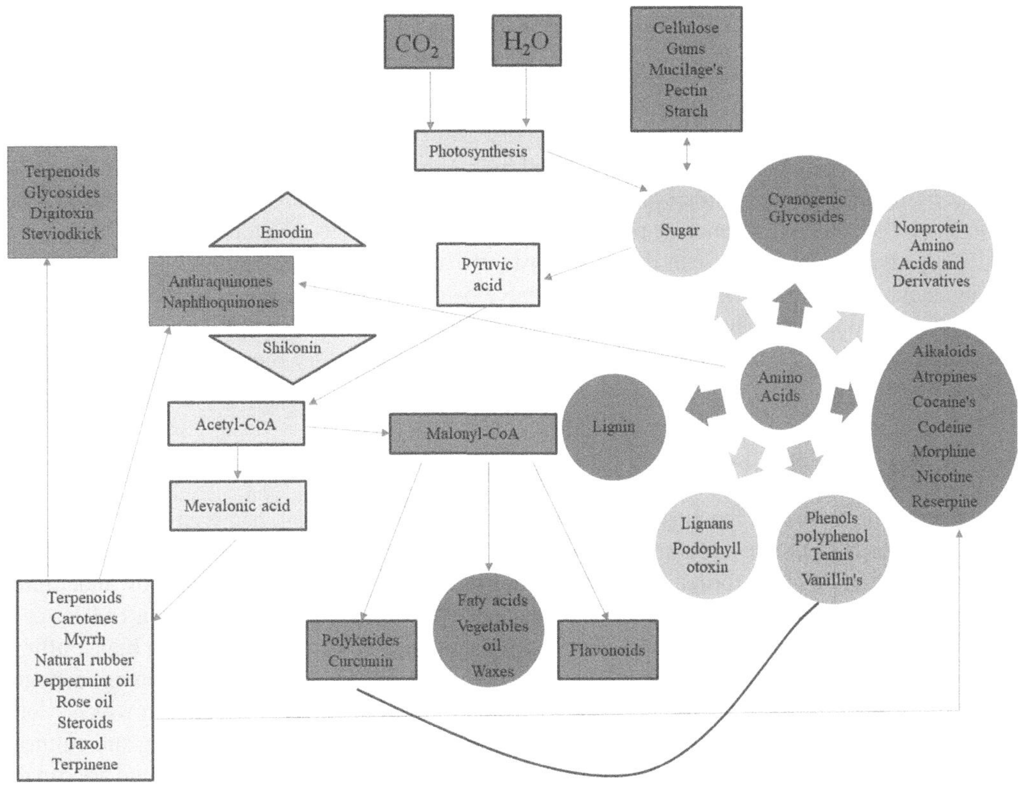

FIGURE 7.1 Biosynthetic origin of some commercially important plant-derived secondary metabolites. (Balandrin et al., 1985; Chawla, 2003.)

interact with other molecules and lead to cell damage (Dilna et al., 2015; Adesulu-Dahunsi et al., 2018c). In *Camptotheca acuminate*, the compound Camptothecin is used as a drug for anti-cancer, antiviral diseases (Lorence et al., 2004). Ginkgolides, a compound found in *Gingko biloba*, is used to treat aging diseases (Ayadi and Tremouillaux-Guiller, 2003). The compound known as verbascoside which is present in *Gmelina arborea* treats fevers, skin issues, and stomach diseases (Dhakulkar et al., 2005). In *Linum flavum* the compound Coniferin is used as a drug for anti-cancer (Lin et al., 2003). In *Papaver somniferum* the compounds present are Morphine, sanguinarine, codeine which are used as a drug for sedative and analgesic diseases (Le Flem-Bonhomme et al., 2004). In *Pueraria phaseoloides* the compound Puerarin is used as a drug for hypothermic, spasmolytic, hypotensive, and anti-arrhythmic diseases (Shi and Kintzios, 2003).

7.1.2 Significance of Hairy Root Culture For Bioactive Compound Production

For numerous reasons that are crucial to biotechnology, hairy roots present a tempting alternative for the synthesis of a valuable secondary compound. Even at higher levels than whole plants or undifferentiated cell cultures, hairy roots can acquire some compounds, such as the parent plant's alkaloids (Gai et al., 2015). Because they don't need phytohormones to grow, they can produce biomass rapidly and have basic growth media needs. In comparison to undifferentiated cell cultures, they have enhanced metabolic production and great genetic stability (Häkkinen and Oksman-Caldentey, 2018).

7.2 BASICS OF HAIRY ROOT CULTURE

Using the natural vector system *Agrobacterium rhizogenes* to transform desired plant species is a novel approach to improving the secondary metabolites production in tissue culture system, which is the result of plant diseases called hairy root diseases (Giri and Narasu, 2000). Usually, dicotyledonous plant species are genetically modified by the gram-negative soil bacteria *Agrobacterium rhizogenes*to make opines, which are a food supply for *Agrobacterium*. According to Shanks and Morgan (1999), this transformation process induces "hairy roots" to develop at the plant's infection site. These transformed roots can develop constantly in a hormone-free culture medium. Several dicotyledonous and monocotyledonous plants have been grown in hairy root cultures, which have been revealed to generate the same secondary metabolites as natural roots. This indicates that hairy root cultures provide an effective approach for producing secondary metabolites (Srivastava and Srivastava, 2007).

7.3 ORIGIN AND APPLICATION OF HAIRY ROOT CULTURES

The interaction between the host plant and the gram-negative soil bacterium *Agrobacterium rhizogenes* leads to the development of hairy roots. "Hairy roots" appear at the plant's infection site as a result of this transformation process. These transformed, hairy roots can grow indefinitely in growth culture medium without growth hormones (Zhou et al., 2011). According to Kim et al. (2002), the major benefit of hairy roots is that they commonly show a similar or even higher affinity for the biosynthesis of secondary metabolites than their parent plants. Hairy root cultures have most likely been found to collect some particular secondary metabolite, even in cases in which it only accumulates in the aerial part of the plant itself. Furthermore, even though ongoing sub culturing and plant regeneration, transformed roots frequently maintain their genetic stability and are capable of regenerating entire healthy plants. It is also known that hairy root cultures produce a variety of secondary metabolites that are absent in the mother plant (Zhou et al., 2011).

7.4 TYPES OF PLANTS SUITABLE FOR HAIRY ROOT CULTURE

Various aromatic and therapeutic plants have had hairy roots produced in them, which are then cultured to produce secondary compounds (Bonhomme et al., 2000). Various susceptible plants, mostly dicotyledon and gymnosperm species, produce simple phenolic substances, like acetosyringone, in response to wounds in natural ecosystems or fields. These substances incite the virulence genes of *Agrobacterium*, which are located on plasmids and are responsible for transferring T-DNA fragments of the Ti- (tumor-inducing) plasmid or Ri- (root-inducing) plasmid (*A. tumefaciens* and *A. rhizogenes*, respectively) to the host plant (Nilsson and Olsson, 1997). DNA fragments that have been transferred play a role in the development of hairy root tissues and malignant crown gall tumours. These tissues produce chemicals known as opines, which are unique amino acids that are used as nutrition by *Agrobacterium* in the soil (Georgiev et al., 2007). According to Abraham and Thomas (2017), these phytochemicals could also be produced by hairy roots, preventing the natural population of therapeutic plants from going extinct.

7.5 EXPLANT SELECTION AND ESTABLISHMENT OF HAIRY ROOT CULTURES

The *rol* genes involved in root initiation and development are inserted into the root cells' genome of the host plant by *Agrobacterium rhizogenes* by the insertion of T-DNA from the Ri plasmid. For this reason, a number of aromatic and therapeutic plants have had hairy roots produced and cultured in order to produce secondary metabolites (Khatodia and Bhatotia, 2013). *A. rhizogenes* plasmids include plant oncogenes, such as *rol* A, *rol* B, *rol* C and *rol* D. These genes are integrated into the plant genomes during *Agrobacterium rhizogenes* infection, resulting in tumour growth and hairy root disease (Bulgakov, 2008). Different explants and media have been used for establishment of hairy root cultures. According to Pawar and Maheshwari (2004), hairy roots in *Solanum surratennes* could be developed from leaf explants and then maintained on MS media without any requirement of an additional phytohormone supply. Sahu et al. (2013), showed that the intermodal explants in *Boerhaavia diffusa* L. could be used for bacterial transformation on standardized MS medium without the use of phytohormones. In *Artemisia vulgaris,* in-vitro shoot tip, leaf, and node explants were selected for transformation and inoculated in the MS medium without growth hormones (Sujatha et al., 2013). The hypocotyls, cotyledon, excised stem, and leaves of *Callerya speciosa* have also been used as explants for genetic transformation using half MS solid media containing 100μM acetosyringone (Yao et al., 2016) in the first stage. In-vitro leaf explants of *Withania somnifera* L. were selected for transformation and inoculated with 0.8% agar in the MS media (Thilip et al., 2015). Grazegorczyk-Karolak et al. (2018) reported that in *Salvia viridis* L., in-vitro leaves and shoots were prefeered explants for bacterial transformation, and inoculated in the MS (Murashige and Skoog, 1962) 0.7% agar medium, either with or without acetosyringone (200 μM). In *Polygala tenuifolia* gene transfer could be achieved with sterile seedling plant leaf explants which were placed in half MS with acetosyringone (0, 100, 200, and 300 μM) for transformation (Ji et al., 2023).

7.6 GENETIC TRANSFORMATION FOR HAIRY ROOT INDUCTION

Plant transformation by *Agrobacterium rhizogenes* has become a significant aspect of plant biotechnology, as it is linked to the production of bioactive compounds with pharmacological importance from the root cultures of a variety of medicinal plants. *A. rhizogenes* interacts with plant cells by expressing a specific segment of its large root-inducing plasmid DNA (pRi T-DNA), which carries 2 auxin biosynthetic genes (iaaH, iaaM) and 4 root-inducing loci (*rol* A to D). The conjugational transfer into the plant genome causes the expression of the pRi T-DNA and leads to the development of a neoplastic "hairy root" disease at the infection site (Sahu et al., 2013). It is widely accepted to apply *Agrobacterium*-mediated gene transfer to produce hairy root lines with high levels of therapeutic production of metabolites (Sujatha et al., 2013).

Different strains of *A. rhizogenes* have been used for genetic transformation of different medicinal plants. Sahu et al. (2013) have used A4, A4T, and LBA9402 strains of *Agrobacterium rhizogenes* for transformation in *Boerhaavia diffusa* L. In *Artemisia vulgaris* explant *A.rhizogenes* A4GUS, ATCC15834, R1601 and R1000 were used to determine the transformation efficiency (Sujatha et al., 2013) while strains MTCC 532 and MTCC 2364 were used for the induction of hairy root in *Withania somnifera* L. (Thilip et al., 2015). Yao et al. (2016) used A4, C58C1, LBA9402, and R1601 strains of *Agrobacterium rhizogenes* for transformation in *Callerya speciosa* while Grazegorczyk-Karolak et al. (2018) applied the A4 strain for genetic transformation in *Salvia viridis* L. (Table 7.1). To confirm successful transformation of explants, hairy roots are analysed for rol C and rol B through DNA (Yao et al., 2016).

Besides selection of suitable *A. rhizogenes* strain and explants types, bacterial growth (OD value), infection time, presence or absence of acetosyringone, co-cultivation period, culture conditions are some of the other factors that influence hairy root culture (Mahakur et al., 2024).

7.7 PRODUCTION OF BIOACTIVE COMPOUNDS AND STRATEGIES FOR ENHANCED YIELD AND QUALITY

Hairy root culture has been established to significantly enhance secondary metabolites in transformed plant tissues. Chromatographic analysis methods are used to analyse the metabolite content from transformed tissue extracts. Besides, The UPLC-PDA-ESI-MS analysis of the hydromethanolic

TABLE 7.1
Production of bioactive compounds through hairy root culture

Plant name	Explants type	Strain used	Bioactive compound	References
Plumbago zeylanica L.	*In-vitro* leaf	R1000, LBA1334	Plumbagin	Kumar et al., 2023
Gymnema sylvestre	Cotyledons and leaf	KCTC 2703	Gymnemic acid	Nagella et al., 2013
Salvia viridis L.	*In-vitro* node and internode	A4	Polyphenolic compounds	Grazegorczyk-Karolak et al., 2018
Callerya speciosa Champ.	Hypocotyls, cotyledons, leaves and excised stem	LBA9402	Polysaccharides	Yao et al., 2016
Artemisia vulgaris	Shoot tip, leaf and node	ATCC15834, A4GUS, R1000, and R1601	Tricyclene, Linalool, Sabinene, Camphene	Sujatha et al., 2013
Withania somnifera L.	*In vitro* leaf	R1000	Withanolide	Thilip et al., 2015
Polygala tenuifolia Wild	Leaf	C58C1, LBA9402, ATCC15834, ACCC10060	Xanthones, Saponins	Ji et al., 2023
Abrus precatorius L.	Seedling	MTCC 532 and MTCC 2364	Glycyrrhizin	Sajjalaguddam and Paladugu, 2016
Raphanus sativus L.	Leaf	MTCC 2364	Flavonoids	Balasubramanian et al., 2018
Bacopa monnieri L.	Leaf and internode	A4, R1000, SA 79, MTCC 2364 and MTCC 532,	Bacoside A	Bansal et al., 2014
Aerva lanata L.	Leaf and callus	MTCC 532 and MTCC 2364	Aervine	Boobalan and Kamalanathan, 2020

extract of the hairy roots of *Salvia viridis* revealed the presence of rosmarinic acid hexoside, caffeic acid, salvianolic acid E, and methyl rosmarinate (Grazegorczyk-Karolak et al., 2018). Similarly, in *Plumbago zeylanica,* the plumbagin content was confirmed by HPLC analysis of hairy root clones (Kumar et al., 2023).

Besides other factors, the role of elicitors has been widely recognized to get a huge volumetric production of secondary metabolites in culture (Wang and Wu, 2013). Elicitors are the extracellular signalling molecules found in plant cells that start or initiate phytoalexin synthesis and plant defensive responses (Ebel and Cosio, 1994). Elicitation has been frequently used to increase the production of secondary metabolites in both plant cell cultures and hairy root cultures. JA (or MeJA), heavy metal ions, and crude extracts or partially purified polysaccharide fractions of fungal cells are the most popular and efficient elicitors for promoting secondary metabolite production in hairy root cultures.

7.8 APPLICATIONS OF HAIRY ROOT CULTURE IN INDUSTRY

Many research investigations on hairy roots have been carried out in shake flasks and bioreactors. Bioreactor production enables the optimization of conditions for enhanced cell growth and secondary metabolite production; allowing for controlled conditions that minimize variations in the amount produced and the quality of the product (Figure 7.2). As confirmed by previous studies, producing hairy roots on a larger scale for industrial use, bioreactor systems must be optimized (Stiles and Liu, 2013).

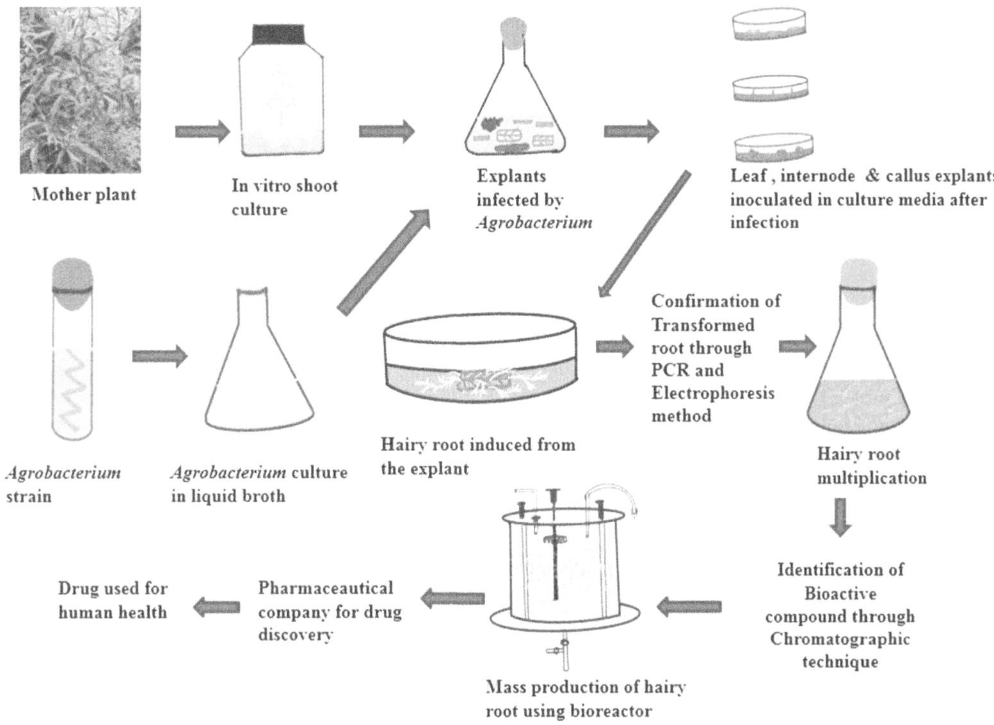

FIGURE 7.2 Diagrammatic representation of bioactive compound production through hairy root culture.

7.9 PHARMACEUTICAL APPLICATIONS

Hairy root cultures can prove to be very beneficial in the pharmaceutical sector as they can provide substantial amounts of subunit vaccines from plants that are both safe and efficient. It was stated the first vaccine candidate created from plants was effective against hepatitis B (Mason et al., 1992). In recent years it has been reported that target proteins can be generated using hairy roots, and the proteins produced exhibit the necessary biological properties. As an alternative to transgenic plants, hairy roots can be utilized to produce proteins in a completely controlled environment, all while removing the risks involved (Gutierrez-Valdes et al., 2020). Because of several beneficial characteristics, including genetic stability, biosynthetic capabilities similar to those of the native plant root, and high biomass production, hairy root cultures induced by *A. rhizogenes* infection have been identified as a promising and popular alternative source of secondary metabolites. All major types of secondary metabolites, including phenolics, flavonoids, alkaloids, and isoprenoids and terpenoids, have been documented to be produced by hairy root cultures (Skarjinskaia et al., 2013). These specific metabolites produce themselves in hairy root cultures through a complex network of multiple unique metabolic pathways that work together effectively. Incomplete understanding exists regarding the genes and metabolic pathways involved in the synthesis of several phytochemicals. The utilization of next-generation sequencing techniques to sequence the transcriptome of hairy root cultures has demonstrated its effectiveness in identifying the genes and metabolic pathways involved in the production of phytochemicals (Halder and Jha, 2018).

7.10 CONCLUSION

This study reports that *Agrobacterium rhizogenes* plays a key role in the induction of hairy roots in different medicinal plants. Different factors like the explants type, OD value, concentration of acetosyringone, culture media, and light influence the induction of hairy roots under *in vitro* conditions. Particular bioactive compounds that are only found in the aerial part of the plant can be identified from the hairy root culture, and the adventitious roots harbour a greater concentration of bioactive compounds. Various elicitors can increase the hairy root's yield, and at the industrial level, bioreactors can help develop a high number of hairy roots and higher metabolite accumulation, which is beneficial to the pharmaceutical industry. Bioactive compounds are used by pharmaceutical companies to produce drugs that are beneficial to human health.

REFERENCES

Abraham, J., and Thomas, T. D. 2017. Hairy root culture for the production of useful secondary metabolites. *Biotechnology and Production of Anti-Cancer Compounds* 1: 201–230.

Adesulu-Dahunsi, A. T., Jeyaram, K., Sanni, A. I., and Banwo, K. 2018c. Production of exopolysaccharide by strains of *Lactobacillus plantarum* YO175 and OF101 isolated from traditional fermented cereal beverage. *Peer J* 6: 1–21.

Ayadi, R., and Tremouillaux-Guiller, J. 2003. Root formation from transgenic calli of *Ginkgo biloba*. *Tree Physiology* 23: 713–718.

Balandrin, M. F., Klocke, J. A., Wurtele, E. S., and Bollinger, W. H. 1985. Natural plant chemicals: Sources of industrial and medicinal materials. *Science* 228(4704): 1154–1160.

Balasubramanian, M., Anbumegala, M., Surendran, R., Arun, M., and Shanmugam, G. 2018. Elite hairy roots of *Raphanus sativus* (L.) as a source of antioxidants and flavonoids. *3 Biotech* 8: 128.

Bansal, M., Kumar, A., and Reddy, B. M. 2014. Influence of *Agrobacterium rhizogenes* strains on hairy root induction and 'bacoside A' production from *Bacopa monnieri* (L.) Wettst. *Acta Physiologiae Plantarum* 36: 2793–2801.

Banwo, K., Olojede, A. O., Adesulu-Dahunsi, A. T., Verma, D. K., Thakur, M., Tripathy, S., ... and Utama, G. L. 2021. Functional importance of bioactive compounds of foods with Potential Health Benefits: A review on recent trends. *Food Bioscience* 43: 101320.

Bonhomme, V., Laurain-Mattar, D., Lacoux, J., Fliniaux, M., and Jacquin-Dubreuil, A. 2000. Tropane alkaloid production by hairy roots of *Atropa belladonna* obtained after transformation with *Agrobacterium rhizogenes* and *Agrobacterium tumefaciens* containing rol A, B, C genes only. *Journal of Biotechnology* 81: 151–158.

Boobalan, S., and Kamalanathan, D. 2020. Tailoring enhanced production of aervine in *Aerva lanata* (L.) Juss. Ex Schult by *Agrobacterium rhizogenes*-mediated hairy root cultures. *Industrial Crops and Products* 155: 112814.

Bulgakov, V. P. 2008. Functions of rol genes in plant secondary metabolism. *Biotechnology Advances* 26: 318–324.

Chawla, H. S. 2003. *Introduction to plant biotechnology*. 2nd edition, Oxford and IBH publishing, Congratulations Pvt Ltd, New Delhi, p. 538.

Dhakulkar, S., Ganapathi, T. R., Bhargava, S., and Bapat, V. A. 2005. Induction of hairy roots in *Gmelina arborea* Roxb. and production of verbascoside in hairy roots. *Plant Science* 169: 812–818.

Dilna, S. V., Surya, H., Aswathy, R. G., Varsha, K. K., Sakthikumar, D. N., Pandey, A., and Nampoothiri, K. M. 2015. Characterization of an exopolysaccharide with potential health-benefit properties from a probiotic *Lactobacillus plantarum* RJF4. *LWT – Food Science and Technology* 64: 1179–1186.

Ebel, J., and Cosio, E. G. 1994. Elicitors of plant defense responses. *International Review of Cytology* 148: 1–36.

Gai, Q. Y., Jiao, J., Luo, M., Wang, W., Ma, W., Zu, Y. G., and Fu, Y. J. 2015. Establishment of high-productive *Isatis tinctoria* L. hairy root cultures: A promising approach for efficient production of bioactive alkaloids. *Biochemical Engineering Journal* 95: 37–47.

Georgiev, M. I., Pavlov, A. I., and Bley, T. 2007. Hairy root type plant in vitro systems as sources of bioactive substances. *Applied Microbiology and Biotechnology* 74: 1175–1185.

Giri, A., and Narasu, M. L. 2000. Transgenic hairy roots: Recent trends and applications. *Biotechnology Advances* 18: 1–22.

Grzegorczyk-Karolak, I., Kuźma, Ł., Skała, E., and Kiss, A. K. 2018. Hairy root cultures of *Salvia viridis* L. for production of polyphenolic compounds. *Industrial Crops and Products* 117: 235–244.

Gutierrez-Valdes, N., Häkkinen, S. T., Lemasson, C., Guillet, M., Oksman-Caldentey, K. M., Ritala, A., and Cardon, F. 2020. Hairy root cultures–A versatile tool with multiple applications. *Frontiers in Plant Science* 11: 33. doi: 10.3389/fpls.2020.00033

Häkkinen, S. T., and Oksman-Caldentey, K. M. 2018. *Progress and prospects of hairy root research. Hairy roots: An effective tool of plant biotechnology*, Springer Nature Singapore Pte Ltd., Singapore: pp. 3–19.

Halder, M., Roychowdhury, D., and Jha, S. 2018. A critical review on biotechnological interventions for production and yield enhancement of secondary metabolites in hairy root cultures. In Srivastava, V., Mehrotra, S., and Mishra, S. (eds.) *Hairy roots* (pp. 21–44). Springer, Singapore.

Hildebrandt, E. M. 1934. Life history of the hairy root organism in relation to its pathogenesis on nursery apple trees. *Journal of Agricultural Research* 10: 857–885.

Ji, H., Yang, B., Jing, Y., Luo, Y., Li, B., Yan, Y., ... and Hu, B. 2023. Trehalose and brassinolide enhance the signature ingredient accumulation and anti-oxidant activity in the hairy root cultures of *Polygala tenuifolia* Willd. *Industrial Crops and Products* 196: 116521.

Khatodia, S., Biswas, K., and Bhatotia, K. 2013. Induction and establishment of hairy root culture of *Solanum xanthocarpum* using *Agrobacterium rhizogenes*. *Journal of Pharmaceutical and BioSciences* 2: 59–63.

Kim, Y. J., Wyslouzil, B. E., and Weathers, P. J. 2002. Secondary metabolism of hairy root cultures in bioreactors. *In Vitro Cellular & Developmental Biology – Plant* 38: 1–10.

Kumar, A., Kumari, A., Demiwal, P., Roy, P., and Sircar, D. 2023. Enhanced production of bioactive plumbagin in hairy root cultures and adventitious root cultures of *Plumbago zeylanica* L. by a novel apocarotenoid elicitor, α-ionone. *Industrial Crops and Products* 203: 117140.

Le Flem-Bonhomme, V., Laurain-Mattar, D., and Fliniaux, M. A. 2004. Hairy root induction of *Papaver somniferum* var. album, a difficult-totransform plant, by *A. rhizogenes* LBA 9402. *Planta* 218: 890–893.

Lin, H. W., Kwok, K. H., and Doran, P. M. 2003. Development of *Linum flavum* hairy root cultures for production of coniferin. *Biotechnol Letter* 25: 521–525.

Lorence, A., Medina-Bolivar, F., and Nessler, C. L. 2004. Camptothecin and 10-hydroxycamptothecin from *Camptotheca acuminata* hairy roots. *Plant Cell Reports* 22: 437–441.

Mahakur, B., Moharana, A., Madkamı, S. K., Naık, S. K., and Barik, D. P. 2024. Optimization of factors affecting *Agrobacterium*-mediated hairy root induction in *Vitex negundo* L. (Lamiaceae). *International Journal of Secondary Metabolite* 11: 244–254.

Mason, H. S., Man-Kit, L. D., and Arntzen, C. J. 1992. Expression of hepatitis b surface antigen in transgenic plants. *Proceedings of the National Academy of Sciences of the United States of America* 89: 11745–11749.

Mehrotra, S., Srivastava, V., Rahman, L. U., and Kukreja, A. K. 2015. Hairy root biotechnology—indicative timeline to understand missing links and future outlook. *Protoplasma* 252: 1189–1201.

Murashige, T. and Skoog, F., 1962. A revised medium for rapid growth and bio assays with tobacco tissue cultures. *Physiologia Plantarum*, 15(3): 473–497.

Nagella, P., Thiruvengadam, M., Jung, S. J., Murthy, H. N., and Chung, I. M. 2013. Establishment of *Gymnemasylvestre* hairy root cultures for the production of gymnemic acid. *Acta Physiologiae Plantarum* 35: 3067–3073.

Nilsson, O., and Olsson, O. 1997. Getting to the root: The role of the *Agrobacterium rhizogenes* rol genes in the formation of hairy roots. *Physiologia Plantarum* 100: 463–473.

Ono, N. N., and Tian, L. 2011. The multiplicity of hairy root cultures: Prolific possibilities. *Plant Science* 180: 439–446.

Pawar, P., and Maheshwari, V. 2004. *Agrobacterium rhizogenes*-mediated hairy root induction in two medicinally important members of family Solanaceae. *Indian Journal of Biotechnology* 3: 414–417.

Ramakrishna, R., Sarkar, D., and Shetty, K. 2020. Functional bioactives from barley for human health benefits. In Shetty, K., and Sarkar, D. (eds.) *Functional foods and biotechnology: Biotransformation and analysis of functional foods and ingredients* (1st ed., p. 454). CRC Press.

Sahu, L., Jena, S., Swain, S. S., Sahoo, S., and Chand, P. K. 2013. *Agrobacterium rhizogenes*-mediated transformation of a multi-medicinal herb, *Boerhaavia diffusa* L.: Optimization of the process and antimicrobial activity against bacterial pathogens causing urinary tract infections. *Frontiers in Life Science* 7: 197–209.

Sajjalaguddam, R. R., and Paladugu, A. 2016. Influence of *Agrobacterium rhizogenes* strains and elicitation on hairy root induction and glycyrrhizin production from *Abrus precatorius*. *Journal of Pharmaceutical Sciences and Research* 8: 1353–1357.

Shanks, J. V., and Morgan, J. 1999. Plant "hairy root" culture. *Current Opinion in Biotechnology* 10: 151–155.

Shi, H. P., and Kintzios, S. 2003. Genetic transformation of *Pueraria phaseoloides* with *Agrobacterium rhizogenes* and puerarin production in hairy roots. *Plant Cell Reports* 21: 1103–1107.

Skarjinskaia, M., Ruby, K., Araujo, A., Taylor, K., Gopalasamy-Raju, V., Musiychuk, K., and Yusibov, V. 2013. Hairy roots as a vaccine production and delivery system. *Biotechnology of Hairy Root Systems* 1: 115–134.

Srivastava, S., and Srivastava, A. K. 2007. Hairy root culture for mass-production of high-value secondary metabolites. *Critical Reviews in Biotechnology* 27: 29–43.

Stewart, F. C. 1900. *A fruit-disease survey of western New York in 1900 (No. 191)*. Agricultural Experiment Station, Geneva, New York.

Stiles, A. R., and Liu, C. Z. 2013. *Hairy root culture: Bioreactor design and process intensification. Biotechnology of hairy root systems*, Springer-Verlag, Berlin Heidelberg: pp. 91–114.

Sujatha, G., Zdravković-Korać, S., Ćalić, D., Flamini, G., and Kumari, B. R. 2013. High-efficiency *Agrobacterium rhizogenes*-mediated genetic transformation in *Artemisia vulgaris*: Hairy root production and essential oil analysis. *Industrial Crops and Products* 44: 643–652.

Thilip, C., Soundar Raju, C., Varutharaju, K., Aslam, A., and Shahjanah, A. 2015. Improved *Agrobacterium rhizogenes*-mediated hairy root culture system of *Withania somnifera* (L.) Dunal using sonication and heat treatment. *3 Biotech* 5: 949–956.

Wang, J. W., and Wu, J. Y. 2013. Effective elicitors and process strategies for enhancement of secondary metabolite production in hairy root cultures. In: Doran, P. (ed.) *Biotechnology of hairy root systems. Advances in biochemical engineering/biotechnology* (Vol. 134: pp. 55–89). Springer, Berlin, Heidelberg.

Yao, S. C., Bai, L. H., Lan, Z. Z., Tang, M. Q., Zhai, Y. J., Huang, H., and Wei, R. C. 2016. Hairy root induction and polysaccharide production of medicinal plant *Callerya speciosa* Champ. *Plant Cell, Tissue and Organ Culture* 126: 177–186.

Zhou, M. L., Zhu, X. M., Shao, J. R., Tang, Y. X., and Wu, Y. M. 2011. Production and metabolic engineering of bioactive substances in plant hairy root culture. *Applied Microbiology and Biotechnology* 90: 1229–1239.

8 Transformation of Medicinal Plants using *Agrobacterium tumefaciens*

Anupriya Chatterjee, Shradha Nirwan and Mukesh Saran*
*Corresponding author

8.1 INTRODUCTION

The goodness of medicinal plants has always nurtured human health. Ancient societies like the Egyptians, Greeks, Chinese, and Indians used medicinal herbs, and they recorded their knowledge in books like the Ebers Papyrus, the Charaka Samhita, and the Shen Nong Ben Cao Jing (Heinrich et al., 2018). The basis for contemporary pharmacology and the methodical investigation of plant-based medications was laid by these early reports. Medicinal plants are still important because of their many therapeutic properties, cultural importance, and economic worth — even in the age of synthetic pharmaceuticals. They are abundant in bioactive substances with a wide range of pharmacological actions, including phenolics, glycosides, alkaloids, and flavonoids. For instance, morphine, an alkaloid originating from the opium poppy (*Papaver somniferum*), is a potent painkiller that is still in high demand today (Heinrich et al., 2018). Comparably, millions of lives have been saved globally by the anti-malarial medication artemisinin, which is derived from *Artemisia annua* (Cragg and Newman, 2013). Atropine, the primary alkaloid of *Atropa belladonna*, is known to have potential bronchodilator, mydriatic, and parasympatholytic property and is in high demand for preparation of anticholinergic drugs (Chaachouay and Zidane, 2024). There is a dire need to enhance the genetic make-up of medicinal plants to make them able to synthesize the desired bioactive compounds and bio-pharmaceuticals for mankind. *Agrobacterium*-mediated genetic transformation of plants is a challenging procedure that only takes place when a plant cell acquires the ability to regenerate by itself. It requires a careful screening of integrated transgene expression. Previous research has revealed that the regeneration efficiency of medicinal plants is frequently found to be low after the transformation procedure. Apocynaceae, Araceae, Araliaceae, Asphodelaceae, Asteraceae, Begoniaceae, Crassulaceae, Fabaceae, Lamiaceae, Papaveraceae, Plantaginaceae, Scrophulariaceae, and Solanaceae are the families of significant medicinal plants from which genetic transformation attempts have been made thus far (Bandurska et al., 2016).

8.1.1 Application of Medicinal Plants

8.1.1.1 In Pharmaceuticals

Drug development and discovery are mostly fueled by medicinal plants, from which many contemporary medications are produced or inspired by phytochemicals. Several commercial drugs are prepared from plant extracts (for example Digitalis drug prepared from extracts of *Digitalis purpurea)* (Chaachouay and Zidane, 2024). Besides, active ingredients in plants form grounds for

DOI: 10.1201/9781003482659-8

other drug development, for example, salicin, a substance present in willow bark (*Salix* spp.), served as the basis for the development of aspirin (acetylsalicylic acid) (Cragg and Newman, 2013).

8.1.1.2 Functional Foods and Nutraceuticals

Nutraceuticals and functional foods are prime examples of the notion that food may be medicine. Because of their ability to promote health, medicinal plants like green tea (*Camellia sinensis*), garlic (*Allium sativum*), and turmeric (*Curcuma longa*) are utilized in dietary supplements and functional foods. Beyond just providing basic nourishment, these items also save us from chronic diseases and improve general health (Singh, 2015).

8.1.1.3 Personal Care and Cosmetics

Medicinal plants are home to plethora of bio-active compounds which are essential to skin care, hair care, and anti-aging. Herbal cosmetic products frequently contain aloe vera, chamomile, and rosemary for their calming, antioxidant, and moisturizing properties (Atanasov et al., 2015).

8.1.1.4 Complementary and Alternative Medicine (CAM)

Complementary and alternative medicine (CAM) modalities, such as aromatherapy, homeopathy, and herbal medicine, primarily depend upon medicinal plants. These therapies provide non-conventional and holistic therapy seekers with alternate options for treatment. Aromatherapy uses essential oils from plants like peppermint and lavender helping people to relax and cope with stress in a better way (Ekor, 2014).

8.1.1.5 Veterinary Medicine and Agriculture

Medicinal herbs are also used as natural growth accelerators and insecticides in agriculture. For instance, neem (*Azadirachta indica*) is utilized in organic farming due to its insecticidal qualities. Plants like calendula and aloe vera are used in veterinary medicine as natural alternatives to synthetic medications for animal wounds and dermatological disease treatment (Singh, 2015).

8.2 OVERVIEW OF *AGROBACTERIUM-MEDIATED* GENETIC TRANSFORMATION

Since the beginning of 20th century, *Agrobacterium tumefaciens* and several related *Agrobacterium* species have been identified as phytopathogens. The ability of *Agrobacterium* to transfer T-DNA to plant cells has been used for plant genetic engineering during the last 40 years. Ever since the first transgenic plant have been generated with AMT in the early 1980s, researchers have enhanced the genetic traits of this "natural genetic engineer" for expanding the scopes in biotechnology. These genetic modifications have broadened the host range of the bacterium. Nevertheless, rather than modifying the genes of the host or bacteria, the majority of significant advancements included modifications to the transformation and regeneration methodology of plants.

8.3 BIOLOGY OF *AGROBACTERIUM TUMEFACIENS*

It is a soil borne, gram negative phytopathogen, member of the Rhizobiaceae family, infecting the majority of dicotyledonous plants causing crown gall/tumor disease at the shoot-root junction. In order to cause tumors, *A. tumefacien* employs a special virulence mechanism in which it transfers the "transfer DNA" or "T-DNA" into host cells, where the T-DNA eventually integrates and becomes a part of plant genome. One major factor influencing the *A. tumefaciens'* pathogenicity is the presence of large Ti (tumor inducing) plasmid. Ti plasmid primarily harbors the majority of the virulence genes or vir genes and the T-DNA. *A. tumefaciens* that lack the Ti plasmid are found to be avirulent. The *A. tumefaciens* strain C58, is the first species of *Agrobacterium* to have a completely sequenced genome, with two different T-DNA sections. It has two incomplete direct repetitions which are referred to as the left and right border sequences, respectively, spanning approximately 25 bp each surround the T-DNA region (Christie and Gordon, 2014).

When *A. tumefaciens'* pathogenicity is induced and activated by the host signal, T-DNA is transmitted in a linear single-stranded form. VirD2 converts the coding strand of the Ti plasmid's T-DNA region into T-strand. Atmakuri et al. (2007) reported that the relaxosome protein complex is formed by the assembly of four Vir proteins, namely VirD2, VirD1, VirC1, and VirC2, at the T-DNA border repeats of the Ti plasmid. This process starts the creation of T-strand. An endonuclease called VirD2 nicks the T-strand from the T-DNA region. The VirD2-T-DNA nucleoprotein complex is formed when the T-strand is cleaved from the border site, leaving VirD2 covalently attached to the T-strand's 5′ end via a phosphotyrosine bond. It is thought that bound VirD2 directs the T-strand's translocation and shields its 5′ end from exonucleolytic assault. To split the superhelical T-border, VirD1 VirD2.

VirC1 and VirC2 jointly, highly boosts the production of T-DNA. Genes in the T-DNA region can be translated in the host and causes a gall once the DNA has been inserted into the host cell and places itself into the host genome. T-DNAs contain genes that code for the enzymes needed to synthesize opiates and phyto hormones (cytokinin and auxin). Auxin and cytokinin buildup in the altered host cells resulting in the hyperproliferation of the cells, which propels the changed tissue's neoplastic expansion. Since the Ti plasmid contains the genes necessary for opine catabolism, the synthesis of opines give *A. tumefaciens* the only nourishment it needs. As a result, the opine-producing tumor gives the infecting *A. tumefaciens* a more specialized and less competitive ecological niche (Guo et al., 2011).

The finding that any desired DNA sequence can substitute native T-DNA without affecting the transfer process of DNA from *A. tumefaciens* to the host was the source of inspiration for AMT. Since then, it was frequently employed as a gene vehicle to introduce the desired DNA fragments into plants.

8.4 MAPPING THE MOLECULAR MECHANISM OF AGROBACTERIUM PATHOGENESIS

Research on the crown gall formation by *Agrobacterium* has been continuing for generations. As a result of this, we now have a fairly good understanding of how the disease works at a molecular level. What makes *Agrobacterium* pathogenic is that it transforms the host genetically by transferring DNA fragments from itself into the host cell. This process takes a long time and has many steps; these include (a) identification and reaction by Agrobacterium to its host; (b) synthesis and transfer of T-DNA plus other substrates by the bacteria to the host cell; (c) movement of T-DNA within the infected cell and its placement into host plant genome; and (d) gene expression from integrated T-DNAs leading to crown gall development. Several detailed reviews of this biological process have been published (Bourras et al., 2015; Guo et al., 2011; Lacroix and Citovsky, 2013; Nester, 2015; Pitzschke, 2013; Tzfira and Citovsky 2006; Ziemienowicz, 2014).

8.4.1 Molecular Perspective *Agrobacterium* Response To The Host Plant

Commonly, *Agrobacterium* can live freely in the soil since most of its virulence genes are dormant. Once it approaches the plant, on receiving signals from the rhizosphere, the virulence genes get activated. This activation uses three two-component systems – VirA/VirG, ChvG/ChvI, and a chemotaxis system – to sense different plant substances, such as organic acids, phenolic compounds, and sugars. These mechanisms aid *Agrobacterium* in initiating its virulence program and adapting to its surroundings. To optimize virulence gene expression, the VirA/VirG system is essential for integrating signals of acidity, phenols, and sugars. The Ti plasmid constitutes the majority of vir genes, and VirA/VirG controls how these genes are expressed. VirA detects signals from plants and triggers VirG, which in turn encourages the transcription of further vir genes.

Precise host recognition and infection are ensured by this cascade. Only virA and virG of the vir operons are constitutively expressed, which allows Agrobacterium to quickly respond to signals from plants and activate its virulence machinery.

8.4.2 Synthesis of T-DNA and Export of Transferred Substrates

The journey of T-strand, in the complex world of *Agrobacterium* starts when the endonuclease VirD2 cuts a single-strand of T- DNA off the Ti plasmid. A group of Vir proteins called the relaxosome complex – VirD2, VirD1, VirC1, and VirC2 — are responsible for coordinating the synthesis of this T-strand. When it is prepared, the T-strand travels toward the host cell under the protection of VirD2. Using either UPP-dependent or independent processes, the *Agrobacterium* reaches the host cell by chemotaxis and attaches itself, creating biofilms for secure attachment. The Type IV secretion system (T4SS), a complex apparatus consisting of 11 VirB proteins and the coupling protein VirD4, facilitates the crucial transport of T-DNA. The VirD2-T-DNA complex along with other virulence proteins are translocated into the host via this mechanism, when VirD4 is present. The T-strand is essentially made into a "hitchhiker" in this intricate protein translocation journey by VirD4, which detects and couples the substrates after this mechanism translocates the VirD2-T-DNA complex and other virulence proteins into the host.

8.4.3 The Journey of T-DNA in Host Cell

The naked T-strand needs protection during its complex trek from *Agrobacterium* to a host cell as portrayed in Figure 8.1. Here comes protein VirE2, which binds to the single-strand DNA and safeguards it from host nucleases. Fascinatingly, *Agrobacterium* enters the host cell by secreting both the VirD2-T-DNA complex and the VirE2 independently. In the host cytoplasm, they combine to form a filamentous T-complex. This T-complex travels in the direction of the host nucleus, occasionally assembling into a larger "super-T-complex" containing more virulence and host proteins. The method takes advantage of nuclear localization signals (NLS) to take control of the host's nuclear import mechanism. Powered by the host's myosin XI-K, the T-complex crosses the cytoplasm of the host along the F-actin network and endoplasmic reticulum, guaranteeing its safe passage to the nucleus where it can exert its influence.

8.4.4 T-DNA Integration in the Host Plant Genome and Tumor Formation

To enable T-DNA integration into the host genome, the T-complex, must shed the VirE2. It is thought that the host's ubiquitin-proteasome system facilitates this disintegration. A virulence protein called VirF targets VIP1 and VirE2 for destructing them by imitating eukaryotic F-box proteins. By stabilizing VirF, VirD5 slows down its rapid deterioration. Though it may include transcription factors or generic chromatin proteins, the precise process by which T-complex targets chromatin is still unknown. Although this is still up for debate, research indicates that the DNA repair machinery of the host plant, frequently helps the T-DNA to integrate and settle itself into the host genome. The integrated T-DNA is split into genes that promote cancer and opine synthesis, along with eukaryotic promoter elements, indicating a eukaryotic origin. These intron-free genes allow the altered plant cells to create opines, which are bacterial nutrients, and hormones that cause tumors.

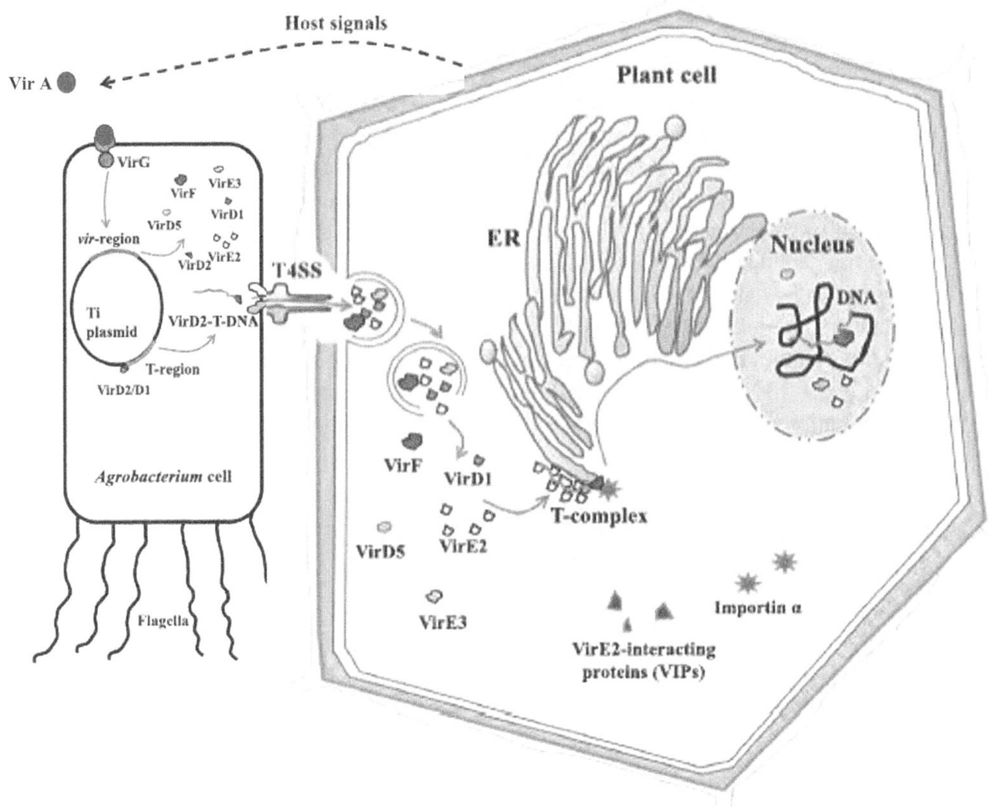

FIGURE 8.1 A schematic diagram portraying the molecular mechanism of *Agrobacterium*-mediated T-DNA transmission process with insights on the virulence protein delivery into the host cell. (Guo et al., 2019.)

8.5 *AGROBACTERIUM TUMEFACIENS* MEDIATED PLANT GENETIC TRANSFORMATION

In today's time, reports convey that *Agrobacterium tumefaciens* mediated genetic transformation technique is an effective, economical and quickest mode of AMT. Using *A. tumefaciens* as a gene vector, the first plant transgenic was developed in the early 1980s, and it was also studied that healthy transgenic plants were able to pass on the transgene to their offspring successfully (Zambryski et al., 1983). Ten years later, *A. tumefaciens* successfully transformed maize, a monocotyledonous plant outside of its "normal host range" (Kado, 2014; Chilton, 1993). It was successfully conducted in the 1980s for transforming firstly dicots, followed by monocots. Its success has opened new avenues for enhancing the medicinal plant's therapeutic properties, producing novel compounds, and improving their cultivation. It is essential to create a common transformation technique to enhance the genomes of therapeutic plants, which are abundant in essential secondary metabolites (Sivanandhan et al., 2016). AMT is the most successful technique available for laboratory settings, out of all the approaches available for genetic transformation (Niazian et al., 2017b). The strain, *Agrobacterium* cell concentration, the concentration of phenolics, such as acetosyringone, plant species/cultivar/genotype, plant growth regulators (PGRs), explant, light and temperature during preculturing and co-cultivation, concentration of antibiotics, wound treatment in plant tissue, and transformed cell selection technique are some of the essential factors that influence the AMT

(Niazian et al., 2017b). Species specific or plant specific optimized transformation protocol will give rise to a higher percentage of regeneration and transformation efficiency. Ten to fourteen days after transformation, stable expression happens because it takes longer for the T-DNA to be incorporated into the host genome (Lacroix and Citovsky, 2013). Transient expression of the transgene typically defined as an expression peak that happens two to four days following host tissue infection. The generation of completely transgenic plants is contingent upon the expression being steady. The stable transgenic cells must be extracted from the agro-infected plant tissues in order to create true plant transformants, and these cells must be able to regenerate an entire plant (Anami et al., 2013). AMT as opposed to steady expression, provides a quick and easy way to express desired genes. *Agrobacterium*-mediated transient expression has garnered significant attention since 20 years as a result of its adoption by a number of biotechnological businesses for the bio-synthesis of medicinal proteins in plants (Kopertekh and Schiemann, 2017). The host range of *Agrobacterium* has also been expanded to non-traditional plant eukaryotes, such as yeast (Bundock et al., 1995), fungi (De Groot et al., 1998; Li et al., 2017a), and human cultured cells (Lacroix and Citovsky, 2018). The uses of genetic transformation mediated by *Agrobacterium* extends well beyond the natural plant hosts of these bacteria.

8.6 *AGROBACTERIUM TUMEFACIENS* MEDIATED GENETIC TRANSFORMATION OF MEDICINAL PLANTS

It's been documented that *A. tumefaciens* is useful in the genetic transformation of plants with medicinal and aromatic significance (Gomez-Gale'ra et al., 2007; Khan et al., 2009). There are surprisingly few transformation reports on medicinally significant plants (Gomez-Gale'ra et al., 2007; Wang et al., 2008, Pandey et al., 2010). *Plumbago zeylanica* (Wei et al., 2006) and *Euphorbia nivulia* (Sunandakumari et al., 2005) are two examples of plants that have been found to be resistant to regeneration following genetic transformation. These plants possess efficiently secondary pathways and a well demonstrated regeneration system. *Mentha citrata* (Spencer et al., 1990) *and Coleus forskohlii* (Mukherjee et al., 2000) have both undergone genetic transformation using the wild-type nopaline strain of *A. tumefaciens* to create transformed shoot cultures and cell suspension cultures, respectively. The reporter gene *gusA* (β-glucuronidase A) and selection marker *nptII* have been widely utilized to develop transgenic plants using mostly leaf and hypocotyl explants in plants for example *Mentha piperita* L. (Niu et al., 1998), *Artemisia annua* (Vergauwe et al., 1998), *Scrophularia buergeriana Miq.* (Park et al., 2003), *Digitalis minor L.* (Sales et al., 2003), *Echinacea purpurea* (Wang and To, 2004), and *Ruta graveolens* (Lievre et al., 2009). An extensive detail is shared in Table 8.1.

8.7 CASE STUDIES OF *AGROBACTERIUM TUMEFACIENS* MEDIATED TRANSFORMATION OF MEDICINAL PLANTS

Bacopa monnieri – Also known as 'Brahmi', *B. monnieri* belongs to the Scrophulariaceae family, and ranked 2nd in a priority species list of the most significant medicinal plants (Rajani, 2008) based on its medicinal importance, shared value, and potential for further research and development. Ramesh et al. (2011) gave a method for *A. tumefaceins* based transformation of the plant using axenic nodal segments as explants. *A. tumefaciens* strain EHA 105, with binary vector pCAMBIA 130 was utilized for genetic transformation through a method given in Figure 8.2. The system so developed ensured generation of stable GUS gene transformants with a high transformation efficiency of 68.8%.

Trachyspermum ammi – This is a useful medicinal industrial plant that belongs to the Apiaceae family. The essential oil contributes to a wide range of medicinal applications, such as antibacterial, antifungal, anti-inflammatory, antioxidant, antibacterial, cytotoxic, anti-lithiasis, nematicidal,

TABLE 8.1
Highlights the *Agrobacterium* mediated transformation of medicinal plant species

Plant Species	Traits/Properties	Explant	Strain and Vector	GOI/SMG/RG	Additive/Treatment	Efficiency (%)	References
African basil (Ocimum gratissimum)	Anti-epileptic, anti-cancerous, antinociceptive	Cotyledonary node	LBA4404, pBI121	—/nptII/gusA	Acetosyringone	6–10	Niazian et al. (2019)
Blue mallee (Eucalyptus polybractea)	Essential oil	Nodal	AGL1, pMDC204	—/hptII/gus	—	6	Niazian et al. (2019)
Cumin (Cuminum cyminum L.)	Essential oil, anti-oxidant	Embryos	EHA105, pCAMBIA1301	—/hptII/gus	Acetosyringone	1.5	Niazian et al. (2019)
Dendrobium catenatum	Alleviate rheumatoid arthritis, diabetes, obesity, etc.	Primary protocorms	GV3101, pCAMBIA1305-P600	—/nptII/gus&gfp	Acetosyringone/SAAT/Tween 20, Silwet-77, and TritonX-100	56.5	Niazian et al. (2019)
Foxglove (Digitalis purpurea)	Cardiotonic glycosides, anti-cancerous	Mature leaf	GV2260/pBI121 and GV3101/pBI121	—/nptII/gusA	Acetosyringone	52.2% and 60%	Niazian et al. (2019)
Gentiana tibetica	Anti-bacterial, anti-oxidant, anti-cancer, and anti-viral	Leaf blade of axenic culture	pCB 301 and pCH32	—/nptII/uidA	l-glutamine	22.5	Niazian et al. (2019)
Greenviolet (Hybanthus enneaspermus L.)	Anti-microbial, aphrodisiac, demulcent, antitussive, anticonvulsant, hepatoprotective	Leaf	LBA4404, pCAMBIA2301	—/nptII/gusA	Acetosyringone	28	Niazian et al. (2019)
Indian ginseng (Withania somnifera L.)	Anti-cancer, anti-microbial, anti-oxidant, anticonvulsant, anti-neurodegenerative, and immune-modulatory agent	Leaf	R1000, -	—/nptII/gusA	Acetosyringone/vacuum infiltration/SAAT	—	Niazian et al. (2019)
Indian ginseng (Withania somnifera L.)	Anti-cancer, osteo-protective, cardioprotective, antibacterial, stress-reliever	Nodal	GV3101, pIG121Hm	—/hptII/nptII/gusA	Acetosyringone	10.6	Niazian et al. (2019)
Macleaya cordata	Anti-bacterial, anti-microbial, anti-cancer	Young leaf and stem segments	GV3101, pCAMBIA2301	—/nptII/uidA	Acetosyringone	15.7	Niazian et al. (2019)

(*continued*)

TABLE 8.1 (Continued)
Highlights the Agrobacterium mediated transformation of medicinal plant species

Plant Species	Traits/Properties	Explant	Strain and Vector	GOI/SMG/RG	Additive/Treatment	Efficiency (%)	References
Orchid (Dendrobium lasianthera)	Anti-microbial, anti-cancer, anti-bacterial, cytotoxic activity	Protocorm	LBA4404, pG35SKNAT1	—/nptII/—	Acetosyringone	70	Niazian et al. (2019)
Pelargonium graveolens	Anti-bacterial, anti-fungal, anti-oxidant, anti-inflammatory, anti-cancer	Petiolar segments, leaf	LBA4404, pBI121	—/nptII/gusA	Acetosyringone	69.5	Niazian et al. (2019)
Scutellaria ocmulgee	Anti-inflammatory, anti-oxidative, anxiolytic, anti-viral, anti-cancer	Shoot tip, nodal segments	EHA105, pq35SGR	—/nptII/gusA&gfp	—	—	Niazian et al. (2019)
Ajowan (Trachyspermum ammi (L.) Sprague)	Antibacterial, antifungal, and antioxidant activities	15 days old hypocotyl	LBA4404, pBI121	BADH/nptII	Acetosyringone	2.94	Niazian et al. (2019)
Veratrum dahuricum	Helps in the treatment of stroke, dysentery, jaundice, headache, and chronic malaria	Fresh embryonic calli	LBA440, pCAMBIA3300, PPT/nptII/GUS	—	Acetosyringone	15	Ma et al. (2020)
Artemisia annua	Anti-plasmodial, anti-hyperlipidaemic, anti-inflammatory, anti-microbial, anti-cholesterolemic, anti-viral	Slightly opened bud	EHA105, Pcambia1300	NptII	Acetosyringone	26.9	Qamar et al. (2022)

GOI – Genes of Interest, SMG – Selection Marker genes, SAAT – Sonication-Assisted Agrobacterium-mediated transformation, and RG – Reporter genes.

The reported transformation efficiency by author(s) are on the basis of the percentage of PCR-positive plants developed –: Data not reported by author(s).

FIGURE 8.2 Nodal explant mediated *B. monnieri* AMT. (a) Direct induction of adventitious shoot induction from wild nodal explants on hygromycin-free culture medium (control); (b) Shoot regeneration of putative transformants on culture media supplemented with hygromycin; (c) Positive transformant (T0) showing the GUS expression; (d) The cross section of nodal cross sections reflecting GUS expression at 40x; (e) The transformed leaf explants showing blue gus spot in; (f) Transgenic *B. monneiri* acclimatized on soil. (Ramesh et al., 2011.)

anthelmintic, and anti-filarial activities. Thymol, a phenolic monoterpenoid derivative of cymene and isomer of carvacrol is the major constituent in ajwain oil. Thymol is known for its strong germicidal, anti-spasmodic, and fungicidal effects. A study by Niazian et al. (2019) reported the successful transformation of ajowan plants by both plant tissue culture and in planta AMT with a transformation percentage of 2.94% and 1.42%, respectively. The transformants exhibited enhanced drought and salinity tolerance as compared to the wild plants with the expression of betaine aldehyde dehydrogenase *(BADH)* gene. Besides, the *BADH*-transformed plants showed a significant increase in thymol content (55.07%) as compared to wild type plants (39.20%) under drought stress condition.

Veratrum dahuricum – Alternatively named as mountain corn, *Veratrum dahuricum* (family:Liliaceae) is used in the traditional Chinese medicine system and drawing attention from pharmaceutical industry for its anticancer product named cyclopamine. A recent study by Ma et al. (2020) describes an efficient AMT transformation protocol with a 56% regeneration efficiency and 15% transformation efficiency. The embryogenic callus of the plant was used for transformation using *A. tumefaciens* strain LBA4404, which carried the binary vector pCAMBIA3300, with bar gene expression cassette that imparts resistance to the herbicide phosphinothricin (PPT). The fast-growing and fresh yellow embryonic calli as exhibited in Figure 8.3, were inoculated in the

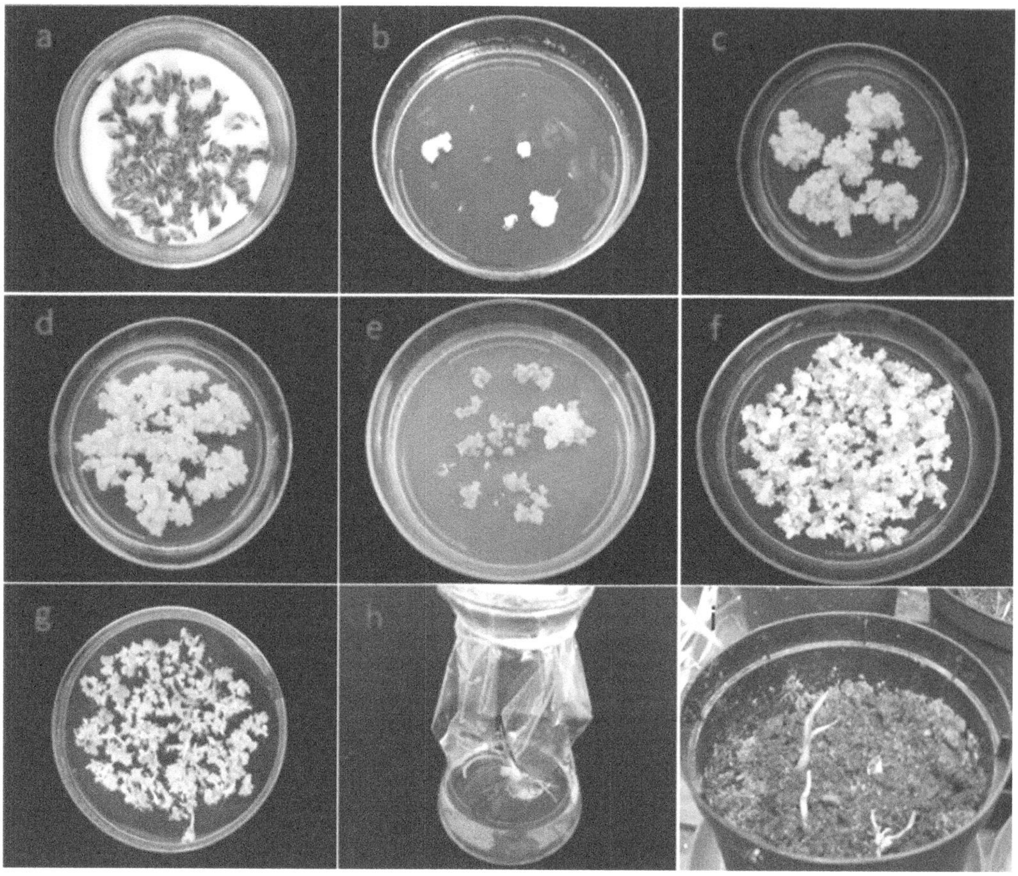

FIGURE 8.3 AMT genetic transformation of *Veratrum dahuricum* with pCAMBIA3300-35S-bar expression vector. (a) Surface-sterilized seeds of *V. dahuricum* (b) induction of embryogenic calli on modified MS-medium containing 8 mg/L picloram; (c,d) proliferation of embryogenic calli on AA medium containing 4 mg/L picloram; (e) selection of transformed callus on AA medium containing 4 mg/L 2,4-D and 16 mg/L PPT; (f) transformed bar calli on AA selection medium; (g) regenerated green plant on R2M medium containing 3 mg/L 6-BA; (h) plantlets with developed roots on hormone-free medium R2M; (i) potted plantlets in peat soil. (Ma et al., 2020.)

recombinant bacterial cell suspension. The PPT resistant calli that showed viability were again sub-cultured on shoot regeneration media. RT-PCR and Quick PAT/bar strips, Southern blot analysis further confirmed bar gene expression in the transgenic plants.

Artemisia annua – Belonging to family Asteraceae, the plant is valued for its essential oils used in perfumery and therapeutics, showing anti-plasmodial, anti-hyperlipidaemic, anti-inflammatory, anti-microbial, anti-cholesterolemic, antioxidant, and hepatoprotective properties (Lubbe et al., 2012; Wang et al., 2020; Shin et al., 2020; Qamar et al., 2022). *A. annua* L. is primarily used to isolate artemisinin which has encouraging anti-malarial properties. For development of transgenic lines of *A. annua*, a floral dipping method using half open, immature bud stage capitula and *A. tumefaciens* strain EHA105, with pCambia1300 as an expression vector has been reported to give high transformation efficiency of 26.9%. (Qamar et al., 2022) (Figure 8.4).

FIGURE 8.4 The steps carried out while dipping flowers of *Artemisia annua*. (A) The inflorescence of *A. annua* L.; (B) A 10x magnification of a bud from immature inflorescence utilized for floral dipping; (C) The green, young, and un-bloomed bud stage inflorescence was dipped in infiltration medium; (D) Paper bags were placed over the dipped (infected) inflorescences; (E) T1 treated plant progenies; (F) Plants with the transgene were chosen and cultivated in a transgenic containment zone. (Qamar et al., 2022.)

8.8 CONCLUSION AND FUTURE REMARKS

The demand for plant-based products has grown significantly across the globe. More than 85% of people in the Middle East, Latin America, Africa, and Asia get the majority of their medical care from traditional medicine, particularly from herbal remedies. In the European Union, about 100 million people still utilize traditional, complementary, or herbal remedies; in certain other countries, this number can reach 90% of the population. *Agrobacterium*-mediated genetic transformation contributes to improving the concentration of bioactive compounds and their therapeutic value through the overexpression of gene of interest involved in the production of secondary metabolites and downregulation of genes producing adverse compounds. Enhanced production of bioactive compounds using advanced tools helps to meet the increasing demand of effective therapeutic drugs to cure life-threatening diseases, like cancer, diabetes, skin care, infertility, etc.

In the present scenario, the development of an efficient *Agrobacterium*-mediated transformation (AMT) protocol for essential medicinal plants can broaden the scope of metabolic engineering or trait enhancement in medicinal plants.

ABBREVIATIONS

A	Absorbance
AMT	Agrobacterium Mediated Transformation
ACTs	Artemisinin-based combination
BADH	Betaine aldehyde dehydrogenase
BAP	6-Benzylaminopurine
CaMV	Cauliflower Mosaic Virus
CAM	Complementary and alternative medicine
2,4-D	2,4-Dichlorophenoxyacetic acid
GUS	β-glucuronidase
hpt II	Hygromycin phosphotransferase
MS	Murashige and Skoog
NAA	1-Naphthaleneacetic
*Npt*II	Neomycin phosphotransferase
PPT	Phosphinothricin
T-DNA	Transfer DNA
Vir	Virulence
YEB	Yeast Extract Broth
YEM	Yeast Extract Media

REFERENCES

Anami, S., Njuguna, E., Coussens, G., and De Buck, S. 2013. Activation tagging in tomato identifies a transcriptional regulator of anthocyanin biosynthesis, modification, and transport. *Plant Physiology*, 163(2): 1056–1071.

Atanasov, A. G., Waltenberger, B., Pferschy-Wenzig, E.-M., Linder, T., Wawrosch, C., Uhrin, P., Temml, V., Wang, L., Schwaiger, S., Heiss, E. H., Rollinger, J. M., Schuster, D., Breuss, J. M., Bochkov, V., Mihovilovic, M. D., Kopp, B., Bauer, R., Dirsch, V. M., and Stuppner, H. 2015. Discovery and resupply of pharmacologically active plant-derived natural products: A review. *Biotechnology Advances*, 33(8): 1582–1614.

Atmakuri, K., Cascales, E., Burton, O. T., Banta, L. M., and Christie, P. J. 2007. Agrobacterium ParA/MinD-like VirC1 spatially coordinates early conjugative DNA transfer reactions. *The EMBO Journal*, 26(10): 2540–2551.

Bandurska, H., Niedziela, J., and Chadzinikolau, T. 2016. Genetic transformation in medicinal plants: Advances and limitations. *Journal of Plant Biotechnology*, 13: 45–56.

Bourras, S., Rouxel, T., and Meyer, M. 2015. Agrobacterium tumefaciens-mediated transformation: State of the art and future prospects. *Molecular Plant Pathology*, 16(4): 823–832.

Bundock, P., Dulk-Ras, A., Beijersbergen, A., and Hooykaas, P. J. J. 1995. Trans-Kingdom T-DNA Transfer from Agrobacterium tumefaciens to Saccharomyces cerevisiae. *The EMBO Journal*, 14: 3206–3214.

Chaachouay, N., and Zidane, L. 2024. Plant-derived natural products: A source for drug discovery and development. *Drugs and Drug Candidates*, 3(1): 184–207. https://doi.org/10.3390/ddc3010011

Chilton, M. D. 1993. Agricultural biotechnology: High-impact activity in university research. In *Perspectives on Agricultural Biotechnology in the 21st Century* (pp. 21–25).

Christie, P. J., and Gordon, J. E. 2014. The agrobacterium Ti plasmids. *Microbiology Spectrum*, 2(6): PLAS-0034-2014.

Cragg, G. M., and Newman, D. J. 2013. Natural products: A continuing source of novel drug leads. *Biochimica et Biophysica Acta (BBA) – General Subjects*, 1830(6): 3670–3695.

De Groot, M. J. A., Bundock, P., Hooykaas, P. J. J., and Beijersbergen, A. G. M. 1998. Agrobacterium tumefaciens-Mediated Transformation of Filamentous Fungi. *Nature Biotechnology*, 16: 839–842.

Ekor, M. 2014. The growing use of herbal medicines: Issues relating to adverse reactions and challenges in monitoring safety. *Frontiers in Pharmacology*, 4: 177.

Gomez-Gale'ra, S., Pelacho, A. M., Gene, A., Capell, T., and Christou, P. 2007. The genetic transformation of medicinal and aromatic plants. *Plant Cell Reports*, 26: 1689–1715.

Guo, M., Ye, J., Gao, D., Xu, N., and Yang, J. 2019. Agrobacterium-mediated horizontal gene transfer: Mechanism, biotechnological application, potential risk and forestalling strategy. *Biotechnology Advances*, 37(1): 259–270.

Guo, M., Gelvin, S. B., and Wise, A. A. 2011. The biology of the Agrobacterium-plant cell interaction. In *The Molecular Biology of Plant Cells* (Vol. 2, pp. 381–416).

Heinrich, M., Barnes, J., Gibbons, S., and Williamson, E. M. 2018. *Fundamentals of Pharmacognosy and Phytotherapy*. Elsevier Health Sciences.

Hiei, Y., Ohta, S., Komari, T., and Kumashiro, T. 1994. Efficient transformation of rice (Oryza sativa L.) mediated by Agrobacterium and sequence analysis of the boundaries of the T-DNA. *Plant Journal*, 6: 271–282.

Kado, C. I. 2014. Historical account on gaining insights on the mechanism of crown gall tumorigenesis induced by Agrobacterium tumefaciens. *Frontiers in Microbiology*, 5: Article 340.

Kado, C. I. 2014. Historical events that spawned the field of plasmid biology. *Microbiology Spectrum*, 2: PLAS-0010-2013.

Khan, T., Shankar, G., and Rajam, M. V. 2009. Agrobacterium-mediated transformation of eggplant (Solanum melongena L.) using root explants: Optimization of factors. *In Vitro Cellular and Developmental Biology – Plant*, 45: 402–411.

Kopertekh, L., and Schiemann, J. 2017. Agrobacterium-mediated transient expression as an efficient tool for plant biotechnology. In J. Schiemann and C. P. Haines (Eds.), *Methods in Molecular Biology* (Vol. 1640, pp. 47–59). Springer.

Lacroix, B., and Citovsky, V. 2013. The roles of bacterial and host plant factors in Agrobacterium-mediated genetic transformation. *International Journal of Developmental Biology*, 57(6–8): 467–481.

Lacroix, B., and Citovsky, V. 2018. Transfer of DNA from bacteria to eukaryotes. *MBio*, 9(5): Article e00908–18.

Li, P., Tao, Z., Dean, R. A., and Meyers, B. C. 2017a. Strain- and treatment-specific genome-wide responses of rice and the rice blast fungus magnaporthe oryzae during early infection. *BMC Genomics*, 18: 741.

Lièvre, K., Tran, T.L., Doerper, S., Hehn, A., Lacoste, P., Thomasset, B., Bourgaud, F., Gontier, E. 2009. Agrobacterium-mediated transformation of Ruta graveolens L. *Methods Mol Biol*. 547: 235–248. doi: 10.1007/978-1-60327-287-2_19. PMID: 19521849.

Lubbe, A., Seibert, I., and Klimkait, T. 2012. *Ethnopharmacology in Overdrive: The Remarkable Anti-HIV Activity of Artemisia annua*. Elsevier. doi:10.1016/j.jep.2012.03.024

Ma, R., Yu, Z., Cai, Q., Li, H., Dong, Y., Oksman-Caldentey, K. M. and Rischer, H. 2020. Agrobacterium-mediated genetic transformation of the medicinal plant Veratrum dahuricum. *Plants*, 9: 191.

Mukherjee, A., Roy, S. C., and Bhadra, S. K. 2000. Agrobacterium-mediated genetic transformation of Coleus forskohlii Briq: An important medicinal plant. *Plant Cell Reports*, 19: 853–858.

Nester, E. W. 2015. Agrobacterium: Nature's genetic engineer. *Frontiers in Plant Science*, 5: Article 730.

Niazian, M., Sadat Noodezh, M., Galuszka, P., and Mortazavian, S. M. M. 2017b. Tissue culture-based agrobacterium-mediated and in planta transformation methods. *Biotech*, 7: 12–23.

Niazian, M., Sadat-Noori, S. A., Tohidfar, M., Galuszka, P., and Mortazavian, S. M. M. 2019. Agrobacterium-mediated genetic transformation of ajowan (Trachyspermum ammi (L.) Sprague): An important industrial medicinal plant. *Industrial Crops and Products*, 132: 29–40.

Niu, Q., He, C., and Zhang, Y. 1998. Transformation of peppermint (Mentha piperita) with Agrobacterium rhizogenes and regeneration of plants with altered morphology and essential oil content. *Plant Cell, Tissue and Organ Culture*, 55: 169–176.

Pandey, B. K., Singh, H., and Sudhakara, R. 2010. Efficient plant regeneration of Saussurea lappa Clarke: An endangered medicinal plant. *In Vitro Cellular and Developmental Biology – Plant*, 46: 476–481.

Park, K., Lee, J., and Kim, K. 2003. Genetic Transformation of the Medicinal Plant Scrophularia buergeriana Miq. (Scrophulariaceae) by Agrobacterium rhizogenes. *Plant Cell Reports*, 21: 1198–1203.

Pitzschke, A. 2013. Agrobacterium infection and plant defense—transformation success hangs by a thread. *Frontiers in Plant Science*, 4: Article 519.

Qamar, F., Khan, S., Ashrafi, K., Iqrar, S., Quadri, S. N., Saifi, M., and Abdin, M. Z. 2022. Germline transformation of Artemisia annuaL. plant via in planta transformation technology "Floral dip". *Biotechnology Reports*, 36: e00761.

Rajani, M. 2008. Brahmi: Priority species of significant medicinal plants. *International Journal of Ayurvedic Medicine*, 12: 87–94.

Ramesh, M., Alagarsamy, K., Vijayakumar, K., Virgin, L. J., and Pandian, S. 2011. Agrobacterium-mediated transformation of pharmaceutically important Indian medicinal herb Bacopa monnieri (L.). *Journal of the Medical Plants Research*, 5: 2316–2321.

Sales, M., Simon, E., and Quiroz, F. 2003. Agrobacterium tumefaciens-Mediated Transformation of Digitalis minor, a Mediterranean Endemic medicinal plant. *Plant Cell Reports*, 21: 1173–1178.

Shen, Q., Zhang, L., Liao, Z., Wang, S., Yan, T., Shi, P., Liu, M., Fu, X., Pan, Q., Wang, Y., Lv, Z., Lu, X., Zhang, F., Jiang, W., Ma, Y., Chen, M., Hao, X., Li, L., Tang, Y., Lv, G., Zhou, Y., Sun, X., Brodelius, P. E., Rose, J. K. C., and Tang, K. 2018. *The genome of Artemisia annua Provides Insight Into The Evolution of Asteraceae Family and Artemisinin Biosynthesis*. Elsevier. doi: 10.1016/j.molp.2018.03.015

Shin, J. Y., Kang, H. J., Cho, B. O., Park, J. H., Che, D. N., Hao, S., Wang, F., Sim, J. S., Sim, D. J., and Jang, S. I. 2020. Ameliorative effects of cirsium jaonicum, artemisia annua and curcuma longa on non-alcoholic fatty liver disease. *Korean Journal of Pharmacognosy*, 51: 178–185. doi: 10.22889/KJP.2020.51.3.178

Singh, R. 2015. Medicinal plants: A review. *Journal of Plant Sciences*, 3(1–1): 50–55.

Sivanandhan, G., Arun, M., Mayavan, S., Rajesh, M., Jeyaraj, M., and Muthuselvam, M. 2016. Chitosan enhances with high frequency plant regeneration efficiency in Drosera burmannii Vahl. *African Journal of Biotechnology*, 15(2): 25–36.

Spencer, A., Towers, G. H. N., and Kirby, E. G. 1990. Agrobacterium-mediated transformation of Mentha citrata Ehrh. *Journal of Plant Physiology*, 136: 194–198.

Sunandakumari, C., Remashree, A. B., and Kiran, S. 2005. Agrobacterium-mediated genetic transformation and regeneration of Euphorbia nivulia Buch-Ham. *Plant Cell, Tissue and Organ Culture*, 82: 123–129.

Tzfira, T., and Citovsky, V. 2006. Agrobacterium-mediated genetic transformation of plants: Biology and biotechnology. *Current Opinion in Biotechnology*, 17(2): 147–154.

Vergauwe, A., Cammaert, R., and Jacobs, M. 1998. Agrobacterium rhizogenes-Mediated Transformation of Artemisia annua L. and Regeneration of Transgenic Plants. *Plant Cell Reports*, 18: 168–173.

Wang, D., Cui, L., Chang, X., and Guan, D. 2020. Biosynthesis and characterization of zinc oxide nanoparticles from Artemisia annua and investigate their effect on proliferation, osteogenic differentiation and mineralization in human osteoblast-like MG-63 Cells. *Journal of Photochemical Photobiology: B*, 202: Article 111652. doi: 10.1016/j.jphotobiol.2019.111652

Wang, Y., and To, K. Y. 2004. Agrobacterium tumefaciens-Mediated Transformation of Echinacea purpurea. *Plant Cell, Tissue and Organ Culture*, 79: 327–335.

Wang, Z. Y., Hobbs, S. L., and Herrin, D. L. 2008. One-step agrobacterium-mediated transformation of the Edible Cyanobacterium Arthrospira platensis C1. *Plant Biotechnology Journal*, 6: 194–207.

Wei, Z. M., Huang, Y. Z., and Yang, Y. F. 2006. Genetic transformation of Plumbago zeylanica L. by Agrobacterium tumefaciens and regeneration of transgenic plants. *Plant Cell, Tissue and Organ Culture*, 85: 297–305.

Zambryski, P., Joos, H., Genetello, C., Leemans, J., van Montagu, M., and Schell, J. 1983. Ti plasmid vector for the introduction of DNA into plant cells without altering their normal regeneration capacity. *The EMBO Journal*, 2: 2143–2150.

Ziemienowicz, A. 2014. Agrobacterium-mediated plant transformation: Factors, applications and recent advances. *Biocatalysis and Agricultural Biotechnology*, *3*(4): 95–102.

9 Chloroplast Engineering for Improvement of Plant Varieties

*Ravinsh Kumar, Ashutosh Singh and Amrita Srivastava**

*Corresponding Author

9.1 INTRODUCTION

The plants possess specialized structures called chloroplasts, enabling them to be self-sufficient in nourishment due to their photoautotrophic nature. The various scientific theory suggests that chloroplasts, a class of plastids specific to plant cells and eukaryotic algae, along with mitochondria, originated from independent, prokaryotic organisms. Each plant cell can have up to 300 chloroplasts (Boffey and Leech, 1982), which are responsible for photosynthesis. Plastids, genome size ~120–150 kb within a single plant cell can vary considerably a range of 1,000 to 10,000 copies per cell (Bendich, 1987). Plant genetic material is located in three different sites within the cell: the nucleus, chloroplasts, and mitochondria. Each of these compartments has its own set of genes responsible for passing on heritable features (Verhounig et al., 2010). Over evolutionary time, chloroplasts underwent genetic material transfer from engulfed photoautotrophic organisms to the host cell's nucleus. This process reduced the chloroplast's genetic autonomy, resulting in its current small genome size. Consequently, the chloroplast encodes only a minimal number of its own proteins. The majority of chloroplast proteins are now encoded in the host nucleus and subsequently imported into the chloroplast after translation (Richardson and Schnell, 2020). It is very well reported that angiosperms exhibit a strong bias towards maternal inheritance of chloroplasts, affecting roughly 80% of all species. This pattern is remarkably stable and independent of processes like gene duplication, polyploidy, and recombination, which frequently occur in plant nuclear genomes (Maliga, 2002; Tangphatsornruang et al., 2010). To date, over 170 chloroplast genomes from various species have been fully sequenced, including many agriculturally significant crops like rice, maize, sugarcane, wheat, tomato, and mungbean. Chloroplast have circular double stranded DNA varies its size from 120 to 220 kb. Its gene organize into four distinct regions in which two inverted repeats (IRs, 20–28 kb) that are separated by a small single copy (16–27 kb) region and a large single copy (80–90 kb) region (Asif et al., 2010). These genomes are valuable for studying molecular evolution and phylogenetic studies (Liò and Goldman, 1998). Additionally, the maternal inheritance of chloroplast genomes is beneficial for genetic engineering since it prevents cross-recombination. Genetic engineering has primarily been focused on the nuclear genome (Gosal et al., 2009). Unlike nuclear genomes, chloroplast genomes are advantageous for genetic engineering due to their maternal inheritance which minimizes the risk of transgenes escaping into the environment through pollen flow (Singh et al., 2010). However, genetic engineering has primarily focused on the nuclear genome due to established techniques. Chloroplast have ability to receive foreign DNA and this makes chloroplast a target for genetic engineering (Daniell et al., 2002). In 1980s plastid engineering was started to manipulate photosynthesis and optimize it and increase its potential by enhancing its photosynthesis process (Grevich and Daniell, 2005). Chloroplast engineered plants are termed as transplastomic (Maliga, 1993). Chloroplast engineering strategically modifies plants by introducing foreign genes directly into the chloroplast's unique circular, double-stranded DNA, bypassing

DOI: 10.1201/9781003482659-9

the nuclear genome. This approach offers several advantages. Diekmann et al. (2008) highlight the importance of having complete chloroplast genome sequences for efficient chloroplast engineering. They achieved this by developing a cost-effective method for isolating pure chloroplast DNA from grasses, adapting existing protocols used for flowering plants. Ultimately, chloroplast engineering allows for the targeted production of specific foreign proteins within the plant itself (Maliga, 2001). Many biotechnological methods equipped with molecular engineering are powerful tools for integrating foreign DNA into chloroplasts. Unicellular green algae, *Chlamydomonas reinhardtii*, is particularly valuable for pioneering chloroplast transformation studies (Nickelsen and Kück, 2000). *C. reinhardtii* offers several advantages: it possesses a single, large chloroplast occupying a significant portion of the cell volume. Moreover, this sole chloroplast harbors numerous copies of its own genome (Boynton et al., 1988). This simplifies the process of introducing and integrating foreign DNA. The successful chloroplast transformation in a higher plant, tobacco, stands as a significant milestone in chloroplast engineering. Cheng et al. (2010) have established a method for transforming plastids in top cultivars of rapeseed (*Brassica napus* L.). Later on, Lin et al. (2014) introduce cyanobacterial functionalities into crop chloroplasts to augment CO_2 fixation and they established tobacco plants expressing crucial enzymes of cyanobacterium *Synechococcus elongatus*. Plastid transformation has been successfully incorporated in ornamental plants such as petunia (Zubko et al., 2004), as well as in legumes (Dufourmantel et al., 2004), fiber crops like cotton, root vegetables like carrot, leafy greens like lettuce, cereals like rice (Kumar et al., 2004; Lelivelt et al., 2005; Lee et al., 2006), trees such as poplar (Okumura et al., 2006), and other Solanaceae family members like tobacco (Ko et al., 2009) and mulberry (Umate, 2010). The successful transformation of eggplant by Singh et al. (2010) further extends this list of plant species.

9.2 CHLOROPLAST ENGINEERING TOOLS AND ITS APPLICATIONS

Plant scientists are increasingly exploring chloroplast genetic engineering as an alternative approach to nuclear transformation for introducing foreign DNA into plant cells. The development of organelle transformation became feasible with the invention of the gene gun. Researchers can employ biolistics, a technique that uses forceful acceleration to introduce purified or synthetic DNA into a wide range of organisms and cell types. Remarkably, this technology is effective for cell organelles, including chloroplasts and even smaller mitochondria, using standard particle sizes of 0.4–1.7 μm. Transforming chloroplasts in seed plants was long considered highly challenging and even deemed impossible by many researchers. A single mesophyll cell can harbor thousands of copies of plastid DNA, while containing a much smaller number of chloroplasts (Golczyk et al., 2014). This abundance of plastid genomes presented an initial challenge, but researchers successfully transformed chloroplasts in the green alga *Chlamydomonas*, followed by tobacco plants, as mentioned earlier. These breakthroughs established foundational principles for plastid genome engineering, and subsequent tools and applications were developed using these early model systems. Here, we have described methods used in chloroplast engineering (Figure 9.1).

9.3 INTEGRATION OF FOREIGN DNA

Several methods exist for delivering DNA to chloroplasts, including biolistic bombardment, polyethylene glycol (PEG) transfection, carbon nanotube delivery and targeted delivery using biorecognition motifs.

Biolistic Alteration – Biolistic transformation is a powerful tool for modifying chloroplasts. This technique involves introducing a plasmid carrying a selectable marker gene and the desired gene of interest directly into the chloroplasts. The foreign genes become integrated into the plastid DNA through a process called homologous recombination. Flanking sequences on the plasmid similar

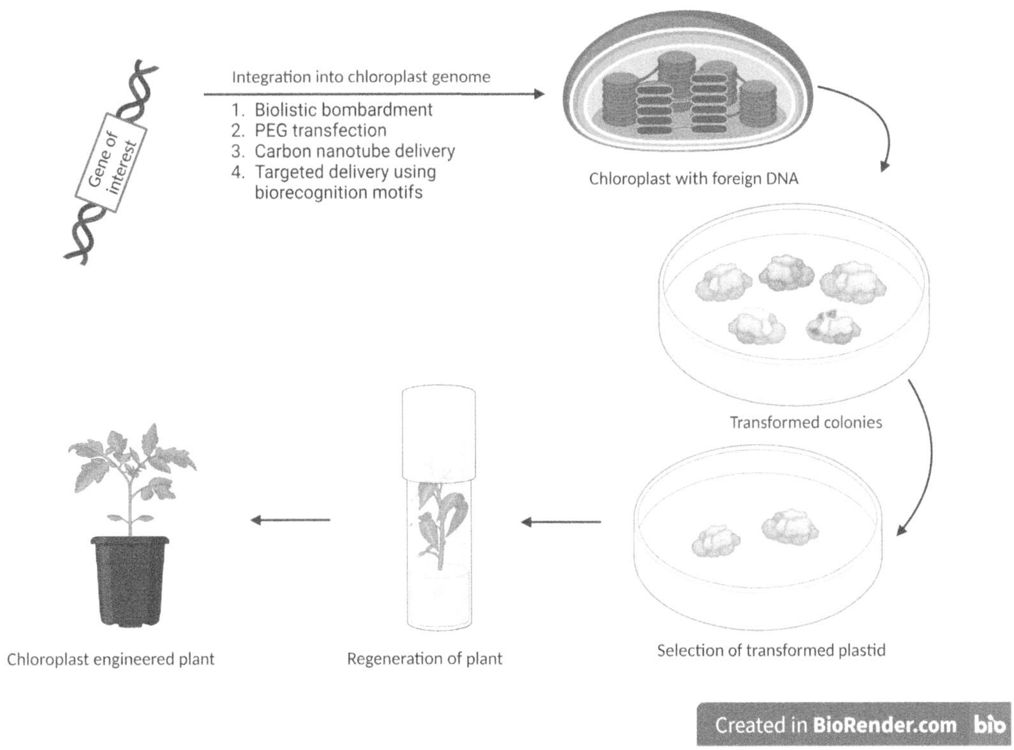

FIGURE 9.1 Production of a chloroplast engineered plant.

to those in the chloroplast genome guide this precise integration (Daniell et al., 2002). A technique called biolistic, also known as gene gun bombardment, utilizes a device to accelerate microscopic particles coated with DNA into cells. These particles, typically composed of tungsten or gold due to their density and inertness, physically pierce the cell wall and membrane, delivering the DNA payload. First successful implementation of this technique took place in unicellular green alga *Chlamydomonas reinhardtii* where functional wild-type *atpB* gene was efficaciously introduced through biolistic into the chloroplast of mutant algal lacking this gene (Boynton et al., 1988). A noteworthy application of biolistic was studied by Cosa et al. (2001) in which *Bacillus thuringiensis* cry2Aa2 operon was introduced in the chloroplast genome of tobacco plant using the biolistic, resulting in the successful transformation. Researchers are using this technique efficiently to improve a wide range of crops, including tomatoes, eggplants, potatoes, rice, bitter squash, cotton, carrots, and soybeans.

PEG-Mediated engineering – Genome of chloroplasts can also be modified through PEG-mediated transformation. It is one of the common methods used in chloroplast transformation. In this method, protoplasts are co-cultured with the desired DNA construct in the presence of PEG. Permeability of the cell membrane is compromised by PEG in the presence of divalent cations. Thus, destabilizing the cell membrane. Foreign DNA construct and divalent cation complex could easily pass through the cell membrane in the form of vesicles and successfully transform the chloroplasts (Low et al., 2018). Chloroplast transformation takes place in three steps – protoplast and chloroplast isolation, introduction of foreign DNA into the chloroplast and regeneration of chloroplast. The initial step of PEG-mediated transformation is protoplast suspension preparation by employing enzymatic treatments on the plant cells. This step is followed by centrifugation and further purification of protoplast suspension to obtain pure chloroplast suspension. So that no other cellular component, other than chloroplast, is involved in transformation. The chloroplast obtained is mixed with the DNA construct and PEG is added to the mixture. The transformed chloroplast with foreign DNA is selected using the selectable marker gene (Kumar et al., 2021). PEG-mediated transformation offers some advantages such as – it shows high transformation efficiency in protoplasts, huge-sized DNA construct can be easily up taken by protoplast without inducing any physical damage to the cell membrane and protoplasts.

Delivery by Carbon Nanotubes – Sumio Iijima discovered carbon nanotubes carriers in 1991. Caron nanotubes possess high aspect ratio, high surface area and are very small in size. Thus, they easily absorb onto or conjugate with therapeutic molecules. Kwak et al. (2019) have demonstrated specifically designed single-walled carbon nanotubes (SWNTs) that can easily traverse the plants cell wall, cell membrane and double membraned lipid bilayer of the chloroplast and get kinetically trapped there. This method is based upon the lipid exchange envelope penetration mechanism. Chitosan wrapped SWNTs own high surface charge and are able to passively transport and irreversibly localize within the chloroplast (Giraldo et al., 2014). Chitosan is used for polycationic DNA delivery for plant transformation because forms complexes with negatively charged DNA through electrostatic attraction and protects it from nuclease degradation and another reason is that it is a biodegradable polysaccharide which is abundantly available and poses no threat to the plants. The success of this method hinges on the distinct pH environments within plant cells. The acidic cytoplasm (pH 5.5) strengthens the association between DNA and SWNTs. However, once inside the chloroplast, the neutral pH of the stroma (fluid-filled region) triggers the release of DNA from the SWNT complexes. For successful transformation it is essential to maintain a proper ratio of DNA and SWNTs (Kumar and Ling, 2021). Since the process is a targeted one, the DNA is expected to be properly integrated into the chloroplast genome. Experiments with yellow fluorescent protein (YFP) as a reporter gene revealed that a 1:6 DNA to SWNT ratio was mildly satisfactory for expression of YFP in the chloroplasts of *Arabidopsis* protoplasts, while a 1:1 ratio did not produce any expression. This indicates that a higher DNA load can neutralize the surface charge of the complexes, compromising their stability and ability to enter within chloroplasts (Body et al., 2019). Successful expression of transgene containing YFP reporter gene via same method has been experienced in arugula (*Eruca vesicaria*), spinach (*Spinacia oleracea*), watercress (*Nasturtium officinale*) and tobacco (*Nicotiana tabacum*). This selective method of chloroplast transformation by nanoparticles is easily accomplished and does not require expensive equipment and most important, it is cost-friendly. As compared to transformation via traditional method such as PEG-mediated and biolistic transformation, minimum amount of DNA is required in nanoparticle mediated gene delivery (Sharma et al., 2023).

Biorecognition Motif-Based Delivery – Although the study on nanoparticles for targeted delivery is continuous, however, precise delivery to sub-cellular compartment is still a challenge. Presently various approaches to improve *in vivo* nanoparticle delivery is being done by modulating the size and charge of nanoparticles but significant result is still not obtained regarding specific subcellular localization (Wu et al., 2017). Santana et al. (2020) have developed a novel approach for the delivery of chemicals specifically to the chloroplast in *Arabidopsis thaliana* by means of engineered nanomaterial guided by peptide recognition motif. A 14-amino acid signal peptide from RUBISCO is utilized for biorecognition motifs that will guide the engineered nanoparticles to the target place. Multifunctional quantum dots are used as carrier to deliver biochemicals. For this, chemicals which have to be delivered in the chloroplast, such as methyl viologen and ascorbic acid, are coupled to the surface of quantum dots with the help of conjugated β-cyclodextrins (β-CD). β-cyclodextrins is a cyclic oligosaccharide that is used for encapsulation of methyl viologen and ascorbic acid and helps in the inclusion complex formation between chemicals and quantum dots.

9.4 INCORPORATING FOREIGN DNA SEQUENCES WITHIN CHLOROPLAST GENOME

For successful transformation, efficient integration of transgene in chloroplast genome is required. To make this possible, transgene is flanked by intergenic sequences on either side that helps in site-specific homologous recombination. Efforts were made for sequencing of crop chloroplast genomes so that through codon optimization and using superlative regulatory sequences, translation and transgene integration could be enhanced (Daniell et al., 2021). The most used site of integration of transgene in chloroplast genome is the transcriptionally active intergenic region between *trnl-trnA* genes of the *rrn* operon. Any integration of foreign genes at this site offers impressive expression (Cosa et al., 2001). *pPRV* (plastid repeat vectors) vector series facilitates the insertion of transgene between *trnV* and *rpsZ12/7* and is uniquely suited to study the plasmid promoter activity (Zoubenko et al., 1994). Another factor that is necessary for regulating the expression of transgene is the presence of regulatory sequences upstream (5' UTR) and downstream of the transgene (3' UTR). Since *psbA* gene encodes the extremely translated protein in the chloroplast and it gets translationally active in the presence of light, the regulatory regions of psbA is used as the best option for use in an expression cassette (Klein and Mullet, 1987).

For successful expression of transgene in chloroplast there must be a chloroplast specific promoter, 5′/3′ regulatory sequences and selectable marker to enhance the efficiency of transcription and translation of the gene. *Prna, PpsbA, PrbcL, psaA* and *atpI* are some commonly used promoters for chloroplast transformation (Adem et al., 2017). However, plastid rrn promoter is found to be more effective for achieving high expression levels (Newell et al., 2003). Recently, MoChlo, a complete chloroplast cloning system has been developed that consist of 47 chloroplast promoters for flexible and rapid plastid genetic engineering (Occhialini et al., 2019). 5'/3' regulatory sequences are equally important for stable integration and enhanced expression. Some commonly used 5'-UTRs are *Ggagg*, T_7G_{10}, *rbcL, atpB, psbA* and *cry2a* while 3'-UTRs commonly used for chloroplast transformation are as follows – *rbcL, rps16, petD* and *psbA* (Adem et al., 2017).

9.5 SELECTION OF TRANSFORMED PLANTS IN CHLOROPLAST ENGINEERING

Following the introduction of foreign DNA, transformation is achieved only in a small number of plastomes (plastid genomes). This results in a heteroplasmic state where the transformed plastomes own foreign DNA and others possess the original DNA. Through repeated culturing (sub-culturing) and selection using specific growth media, this heteroplasmic state progresses towards a uniform state (homoplasmy) where all plastomes within a cell are transformed with the desired DNA. Therefore, for efficient and stable transformation in crop species, the chosen plant should be compatible with tissue culture techniques and demonstrate a high capacity for regeneration. The *aadA* gene, encoding *Aminoglycoside 3'-Adenylyltransferase*, is a widely used tool for selecting chloroplast transformants. Bacterial *aadA* gene that confers resistance to the streptomycin and spectinomycin is used as selectable marker in plastid transformation. It binds to the 30S ribosomal subunit and inhibits translation process in chloroplasts (Carter et al., 2000; Sváb and Maliga, 1993). It was first used in *Chlamydomonas reinhardtii* (Goldschmidt-Clermont, 1991; Fischer et al., 1996). Following this, use of *aadA* for selection has been successfully used in other plants like *Arabidopsis* and poplar (Sikdar et al., 1998; Okumura et al., 2006). This technique has facilitated the introduction of over 100 transgenes into higher plant plastids.

Transplastomic technology has several advantages over the concerns associated with genetically modified crops such as environmental persistence, reduced pleiotropic effects and maternal inheritance. Other than *aadA*, kanamycin/neomycin resistance gene *nptII* (neo) and the chloramphenicol resistance gene *cat* (Carrer et al., 1993) is also used as selectable marker for chloroplast transformation. But the use of antibiotic resistance genes as a selectable marker in transgenic crops for commercial use requires approval from concerned regulatory bodies. So, alternative selectable markers based on metabolic processes have been explored. Betaine aldehyde dehydrogenase gene (*BADH*) found within chloroplast is involved in the synthesis of glycine betaine which protects the cells during osmotic stress. It catalyzes 3-aminopropionaldehyde to produce β-Ala which after methylation form β-Ala betaine (Rathinasabapathi et al., 2001). Some other genes related to metabolism that are used as a selectable marker in plastid transformation are *bar, codA, ARG9, ASA2* (Day and Goldschmidt-Clermont, 2011; Cui and Jiang, 2014). In a way the use of selectable markers ease the way of selecting transformed plants but on the other hand it poses a threat due to its inclusion and persistence in transformed plants.

Therefore, creating marker-free GM plants is crucial to address health and environmental concerns and ensure the safe use of GM plants in agriculture and industry. Advanced techniques like the Cre-*lox* recombination system have shown promise in removing selectable marker genes following chloroplast transformation (Shao et al., 2014). Recent advancements in producing foreign protein drugs within marker-free lettuce plants through chloroplast engineering highlight the promising potential for this approach in biopharmaceutical development (Tanwar et al., 2023).

9.6 IMPROVED PLANT VARIETY USING CHLOROPLAST ENGINEERING

Plastid transformation offers advantages over nuclear methods for crop engineering and developing plant varieties but also faces some limitations. Tobacco remains the main research model, while success in major cereals (rice, wheat, maize) is limited. The regeneration method is a significant hurdle; most cereals rely on somatic embryogenesis, hindering efficient plastid transformation. Targeting non-green plastids, such as amyloplasts in fruits and grains, is challenging due to reduced gene activity in these plastids. Additionally, species-specific vectors are needed for stable protein expression. Despite these challenges, progress has been made in tomato, potato, eggplant, soybean, cauliflower, carrot and, cabbage. Future advancements in overcoming regeneration and non-green plastid limitations, along with the development of universal vectors, are crucial for wider adoption in major food crops. Here, we have summarized major developments in chloroplast engineering for the development of plant varieties.

9.6.1 Chloroplast Engineering In Tomato (*Solanum lycopersicum*)

As a widely cultivated and consumed vegetable, the popularity of tomato is steadily rising due to the presence of lycopene, a compound with well-documented antioxidant and anticancer properties. Recognizing this potential, researchers are increasingly focusing on genetically engineering tomato plants to enhance the levels of health-promoting compounds within the fruit. Additionally, the complete plastid genome sequence for tomatoes is available; facilitating these advancements. Researchers achieved a significant advancement in plant transformation in 2001 (Ruf et al., 2001) successfully produced fertile tomatoes with modified chloroplasts (transplastomic). Their approach involved engineering vectors specifically designed to target tobacco chloroplasts, such as pRB70, pRB94, and pRB95. These vectors were then introduced into leaves of tomato using a biolistic bombardment technique. The pRB70 vector included an *aadA* gene, which was flanked by a ribosomal RNA operon promoter (Prrn) and located in the intergenic space between the *trnfM* and *trnfG* genes. The *psbA* gene functioned as the terminator in this configuration. This strategy effectively led to the stable transformation of the tomato plastid genome and achieved homoplasmy in transplastomic lines, ensuring that the plastid-encoded trait was passed on to the F1 generation. Fruits may hold promise for the production of beneficial compounds. This study found that the accumulation of foreign proteins was significantly greater in fruits compared to leaves. In fact, fruit protein levels reached nearly half of those seen in leaf chloroplasts, the sites of protein production in leaves. This suggests that engineering plant chloroplasts (transplastomic technology) could be a viable approach to develop new plant varieties that produce higher levels of nutraceuticals and biopharmaceuticals within edible parts, potentially improving safety for consumers. Wurbs et al. (2007) developed a more efficient chloroplast transformation system for tomatoes. They modified a previously reported method and used it on tomato (cultivar IPA-6) for modification in the carotenoid biosynthesis pathway. The experiment involved a chloroplast transformation vector derived from another vector (pRB94). This new vector, pRB96, contained an inserted expression cassette with a specific promoter (plastid atpI promoter), an untranslated region (3' UTR) from a gene (*rps16*), *crtY* gene encoding lycopene beta-cyclase and a site for transgene integration. The *crtY* gene was of *Erwinia herbicola* origin. This vector was then bombarded into tomato leaves using a particle bombardment technique. Transformed tomato lines were identified using resistance to two antibiotics (spectinomycin and streptomycin). The resulting transplastomic tomato plants expressed the bacterial lycopene beta-cyclase gene, leading to herbicide resistance and a fourfold increase in pro-vitamin A content within the fruits. The research showcased the success of plastid transformation in altering metabolic pathways to yield nutritional advantages, including in plastids that do not perform photosynthesis. Additionally, the scientists improved carotenoid production in genetically modified tomato plants with transformed plastids by introducing lycopene beta-cyclase genes from two sources: A prokaryotic source (*Erwinia herbicola*), A higher plant source (*Narcissus pseudonarcissus*) (Apel and Bock, 2009). Two novel chloroplast vectors were engineered: pEcrtY harboring the *crtY* gene from the bacterium *Erwinia* and pNLyc harboring the *Lyc* gene sourced from the plant *Narcissus*. The expression cassettes incorporated a promoter from a ribosomal RNA operon fused with a leader sequence and a 3' untranslated region from another gene (*rps16*). These foreign genes were inserted into a specific region between two other genes. An additional gene provided resistance to a herbicide and was flanked by specific sequences for potential removal using a targeted recombination technique. Both vectors were delivered into tomato leaves via particle bombardment. Analysis using restriction fragment length polymorphism (RFLP) confirmed the stable integration of the transgenes and homoplasmy (uniformity) of the chloroplast genome in the transplastomic tomato lines. As expected with maternally inherited plastid traits, no segregation of antibiotic resistance was observed in the first generation (T_1) of transformed plants. Transplastomic tomatoes expressing a plant-derived lycopene beta-cyclase displayed significantly enhanced enzyme activity compared to those harboring the bacterial gene. The conversion of lycopene to beta-carotene, which changed the fruit color from red to orange, led to a 50% increase in provitamin A. This study suggests that plastid transformation for metabolic engineering in non-green tissues can be highly effective when the right combinations of transgenes and expression elements are chosen. Zhou et al. (2008) demonstrated the expression of HIV antigens (P24-nef) in the plastid genome of tomato (cv. IPA-6), marking a significant advancement in developing edible vaccines through tomato chloroplast engineering. The chloroplast transformation vector pZF7lox includes the *p24-nef* fusion gene, regulated by *Prrn* and the phage T7-derived gene 10 leader sequence, with an *rbcL* terminator positioned between the *trnfM* and *trnG* regions. The selectable marker gene *aadA*, controlled by *Prrn* and flanked by loxP sites, enables its removal through Cre-mediated site-specific recombination. The plastid transformation was performed by biolistically bombarding technique in tomato leaves with the pZF7lox vector. The resultent homoplasmic transplastomic tomato plants produced approximately 40% antigen protein. This study revealed the potential of plastids as cost-effective bioreactors for high-yield production of HIV antigens in edible crops. The pTop2 vector, specifically designed as a polycistronic construct, incorporated two identical intercistronic expression elements to separate the genes. The transgene expression cassettes were under the control of a promoter sourced from the tobacco ribosomal RNA operon. This promoter was flanked by the 5' UTR of phage T7 gene 10 and the 3' UTR derived from a plant-specific gene, such as the plastid *rbcL* gene or the tobacco *rbcL* or *rps16* gene. Interestingly, the pTop2 construct employed a 3' UTR from a separate organism, *Chlamydomonas reinhardtii,* to terminate the operon. In contrast, the other constructs utilized tobacco genes' 3' UTRs to enhance the stability of the mature, single-cistron messenger RNAs. The *aadA* gene, employed for selection, was controlled by a chimeric ribosomal RNA operon promoter and fused to the 3' UTR of the *NtTpsbA* gene. Biolistic bombardment with these vectors (pSyHPT and pTop2) on young tomato leaves generated transplastomic tomato plants. Molecular analyses employing RFLP verified the stable integration of the transgene and the uniform presence of the transgene across all chloroplast genomes (homoplasmy) within the transformed plants. These plant lines demonstrated a substantial boost, up to tenfold, in their total tocochromanol content.

9.6.2 Chloroplast Engineering In Eggplant (*Solanum melongena*)

Eggplant is a staple food in many developing countries, offering affordability and essential nutrients. It's a good source of minerals, vitamins, fiber, and boasts phytonutrients with antioxidant properties. Beyond its nutritional value, eggplant holds significance in traditional medicine for treating various ailments. While transgenic research has made strides in crop improvement through genetic engineering, alternative methods like chloroplast genome engineering might be a promising approach for developing new eggplant varieties. This technique could offer advantages over traditional genetic modification, potentially mitigating concerns about cross-pollination between wild relatives and genetically modified eggplant. Singh et al. (2010) used a procedure in which green stem segments were bombarded with a particular vector, pPRV111A, that was intended to integrate into the eggplant's plastid genome. This vector had areas for homologous recombination with eggplant plastid DNA and was bordered by tobacco plastid sequences (*trnV* and *rps7/12*). The *aadA* gene, which encodes the *aminoglycoside adenyltransferase* (AAD) enzyme, was introduced to provide resistance to spectinomycin and streptomycin. This strategy allowed for the successful regeneration of homoplasmic transplastomic eggplants using spectinomycin selection. Subsequent examination of T1 and F1 generations confirmed the *aadA* gene's stable integration, as shown by persistent spectinomycin resistance in the plant population. Eggplant plastid transformation is a revolutionary method of modifying plant genes, perhaps boosting crops and allowing plants to produce.

9.6.3 Chloroplast Engineering In Lettuce (*Lactuca sativa*)

Lactuca sativa L., also known as lettuce, is a cool-season vegetable from the *Asteraceae* family that is grown for its nutritional value and economic relevance. Lettuce is ideal for genetic study because of its short life cycle, full self-fertility, and low cultivation requirements. Biotechnology, which includes chloroplast engineering, has considerable potential to improve the nutritional value of lettuce. Lelivelt et al. (2005) successfully integrated foreign genes into lettuce's plastid genome by employing PEG to mediate plastid transformation. A plastid transformation vector containing a 2,253-bp stretch of lettuce chloroplast DNA that particularly targets the *trnI/trnA* intergenic region. This vector was delivered into lettuce protoplast cells using PEG-mediated transformation, with the *aadA* gene for spectinomycin resistance serving as a selectable marker. Analysis of the first generation (T_1) transplastomic plants revealed spectinomycin resistance and expression of the foreign gene, confirming successful transgene transmission. Additionally, these plants displayed detectable levels of recombinant green fluorescent protein within their plastids. Researchers have successfully used vectors containing regulatory elements from tobacco and lettuce to achieve gene expression in the chloroplasts of lettuce. Through biolistic bombardment, foreign DNA was introduced into lettuce leaf segments, resulting in transplastomic plants, laying the groundwork for future applications in lettuce plastid transformation. Likewise, another group expressed human thioredoxin 1 (*hTrx1*) in lettuce using a vector (pRL200-*hTrx1*) designed with lettuce chloroplast DNA sequences. The transplastomic plants stably integrated the *hTrx1* gene and accumulated the protein at detectable levels (around 1% of total soluble protein), demonstrating the feasibility of using lettuce chloroplasts for producing therapeutic proteins. Further research demonstrated the production of a therapeutic fusion protein (CTB-Pins) in lettuce chloroplasts. Using a custom plasmid (pLS-LF-CTB-Pins) with long flanking sequences homologous to lettuce chloroplast genes *trnI* and *trnA*, the researchers achieved high-level expression (2.5% of total soluble protein) and confirmed maternal inheritance of the transgenes. This study highlighted the potential of edible plants for therapeutic protein production. Additionally, researchers have explored the expression of Dengue virus antigens and cyanobacterial *FBP/SBPase* in lettuce chloroplasts, enhancing growth rates and photosynthetic activity. Japanese researchers also engineered lettuce to produce higher levels of Vitamin E, showcasing the versatility of chloroplast transformation for various biotechnological applications.

Despite these advancements, challenges remain in commercializing transplastomic lettuce due to public acceptance of GMOs and intellectual property issues. However, the high protein accumulation capacity and status as a leafy green vegetable make lettuce a promising platform for biopharmaceutical production.

9.6.4 Chloroplast Engineering In Potato (*Solanum tuberosum*)

Millions rely on the potato, the world's fourth-largest vegetable tuber crop, with annual production, nearing 400 million tons. But this abundant food source is more than just a filler. Packed with essential nutrients like carbohydrates, fiber, vitamin C, and minerals like potassium and iron, potatoes are true nutritional champions. Notably, it was among the first plants targeted for genetic modification, with the potential to revolutionize global food production.

Plastid transformation, pioneered in potatoes by Sidorov et al. (1999), presents a compelling alternative for crop improvement, including the development of edible vaccines. Their approach involved bombarding leaf explants with tobacco-derived chloroplast transformation vectors, achieving the integration of *aadA* and GFP genes into the chloroplast genome. This resulted in transplastomic plants with concentrated GFP accumulation in green tissues and minimal presence in tubers. Building upon this work, Nguyen et al. (2005) successfully transformed the popular potato cultivar Desiree using similar vectors. The resulting lines displayed consistent GFP expression in both tubers and leaves. The subsequent unveiling of the complete potato plastid genome sequence paved the way for enhanced transformation efficiency through optimized vectors and regeneration media. After obtaining the complete plastid genome sequence of potato, advancements in potato plastid transformation have been achieved through optimization of flanking sequences and regeneration medium. This has facilitated the design and testing of various construct vectors using different promoters, showcasing chloroplast engineering's potential for precise protein expression control across plant tissues.

9.6.5 Chloroplast Engineering In Carrot (*Daucus carota*)

Carrot (*Daucus carota* L.) is globally recognized as a vital vegetable crop due to its valuable nutritional benefits, including sugars, vitamins A and C, and dietary fiber. It is sensitive to salt levels, with growth declining by 7% for each 10 mM increase in salinity beyond 20 mM salt (Gibberd et al., 2002; Khan, 2012). This sensitivity makes carrot a promising candidate for developing new crop varieties aimed at enhancing salt tolerance. Salt stress poses a significant abiotic challenge in plant agriculture, exacerbated by excessive irrigation and fertilizer application. Vivek et al. (1999) demonstrated that introducing a spinach gene, betaine aldehyde dehydrogenase (*badh*), into carrots enhanced their salt tolerance through plastid engineering, predating the sequencing of the carrot plastid genome. Their method involves designing a chloroplast transformation vector to target and integrate into a specific region, referencing the tobacco plastid genome for guidance. Ruhlman et al. (2006) offer a significant resource for advancing chloroplast engineering in carrots and related species. Despite the current absence of published research on transplastomic technology development in this crucial edible crop, their work remains pivotal for future exploration in the field.

9.6.6 Chloroplast Engineering In Cabbage (*Brassica oleracea*)

Cabbage, a leafy green vegetable from the *Brassicaceae* family, is widely distributed and a key ingredient in salads worldwide, rich in nutrients. It contains vitamins A, C, K, and B6, as well as iron, folic acid, calcium, glucosinolates, both soluble and insoluble fiber, and is low in sodium. However, its cultivation faces challenges due to various environmental stresses. Genetic engineering, specifically through plastid transformation, allows for the targeted introduction of desirable traits into existing cabbage varieties. By combining advancements in plastid transformation with optimized regeneration protocols, scientists hold significant promise for enhancing cabbage production and resilience.

Liu et al. (2007) successfully transformed cabbage (K-Y cross cultivar) chloroplasts using particle bombardment. Their vector contained flanking regions (*trnV-rrn16S* and *trnI-trnA-rrn23S*) to promote transgene integration, *aadA* gene as marker and *uidA* gene as reporter. The resulting cabbage plants showed protein accumulation of 3.2% to 5.2% of total soluble protein. Following this, they employed a new vector (pASCCBT) including the *aadA* gene for selection and the *cry1Ab* foreign gene under the control of the *Prrn* promoter and a *psbA* terminator for transformation of cabbage chloroplasts. Plants with complete plastid transformation (homoplasmic) accumulated Cry1Ab protein at levels ranging from 4.8% to 11% of TSP and demonstrated significant insecticidal activity against *Plutella xylostella* larvae.

9.6.7 Chloroplast Engineering In Cauliflower (*Brassica oleracea*)

Cauliflower, a prominent member of the Brassicaceae family, shares the global consumption market with cabbage in terms of cultivation and production. Prized for its large, edible flower head (curd), it matures rapidly, forming substantial curds within a month of being transplanted into soil. Genetic transformation techniques, particularly those targeting plastids, hold promise for enhancing cauliflower's resilience against herbicides, insects, diseases, and environmental challenges. This approach additionally paves the way for the development of edible vaccines, making cauliflower transformed using plastids a compelling avenue for genetic improvement in this crop. Cauliflower (cv. Thalassa) underwent plastid transformation using a PEG-mediated protoplast approach. The experiment involved a vector containing flanking sequences (*accD-rbcL*) obtained from *Brassica napus*. This vector housed an expression cassette driven by *Prrn*, the tobacco plastid promoter and a tobacco terminator (psbA 3'UTR). The cassette encoded the aadA gene, conferring spectinomycin and streptomycin resistance, and was integrated in between *accD* and *rbcL* genes of *B. napus*. Achieving uniformity (homoplasmy) in cauliflower plastids through extended protoplast culturing remains difficult due to the low efficiency of plastid transformation. Deciphering the cauliflower plastid genome sequence is essential to improve this crop using plastome engineering techniques.

9.6.8 Chloroplast Engineering In Soybean (*Glycine max*)

The soybean, a leguminous crop, is a global agricultural crop cultivated for its versatile edible beans valued for their protein and oil content. Scientists are exploring genetic engineering techniques to improve soybean, a vital legume crop. Plastid transformation, a method pioneered by Dufourmantel et al. (2004), holds promise for enhancing soybean genetics. Their research created soybean plants (*Glycine max* cv. Jack) with modified chloroplasts (transplastomic) by bombarding embryonic tissues with a custom-designed chloroplast transformation vector. This vector, named pCLT312, contained a gene for selecting transformed cells (*aadA*) and a segment of soybean's own chloroplast DNA. Notably, the resulting T1 generation displayed resistance to spectinomycin. Dufourmantel et al. (2005), achieved a breakthrough in soybean genetic engineering by creating fertile, homoplasmic soybean lines expressing the insecticidal protein *Cry1Ab* from *Bacillus thuringiensis*. This achievement involved a custom-made chloroplast transformation vector derived from *pCLT312*, named *pCLT327*. The vector incorporated a promoter sequence from tobacco's 16S *rDNA* gene, along with regulatory elements from bacteriophage T7 and tobacco's plastid *rbcL* gene. The transgene, encoding *Cry1Ab*, was inserted between the *rps12/7* and *trnV* genes within the soybean plastid genome. Particle bombardment delivered the *pCLT327* vector into embryogenic cotyledon tissues. The resulting T1 generation displayed robust *Cry1Ab* expression in leaves, stems, and seeds, effectively protecting the plants against the destructive velvet bean caterpillar (*Anticarsia gemmatalis*). Scientists engineered the first herbicide-tolerant plants by incorporating a bacterial gene (likely involved in herbicide metabolism) from *Pseudomonas fluorescens* into the soybean's chloroplast DNA (plastome). A custom vector, designated *pCLT323*, facilitated the insertion of gene cassettes flanked by specific plastome sequences. To achieve expression, the bacterial gene was positioned under the influence of regulatory components derived from the tobacco chloroplast. These elements encompassed 16S *rDNA* promoter and the ribosome binding site sourced from the bacteriophage T7 gene 10. Soybean cotyledon-derived embryogenic calli were bombarded with these engineered vectors. Employing Southern blot analysis, researchers verified the successful integration and inheritance of the foreign gene in the first generation (T1) of transplastomic plants, signifying homoplasmy. These transplastomic soybeans displayed considerably higher levels of the introduced protein in their leaves and exhibited enhanced tolerance to herbicides compared to unmodified plants, while maintaining a normal phenotype.

9.7 ADVANTAGES OF CHLOROPLAST ENGINEERING

Chloroplast engineering has emerged as a compelling alternative to nuclear transformation due to several beneficial features. The high accumulation of foreign proteins, ranging from 5% to 25% of the total soluble protein, is facilitated by the polyploid characteristics of the plastid genome and the stability it offers to transgenes. Additionally, expression is often tissue-specific, offering greater control. Plastids possess a well-defined set of promoters and expression cassettes enabling researchers to finely regulate transgene expression. The natural polycistronic translation mechanism of the plastid genetic system enables the simultaneous introduction of multiple advantageous traits through co-transformation, a process known as "transgene stacking" (Krichevsky et al., 2010). Unlike nuclear DNA, plastid DNA does not possess a complex chromatin structure. This eliminates unwanted position effects often associated with random transgene integration in the nucleus. Moreover, homologous recombination ensures the generation of plants with a single type of transplastome. Plastid genes are not susceptible to DNA methylation and epigenetic gene silencing processes such as co-suppression, which are often obstacles in nuclear transformation. Recent developments in methods for recycling or removing selectable markers have also effectively tackled biosafety issues linked with GM crops. The stringent maternal inheritance of plastid genes in most agricultural plants helps mitigate environmental risks (Huang et al., 2003). Although isolated cases of plastid-to-nucleus gene transfer and pollen-mediated transmission to wild relatives have been observed, plastids generally provide better containment of transgenes compared to nuclear transformation. Chloroplast transformation offers a powerful tool for plant biotechnology, enabling the development of transgenic plants with desirable traits. These traits exhibit greater tolerance towards abiotic and biotic stresses and produce valuable products such as improved plant nutritional value, production of vaccines and biopharmaceuticals. Advantages include enhanced control, efficiency, and safety over traditional nuclear transformation methods for researchers.

9.8 CONCLUSION

To fulfill agricultural and pharmaceutical, it is crucial for biotechnologists to develop new crop varieties that meet the required production and other demands without impacting the environment. Chloroplast engineering, growing over genetic nuclear engineering, is becoming a more effective and promising tool for developing new crop varieties. This method is becoming an attractive tool for biotechnologists. A key limitation of chloroplast engineering is the insufficient knowledge or methods of the regulatory mechanisms controlling gene expression in chloroplasts. Further research is needed to understand chloroplast regulatory mechanisms and their interaction with the nucleus, which are crucial for enhancing foreign protein expression and stability, thus advancing the development of new plant varieties.

ABBREVIATIONS

PEG	polyethylene glycol
RFLP	restriction fragment length polymorphism
SWNTs	single-walled carbon nanotubes
UTR	untranslated regions
YFP	yellow fluorescent protein

REFERENCES

Adem, M., Beyene, D. and Feyissa, T., 2017. Recent achievements obtained by chloroplast transformation. *Plant Methods* 13: 1–11.

Apel, W. and Bock, R., 2009. Enhancement of carotenoid biosynthesis in transplastomic tomatoes by induced lycopene-to-provitamin A conversion. *Plant Physiology* 151: 59–66.

Asif, M.H., Mantri, S.S., Sharma, A., Srivastava, A., Trivedi, I., Gupta, P., Mohanty, C.S., Sawant, S.V. and Tuli, R., 2010. Complete sequence and organisation of the Jatropha curcas (*Euphorbiaceae*) chloroplast genome. *Tree Genetics & Genomes* 6: 941–952.

Bendich, A.J., 1987. Why do chloroplasts and mitochondria contain so many copies of their genome?. *BioEssays* 6: 279–282.

Body, M.J., Dave, D.F., Coffman, C.M., Paret, T.Y., Koo, A.J., Cocroft, R.B. and Appel, H.M., 2019. Use of yellow fluorescent protein fluorescence to track OPR3 expression in Arabidopsis thaliana responses to insect herbivory. *Frontiers in Plant Science* 10: 1586.

Boffey, S.A. and Leech, R.M., 1982. Chloroplast DNA levels and the control of chloroplast division in light-grown wheat leaves. *Plant Physiology* 69: 1387–1391.

Boynton, J.E., Gillham, N.W., Harris, E.H., Hosler, J.P., Johnson, A.M., Jones, A.R., Randolph-Anderson, B.L., Robertson, D., Klein, T.M., Shark, K.B. and Sanford, J.C., 1988. Chloroplast transformation in Chlamydomonas with high velocity microprojectiles. *Science* 240: 1534–1538.

Carrer, H., Hockenberry, T.N., Svab, Z. and Maliga, P., 1993. Kanamycin resistance as a selectable marker for plastid transformation in tobacco. *Molecular and General Genetics MGG* 241: 49–56.

Carter, A.P., Clemons, W.M., Brodersen, D.E., Morgan-Warren, R.J., Wimberly, B.T. and Ramakrishnan, V., 2000. Functional insights from the structure of the 30S ribosomal subunit and its interactions with antibiotics. *Nature* 407: 340–348.

Cheng, L., Li, H.P., Qu, B., Huang, T., Tu, J.X., Fu, T.D. and Liao, Y.C., 2010. Chloroplast transformation of rapeseed (Brassica napus) by particle bombardment of cotyledons. *Plant Cell Reports* 29: 371–381.

Cosa, B.D., Moar, W., Lee, S.B., Miller, M. and Daniell, H., 2001. Overexpression of the Bt cry2Aa2 operon in chloroplasts leads to formation of insecticidal crystals. *Nature Biotechnology* 19: 71–74.

Cui, Y., Qin, S. and Jiang, P., 2014. Chloroplast transformation of Platymonas (*Tetraselmis*) subcordiformis with the bar gene as selectable marker. *PloS One* 9: 98607.

Daniell, H., Jin, S., Zhu, X.G., Gitzendanner, M.A., Soltis, D.E. and Soltis, P.S., 2021. Green giant—a tiny chloroplast genome with mighty power to produce high-value proteins: History and phylogeny. *Plant Biotechnology Journal* 19: 430–447.

Daniell, H., Khan, M.S. and Allison, L., 2002. Milestones in chloroplast genetic engineering: An environmentally friendly era in biotechnology. *Trends in Plant Science* 7: 84–91.

Day, A. and Goldschmidt-Clermont, M., 2011. The chloroplast transformation toolbox: Selectable markers and marker removal. *Plant Biotechnology Journal* 9(5): 540–553.

Diekmann, K., Hodkinson, T.R., Fricke, E. and Barth, S., 2008. An optimized chloroplast DNA extraction protocol for grasses (*Poaceae*) proves suitable for whole plastid genome sequencing and SNP detection. *PloS One* 3: 2813.

Dufourmantel, N., Pelissier, B., Garcon, F., Peltier, G., Ferullo, J.M. and Tissot, G., 2004. Generation of fertile transplastomic soybean. *Plant Molecular Biology* 55: 479–489.

Dufourmantel, N., Tissot, G., Goutorbe, F., Garcon, F., Muhr, C., Jansens, S., Pelissier, B., Peltier, G. and Dubald, M., 2005. Generation and analysis of soybean plastid transformants expressing Bacillus thuringiensis Cry1Ab protoxin. *Plant Molecular Biology* 58: 659–668.

Fischer, N., Stampacchia, O., Redding, K. and Rochaix, J.D., 1996. Selectable marker recycling in the chloroplast. *Molecular and General Genetics MGG* 251: 373–380.

Gibberd, M.R., Turner, N.C. and Storey, R., 2002. Influence of saline irrigation on growth, ion accumulation and partitioning, and leaf gas exchange of carrot (Daucus carota L.). *Annals of Botany* 90(6): 715–724.

Giraldo, J.P., Landry, M.P., Faltermeier, S.M., McNicholas, T.P., Iverson, N.M., Boghossian, A.A., Reuel, N.F., Hilmer, A.J., Sen, F., Brew, J.A. and Strano, M.S., 2014. Plant nanobionics approach to augment photosynthesis and biochemical sensing. *Nature Materials* 13(4): 400–408.

Golczyk, H., Greiner, S., Wanner, G., Weihe, A., Bock, R., Börner, T. and Herrmann, R.G., 2014. Chloroplast DNA in mature and senescing leaves: A reappraisal. *The Plant Cell* 26: 847–854.

Goldschmidt-Clermont, M., 1991. Transgenic expression of aminoglycoside adenine transferase in the chloroplast: A selectable marker for site-directed transformation of *Chlamydomonas*. *Nucleic Acids Research* 19: 4083–4089.

Gosal, S.S., Wani, S.H. and Kang, M.S., 2009. Biotechnology and drought tolerance. *Journal of Crop Improvement* 23: 19–54.

Grevich, J.J. and Daniell, H., 2005. Chloroplast genetic engineering: Recent advances and future perspectives. *Critical Reviews in Plant Sciences* 24(2): 83–107.

Huang, C.Y., Ayliffe, M.A. and Timmis, J.N., 2003. Direct measurement of the transfer rate of chloroplast DNA into the nucleus. *Nature* 422: 72–76.

Khan, F., 2012. Morphological characterization of in vitro salt tolerant cell lines and regenerated plants of Daucus carota var. sativus. *Pakistan Journal of Science* 64: 2.

Klein, R.R. and Mullet, J.E., 1987. Control of gene expression during higher plant chloroplast biogenesis. Protein synthesis and transcript levels of psbA, psaA-psaB, and rbcL in dark-grown and illuminated barley seedlings. *Journal of Biological Chemistry* 262(9): 4341–4348.

Ko, S.M., Yoo, B.H., Lim, J.M., Oh, K.H., Liu, J.I., Kim, S.W., Liu, J.R., Choi, K.S. and Yoon, E.S., 2009. Production of fibrinolytic enzyme in plastid-transformed tobacco plants. *Plant Molecular Biology Reporter* 27: 448–453.

Krichevsky, A., Meyers, B., Vainstein, A., Maliga, P. and Citovsky, V., 2010. Autoluminescent plants. *PloS One* 5: 15461.

Kumar, A.U. and Ling, A.P.K., 2021. Gene introduction approaches in chloroplast transformation and its applications. *Journal of Genetic Engineering and Biotechnology* 19: 1–10.

Kumar, S., Dhingra, A. and Daniell, H., 2004. Plastid-expressed betaine aldehyde dehydrogenase gene in carrot cultured cells, roots, and leaves confers enhanced salt tolerance. *Plant Physiology* 136: 2843–2854.

Kwak, S.Y., Lew, T.T.S., Sweeney, C.J., Koman, V.B., Wong, M.H., Bohmert-Tatarev, K., Snell, K.D., Seo, J.S., Chua, N.H. and Strano, M.S., 2019. Chloroplast-selective gene delivery and expression in planta using chitosan-complexed single-walled carbon nanotube carriers. *Nature Nanotechnology* 14: 447–455.

Lelivelt, C.L., McCabe, M.S., Newell, C.A., desnoo, C.B., Van Dun, K.M., Birch-Machin, I., Gray, J.C., Mills, K.H. and Nugent, J.M., 2005. Stable plastid transformation in lettuce (*Lactuca sativa* L.). *Plant Molecular Biology* 58: 763–774.

Lee, S.M., Kang, K., Chung, H., Yoo, S.H., Xu, X.M., Lee, S.B., Cheong, J.J., Daniell, H. and Kim, M., 2006. Plastid transformation in the monocotyledonous cereal crop, rice (Oryza sativa) and transmission of transgenes to their progeny. *Molecules and Cells* 21: 401–410.

Lièvre, K., Tran, T.L., Doerper, S., Hehn, A., Lacoste, P., Thomasset, B., Bourgaud, F. and Gontier, E. 2009. Agrobacterium-mediated transformation of Ruta graveolens L. *Methods in Molecular Biology* 547: 235–248. doi: 10.1007/978-1-60327-287-2_19. PMID: 19521849.

Lin, M.T., Occhialini, A., Andralojc, P.J., Parry, M.A. and Hanson, M.R., 2014. A faster Rubisco with potential to increase photosynthesis in crops. *Nature* 513: 547–550.

Liò, P. and Goldman, N., 1998. Models of molecular evolution and phylogeny. *Genome Research* 8: 1233–1244.

Liu, C.W., Lin, C.C., Chen, J.J. and Tseng, M.J., 2007. Stable chloroplast transformation in cabbage (*Brassica oleracea* L. var. capitata L.) by particle bombardment. *Plant Cell Reports* 26: 1733–1744.

Low, L.Y., Yang, S.K., Kok, D.X.A., Ong-Abdullah, J., Tan, N.P. and Lai, K.S., 2018. Transgenic plants: Gene constructs, vector and transformation method. *New Visions in Plant Science* 19: 41.

Maliga, P., 1993. Towards plastid transformation in flowering plants. *Trends in Biotechnology* 11: 101–107.

Maliga, P., 2001. Plastid engineering bears fruit. *Nature Biotechnology* 19: 826–827.

Maliga, P., 2002. Engineering the plastid genome of higher plants. *Current Opinion in Plant Biology* 5: 164–172.

Newell, C.A., Birch-Machin, I., Hibberd, J.M. and Gray, J.C., 2003. Expression of green fluorescent protein from bacterial and plastid promoters in tobacco chloroplasts. *Transgenic Research* 12: 631–634.

Nguyen, T.T., Nugent, G., Cardi, T. and Dix, P.J., 2005. Generation of homoplasmic plastid transformants of a commercial cultivar of potato (*Solanum tuberosum* L.). *Plant Science* 168: 1495–1500.

Nickelsen, J. and Kück, U., 2000. The unicellular green alga *Chlamydomonas reinhardtii* as an experimental system to study chloroplast RNA metabolism. *Naturwissenschaften* 87: 97–107.

Occhialini, A., Piatek, A.A., Pfotenhauer, A.C., Frazier, T.P., Stewart Jr, C.N. and Lenaghan, S.C., 2019. MoChlo: A versatile, modular cloning toolbox for chloroplast biotechnology. *Plant Physiology* 179: 943–957.

Okumura, S., Sawada, M., Park, Y.W., Hayashi, T., Shimamura, M., Takase, H. and Tomizawa, K.I., 2006. Transformation of poplar (Populus alba) plastids and expression of foreign proteins in tree chloroplasts. *Transgenic Research* 15: 637–646.

Rathinasabapathi, B., Fouad, W.M. and Sigua, C.A., 2001. β-Alanine betaine synthesis in the Plumbaginaceae. Purification and characterization of a trifunctional, S-adenosyl-L-methionine-dependent N-methyltransferase from Limonium latifolium leaves. *Plant Physiology* 126(3): 1241–1249.

Richardson, L.G. and Schnell, D.J., 2020. Origins, function, and regulation of the TOC–TIC general protein import machinery of plastids. *Journal of Experimental Botany* 71(4): 1226–1238.

Ruf, S., Hermann, M., Berger, I.J., Carrer, H. and Bock, R., 2001. Stable genetic transformation of tomato plastids and expression of a foreign protein in fruit. *Nature Biotechnology* 19: 870–875.

Ruhlman, T., Lee, S.B., Jansen, R.K., Hostetler, J.B., Tallon, L.J., Town, C.D. and Daniell, H., 2006. Complete plastid genome sequence of Daucus carota: Implications for biotechnology and phylogeny of angiosperms. *BMC Genomics* 7: 1–13.

Santana, I., Wu, H., Hu, P. and Giraldo, J.P., 2020. Targeted delivery of nanomaterials with chemical cargoes in plants enabled by a biorecognition motif. *Nature Communications* 11: 2045.

Shao, M., Kumar, S. and Thomson, J.G., 2014. Precise excision of plastid DNA by the large serine recombinase Bxb1. *Plant Biotechnology Journal* 12: 322–329.

Sharma, A., Sharma, A., Kashyap, P., Dhyani, P., Pant, M., 2023. Wheat straw waste utilization for nanoparticles synthesis and their various applications. In: Husen, A. (ed.) *Nanomaterials from agricultural and horticultural products. Smart nanomaterials technology*. Springer, Singapore. https://doi.org/10.1007/978-981-99-3435-5_8

Sidorov, V.A., Kasten, D., Pang, S.Z., Hajdukiewicz, P.T., Staub, J.M. and Nehra, N.S., 1999. Stable chloroplast transformation in potato: Use of green fluorescent protein as a plastid marker. *The Plant Journal* 19: 209–216.

Sikdar, S.R., Serino, G., Chaudhuri, S. and Maliga, P., 1998. Plastid transformation in Arabidopsis thaliana. *Plant Cell Reports* 18: 20–24.

Singh, A.K., Verma, S.S. and Bansal, K.C., 2010. Plastid transformation in eggplant (*Solanum melongena* L.). *Transgenic Research* 19: 113–119.

Sváb, Z. and Maliga, P.A.L., 1993. High-frequency plastid transformation in tobacco by selection for a chimeric aadA gene. *Proceedings of the National Academy of Sciences* 90: 913–917.

Tangphatsornruang, S., Sangsrakru, D., Chanprasert, J., Uthaipaisanwong, P., Yoocha, T., Jomchai, N. and Tragoonrung, S., 2010. The chloroplast genome sequence of mungbean (*Vigna radiata*) determined by high-throughput pyrosequencing: Structural organization and phylogenetic relationships. *DNA Research* 17: 11–22.

Tanwar, N., Arya, S.S., Rookes, J.E., Cahill, D.M., Lenka, S.K. and Bansal, K.C., 2023. Prospects of chloroplast metabolic engineering for developing nutrient-dense food crops. *Critical Reviews in Biotechnology* 43: 1001–1018.

Umate, P., 2010. Mulberry improvements via plastid transformation and tissue culture engineering. *Plant Signaling & Behavior* 5: 785–787.

Verhounig, A., Karcher, D. and Bock, R., 2010. Inducible gene expression from the plastid genome by a synthetic riboswitch. *Proceedings of the National Academy of Sciences* 107: 6204–6209.

Vivek, B.S., Ngo, Q.A. and Simon, P.W., 1999. Evidence for maternal inheritance of the chloroplast genome in cultivated carrot (*Daucus carota* L. ssp. sativus). *Theoretical and Applied Genetics* 98: 669–672.

Wu, H., Tito, N. and Giraldo, J.P., 2017. Anionic cerium oxide nanoparticles protect plant photosynthesis from abiotic stress by scavenging reactive oxygen species. *ACS Nano* 11(11): 11283–11297.

Wurbs, D., Ruf, S. and Bock, R., 2007. Contained metabolic engineering in tomatoes by expression of carotenoid biosynthesis genes from the plastid genome. *The Plant Journal* 49: 276–288.

Zhou, F., Badillo-Corona, J.A., Karcher, D., Gonzalez-Rabade, N., Piepenburg, K., Borchers, A.M.I., Maloney, A.P., Kavanagh, T.A., Gray, J.C. and Bock, R., 2008. High-level expression of human immunodeficiency virus antigens from the tobacco and tomato plastid genomes. *Plant Biotechnology Journal* 6: 897–913.

Zoubenko, O.V., Allison, L.A., Svab, Z. and Maliga, P., 1994. Efficient targeting of foreign genes into the tobacco plastid genome. *Nucleic Acids Research* 22(19): 3819–3824.

Zubko, M.K., Zubko, E.I., Zuilen, K.V., Meyer, P. and Day, A., 2004. Stable transformation of petunia plastids. *Transgenic Research* 13: 523–530.

10 DNA Barcoding of Plants Used for Herbal Drug Preparation

*Bhaskar Pal, Rajdip Goswami, Madhurima Kotal, Biswajit Basu, Sourav Ghosh and Saikat Santra**

*Corresponding Author

10.1 INTRODUCTION

DNA barcoding has transformed herbal medicine sector by ensuring authenticity and quality of drugs. Traditional healing systems, which rely on herbal medicine, and with the worldwide herbal market growing, concerns regarding adulteration, misidentification, and quality control have intensified (Gao et al., 2010). DNA barcoding, a technique that sequences DNA to accurately identify species, offers a solution by allowing for precise identification of plant species (Lal et al., 2024). This method is more reliable and scalable compared to traditional methods (Ali and Khan, 2009). Plant authentication Plant identification DNA barcoding Herbal drugs Fingerprinting (Figure 10.1) illustrates the common plant products and their challenges in herbal drug preparation, which can be addressed through molecular identification tool like DNA Barcoding.

The process of DNA barcoding involves sequencing a small, standardized fragment from previously sequenced regions in the DNA and identifying species compared with an online reference database. The use of DNA barcoding offers several advantages over traditional methods (Table 10.4) such as morphological identification and chemical profiling, including accuracy, reliability, and scalability (Han et al., 2016). The key basis of the DNA bar-coding approach is choosing a uniform DNA segment, typically called a "barcode," that shows constant variation in different species and uniformity within a particular species. Some bar-coding regions used for plants include the matK gene and the rbcL gene, as well as barcodes created by including intergenic spacer regions from the chloroplast genome. Such barcodes are "molecular fingerprints" that help differentiate between closely related plant species. Thus, DNA bar-coding has been efficiently used in herbal drug authentication at all stages of the supply chain, starting from plant materials to finished products.

At the pre-harvest stage, DNA bar-coding can work for authenticating plant species in the wild or cultivated settings, ensuring the use of correct botanical sources (Hollingsworth et al., 2011). During the developing process, DNA bar-coding enables the verification of plant ingredients, detection of adulterants, and monitoring of product quality and consistency. Moreover, DNA barcoding can facilitate regulatory compliance by providing robust scientific evidence for product labelling and quality standards. Numerous case studies illustrate the effectiveness of DNA barcoding in herbal drug authentication. A research investigation, on items sold in the United States found adulteration and replacement where DNA barcoding revealed inconsistencies between stated contents and real components. Another instance saw DNA barcoding crucial in detecting plant species in teas emphasizing the significance of precise species identification, for consumer wellbeing.

Despite its promise the broad application of DNA barcoding in verifying drugs encounters various obstacles (Palhares et al., 2015). To ensure the authenticity and safety of medicine it is essential to establish protocols reference databases and bioinformatics tools. Additionally, incorporating DNA barcoding into frameworks and industry practices is crucial. The evolving nature of plant genomes, genetic introgression and hybridization adds complexity to species identification through DNA barcoding. Overcoming these obstacles will require cooperation, among scientists, regulators,

 DOI: 10.1201/9781003482659-10

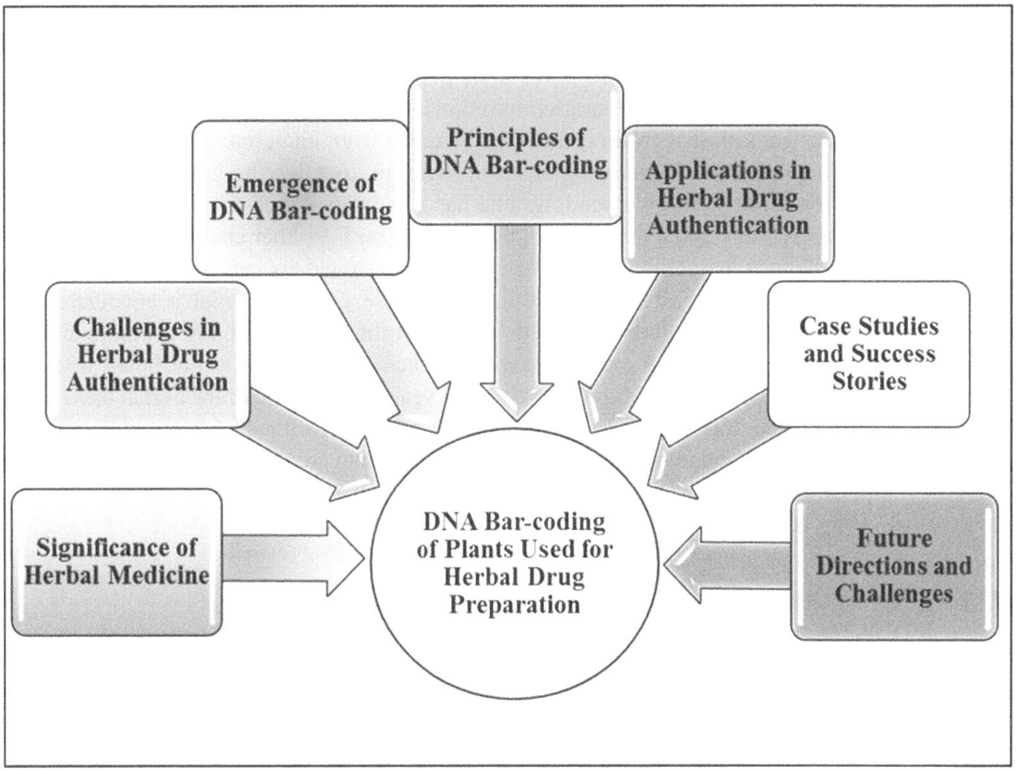

FIGURE 10.1 Overview on DNA bar-coding of plants used for herbal drug.

industry players and policymakers to leverage the advantages of DNA barcoding in safeguarding medicine. In essence DNA barcoding signifies a change in how we authenticate plants utilized in making treatments. Through utilizing the advancements of genetics DNA barcoding provides an adaptable answer to the difficulties of identifying species, adulteration and ensuring quality control in the field of herbal medicine. As this sector progresses DNA barcoding shows potential for improving openness, confidence and sustainability in the market, ultimately benefiting both consumers and those involved in the industry (Parveen et al., 2016).

10.2 BASICS OF DNA BARCODING OF PLANTS USED FOR HERBAL DRUG AUTHENTICATION

DNA barcoding for plants heralds a revolutionizing approach – one that guarantees the authenticity, quality and safety of herbal drugs. This technique delves into molecular genetics to pin down plant species' identities spot-on by reading short, standardized segments of their DNA sequences, a facility and a reference to enormous reference bases, at least (Pretorius et al., 2017). Biochemical integrity of the DNA barcode is made up of the whole and every chain of the herbal drugs. At the pre-harvest level, it can ascertain what species of the plant is – whether it is in the wild type or grown – before including correct botanical inputs into its products; it also inspects plant parts for any contaminants and sets the essential grade and quality of the products. DNA barcoding provides scientific evidence strong enough to ensure regulatory compliance, which promotes transparency in labelling and quality standards; all this is done without any compromise to any level of the supply chain with regards to its applications (Techen et al., 2014; Yu et al., 2021).

The adoption of DNA barcoding technology in the field of herbal medicine on a large scale is no easy task. Strong protocol standardization, well-built reference databases, and availability of bioinformatics tools are the keystones that have to be put in place if this innovative approach is to work effectively. Regulation and incorporation into the corresponding frameworks and day-to-day industrial operations is a task that requires a collective effort from numerous scientists, regulators, industries, and policymakers in various regions. Furthermore, to solve the issues associated with species identification through DNA barcoding, one has to consider such factors as genome variability, genetic introgression, and hybridization, thus dealing with a rather challenging problem.

Despite the challenges, barcoding has proactively demonstrated its capacity to guarantee the safety and wholesomeness of herbal medicine amid all these challenges that it encounters. For instance, research carried out on herbal products has shown that they are highly contaminated and substituted in the market with DNA barcodes exposing such commodities as not being what they advertise or misleading. DNA barcoding was useful in revealing that some of the herbal teas contain toxic plant species, hence stressing on the correctness of species identification to avoid harming the consumers. The promotion and implementation of DNA barcoding are envisaged as being beneficial for improving credibility and the sustainability of the international herbal market among consumers and investors. (Zhu et al., 2022).

10.3 DNA MARKERS SELECTION CRITERIA FOR HERBAL DRUG INVESTIGATION

In selecting DNA markers for investigations of herbal drugs the following points are considered in a bid to obtain species identification that is accurate and inertial to the plant species of interest. These are genetic heterogeneity, interbreeding, non-substitutability and the possibilities of increasing the concentration of and sequencing these flows. Firstly, genetic variability has to be considered because the DNA markers require sufficient differentiation in genetic sequences between various species of plants. Such variation ensures that those markers will effectively distinguish the plants that are genetically closely related – very important when affirming the results. The other criterion is universality since the DNA markers that have been chosen should also be present in as many plant taxonomical divisions as possible. This is useful in establishing that no matter the classifications of the plants that people for instance use in traditional medicine, the selected markers can be employed on many of the plants. Functionality is equated with the capacity of DNA markers to be utilized on the specific plant genotype, in a sense that they can be successfully amplified and sequenced. As a result, the selected markers should indicate high PCR intensification success rates; and the obtained data should be quite clear and easy to interpret. In the context of this work, the application is made blunt and efficient for the identification of species using DNA markers of herbal drugs. The flexibility of amplification and sequencing of the DNA work is also important in the use of DNA markers in the herbal medicine practice. The selected markers should be easy to extend and the basic DNA sequencing protocols which are to be used to analyze these markers should not be tedious. Because of this, the processes that are needed in the DNA barcoding become simpler in the laboratory and few appendages and/or knowledge are needed. Hence, genetic variability or polymorphism, prevalence in both exotic and standard populations, utility, as well as the applicability of the PCR and sequencing techniques should be given consideration when selecting the DNA markers to be used in the study of herbal drug preparation. Thus, meeting these criteria, DNA markers can serve as proofs of plant species and insurance of the quality and non-hazardous nature of herbal supplements.

The key aspects of selection criteria for DNA markers can be summarized as below

Genetic Variability- DNA markers ought to be of sufficient level of polymorphism to enable clear distinction on different plant species (Palhares et al., 2015).

Universality- Markers must be universal. To be able to use markers in various herbal species it is mandatory to conserve markers at the multiple taxonomic levels of plants (Tnah et al., 2019).

Functionality- Guidelines have been enlisted regarding the marker functionality criteria for specificity, the plant species, PCR efficiency, and result distinguishability(Gao et al., 2010).

Ease of Amplification and Sequencing- The markers to be used should pose no issue being retained by the common DNA sequencing and are characterized by relatively simple amplification processes (Hollingsworth et al., 2011).

10.4 METHODOLOGICAL/ TECHNOLOGICAL ADVANCEMENTS IN DNA BARCODING

Recent times have seen major improvements in methods and technology for DNA barcoding of herbal plants, making it more accurate, efficient, and practical. Key developments are:

Conventional DNA barcodes - In a typical approach, these are selected on the basis of being generally highly conserved within the species of interest, but sufficiently variable between species to enable discrimination. Such regions are developed for identification of particular species or groups of species. The genes matK (maturase K) and rbcL (ribulose-1,5-bisphosphate carboxylase/oxygenase large subunit) for plants, COI (Cytochrome c oxidase subunit I) for animals, and the ITS (Internal Transcribed Spacer) region for fungi are examples of conventional DNA barcodes. These markers are widely applied and helpful in fields including forensic science, herbal product authenticity, and biodiversity evaluation (Chen et al., 2010).

Next-Generation Sequencing - This breakthrough also made way for high-throughput DNA barcoding of many samples simultaneously, as well as improved efficiency in building comprehensive reference databases and facilitated large-scale biodiversity studies in the field of herbal plant (Techen et al., 2014).

Mini-barcodes and Mini-multiplexing - Mini-barcodes are specific markers for species identification, they require a small amount of DNA. For instance, the compatibility between these mini-multiplexing techniques and simultaneous amplification and sequencing of many mini-barcodes per reaction make them ideal for efficient and reagent saving protocols during DNA barcoding (Tnah et al., 2019).

Metabarcoding - Metabarcoding is combining high-throughput sequencing with DNA barcoding to unravel complex assemblages of mixed environmental DNA samples. In this regard, metabarcoding allows plants diverse species in herbal products to be identified; it also assists in the detection of adulterants or contaminants that provide information on product composition and quality control (Yu et al., 2021).

Machine Learning and Bioinformatics Tools - Bioinformatics tools and machine learning algorithms have been in use to optimize plant growth (Pant et al., 2023). Such algorithms can be created to improve the accuracy of species identification, develop predictive models and automate data analysis in DNA barcoding. With their help, sequencing data can be better interpreted, unknown samples classified and DNA barcoding workflows made more robust (Pretorius et al., 2017).

Portable and Field-Deployable Technologies - The diminutive size and portability of DNA sequencing instruments now allow for species identification in resource-limited field situations. Regulators, producers, and researchers can deploy portable technology at various points throughout the supply chain to undertake real-time verification and quality control on herbal plants (Han et al., 2016).

Integration with Traditional Knowledge Systems - DNA barcoding can be integrated with traditional knowledge systems for the holistic authentication of herbal medicine. Combining molecular data with indigenous knowledge about plant name, identification and medicinal use patterns makes species identification more accurate and culturally meaningful in the preparation of herbal drugs (Hollingsworth et al., 2011).

These advances in methodology and technology have effectively boosted the capability of DNA barcoding for herbal plant medications, providing stronger assurance of the safety, authenticity, and quality control of herbal medicine(Parveen et al., 2016; Chen et al., 2010; Hollingsworth et al., 2011).

10.5 WORKFLOW OF DNA BARCODING

The process of DNA barcoding involves several stages of DNA extraction and further analysis.

(i) Sample Collection: The proper sampling for DNA barcoding species of herbal plants is a step-by-step process ensuring high-quality DNA to be used by taxonomists for species identification (Han et al., 2016). The selection of sampling sites is determined in such a way that it covers a geographical area and the ecological zone and elevation of the target plant species (Mishra et al., 2016). A reliable and correct system of plant identification is established by the application of taxonomic keys or the help of a professional. Plant samples are preferably taken each year at the beginning of the season of their flower or fruit with the use of neat, sharp tools, the use of which ensures the growth of better-quality specimens (Figure10.2). More than a single sample per species is taken to cover the wide range of the genetic variations detected (Techen et al., 2014) . This diversity is needed to ensure that there is the barcode library that can be used in biological and ecological projects (Kress and Erickson, 2008). The detailed records include the recording of GPS coordinates, date, type of habitat, and the name of collector, etc., and must be strictly followed for development of barcode libraries (Raja et al., 2015; Parveen et al., 2016). Proper packaging ensures high-quality samples for DNA barcoding (Howard et al., 2020). Specimens must be obtained from confirmed and legit sources; this could be in consultation with relevant bodies such as museums, herbariums or accredited research institutes. These organizations provide neatly documented samples that ensure the credibility of the DNA samples. All the DNA samples are associated with a museum or herbarium

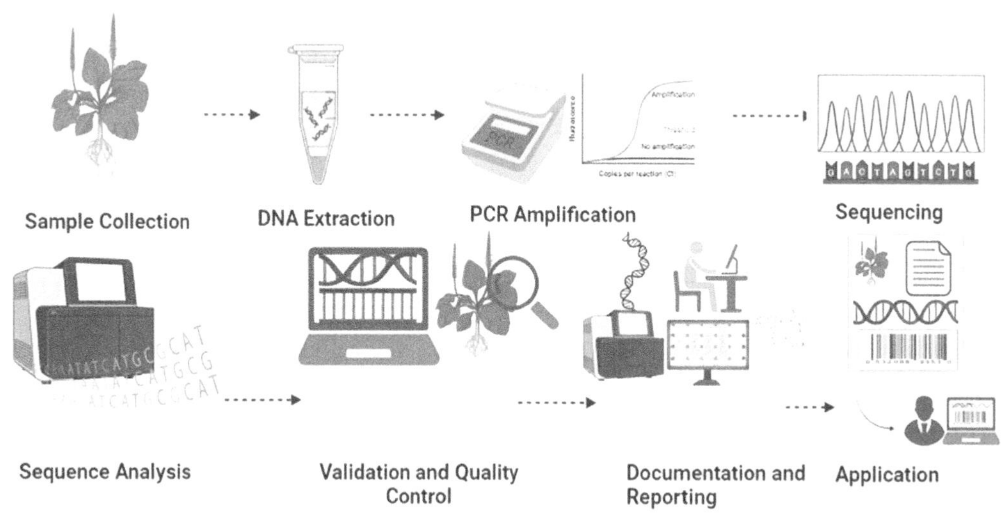

FIGURE 10.2 Illustration of DNA bar-coding for herbal drug authentication.

physical sample stock called voucher, which can be reviewed at any time to assess the identity of the DNA sample contained in the barcode library (Kress, 2017; Hebert et al., 2003; Gostel and Kress, 2022). This way, all possible aspects of a particular specimen's context are described in detail, contributing to the construction of a rich metadata framework for the barcode library (Mishra et al., 2016). After that, samples are processed in a laboratory setting where they are washed, dried and kept into storage in a manner that reduces the degradation and or contamination that can affect the samples. The latter suggests that laboratory processes have be very regulated in order to maintain control over the specimen, in this case the DNA. The metadata provide what was mentioned earlier, the necessary context to the DNA sequences, which greatly enhances the usage of the barcode library (Raja et al., 2015).

(ii) DNA Extraction: The process of DNA separation for medicinal plant species furnishes a number of vital steps needed to obtain the best quality DNA for the barcoding application. Apart from others, only healthy, clean, and ground plant materials, which are ground in liquid nitrogen to destroy the cell walls can be used. Lysis buffer is the means of DNA release and then we will get rid of the DNA by eliminating the contaminants that stick to it (Liu et al., 2015). DNA was precipitated using a solvent, cleaned, and put into a buffer for further use (Mendonça-Filho, 2006). Quality and quantity are assessed using spectrophotometry and gel electrophoresis. Various methods are used for DNA extraction (Abubakar et al., 2017). The Cetyltrimethylammonium Bromide (CTAB) method is effective for plants with high polysaccharide content but is time-consuming (Table 10.1). Silica column-based kits are fast and convenient but expensive (Zhang et al., 2012). The Chelex resin method is simple and quick but yields less DNA. Phenol-chloroform extraction provides high purity but uses hazardous chemicals. SDS-based methods are effective for difficult tissues but require careful handling to avoid DNA shearing (Sahu et al., 2012). Standard techniques ensure the uniformity and credibility among samples (Al'Khafaji et al., 2018).

(iii) PCR Amplification: The PCR (Polymerase Chain Reaction) amplification process is essential for DNA barcoding of herbal plant species (Table 10.2). Initially, the preparation of the PCR mix involves using a small amount of extracted DNA as the template, adding specific forward and

TABLE 10.1
DNA Extraction Methods for Plants

Method	Principle	Advantages	Disadvantages	References
CTAB Method	Uses CTAB buffer to lyse cells and precipitate polysaccharides	Effective for plants with high polysaccharide content	Time-consuming, multiple steps	(Tnah et al., 2019)
Silica Column-Based Kits	DNA binds to silica column in the presence of chaotropic salts	Fast, convenient, and high purity	Expensive, limited capacity	(Huynh et al. 2017
Chelex Resin Method	Uses chelating resin to remove metal ions that inhibit PCR	Simple, quick, and inexpensive	Not suitable for all plant types, lower DNA yield	(Singh, Kumari, and Iyengar 2018)
Phenol-Chloroform Extraction	Organic solvents separate DNA from proteins and other contaminants	High purity and yield	Hazardous chemicals, labor-intensive	(Tan and Yiap, 2009)
SDS-Based Method	Uses SDS detergent to lyse cells and release DNA	Effective for difficult tissues	Requires careful handling of chemicals, potential DNA shearing	(Natarajan et al., 2016)

TABLE 10.2
PCR Methods for DNA Amplification

Method	Principle	Advantages	Disadvantages	References
Conventional PCR	Amplifies DNA using thermal cycling	Widely used, simple, and cost-effective	Limited sensitivity, endpoint analysis only	(Wang et al., 2023)
Real-Time PCR (qPCR)	Quantifies DNA in real-time using fluorescent markers	High sensitivity, quantitative, real-time monitoring	Expensive equipment, requires fluorescent probes	(Watzinger et al., 2006)
Touchdown PCR	Gradually decreases annealing temperature	Increases specificity and yield	More complex optimization	(Korbie and Mattick, 2008)
Nested PCR	Uses two sets of primers in two successive PCRs	Increases specificity, reduces non-specific binding	Labor-intensive, risk of contamination	(Maheaswari, Kshirsagar, and Lavanya, 2016)
Multiplex PCR	Amplifies multiple targets in one reaction	Time-saving, efficient	Complex optimization, primer interactions	(Elnifro et al., 2000)
High-Fidelity PCR	Uses high-fidelity DNA polymerases	High accuracy, reduces error rates	Expensive, slower reaction times	(Ishino and Ishino, 2014)

reverse primers targeting the barcode region, including a mix of deoxynucleotide triphosphates (dNTPs), and using an appropriate PCR buffer for enzyme activity. Additionally, a heat-stable DNA polymerase (Taq polymerase) and MgC12 are optimized for enzyme activity (Ngai et al., 2023). The PCR cycling conditions start with an initial denaturation at 94–95°C to separate the DNA strands. This is followed by repeated cycles of denaturation at 94–95°C for 30–60 seconds, annealing at 50–65°C to allow primers to bind to the DNA template, and extension at 72°C for Taq polymerase to synthesize new DNA strands. The final extension step at 72°C for 5–10 minutes ensures complete DNA synthesis, with the cycle steps repeated for 25–35 cycles. After amplification, PCR product verification is conducted by running the PCR products on an agarose gel for gel electrophoresis. A UV transilluminator is used to see the DNA bands in order to verify successful amplification traceability (Hyder et al., 2024).

(iv) Sequencing: Sequencing is one of the most vital elements for the DNA barcoding technique, in which the nucleotide order of DNA fragments helps to identify the species correctly. This is can be accomplished by techniques such as Sanger sequencing which entails chain-termination and capillary electrophoresis to derive high-accuracy sequences of individual genes (Satam et al., 2023). On the other hand, new technologies like illumina use NGS with high throughput by synthesising new DNA and detecting incorporated nucleotides based on fluorescence signals. Compared to WGS, NGS is less expensive for large scale studies although it does call for special equipment and bioinformatics interpretation. The obtained sequencing data can also be compared to reference databases to corroborate the species identification and therefore plays the crucial role of confirming the identity of herbal plant species (Stranneheim and Lundeberg, 2012). Particular barcoding regions, like COI, rbcL and matK are boosted by PCR. Sequencing technologies like SS or NGS are then used to produce sequence data with high accuracy (Baym et al., 2016).

(v) Sequence Analysis: Sequence analysis entails the use of bioinformatics to align the derived DNA sequences with a reference species and obtain species identification. In this method, molecular

TABLE 10.3
Bioinformatics Tools for Sequence Analysis

Method	Description	Advantages	Disadvantages	References
BLAST (Basic Local Alignment Search Tool)	Compares query sequences to sequences in public databases using algorithms to identify homologous sequences.	Fast, widely used, user-friendly interface.	Limited sensitivity for detecting distant homologs.	(Saeed and Usman, 2019)
Clustalw/Muscle	Multiple sequence alignment tools that align multiple sequences to identify conserved regions and variations.	Suitable for aligning large datasets, scalable.	Computationally intensive for large datasets.	(Edgar, 2004)
Phylogenetic Analysis	Constructs phylogenetic trees to visualize evolutionary relationships between sequences.	Provides insights into evolutionary history.	Requires computational resources and expertise.	(Pavlopoulos et al., 2010)
Hidden Markov Models (HMMs)	Statistical models used to identify sequence patterns and motifs, aiding in functional annotation.	Highly sensitive, effective for motif detection.	Complex to construct and interpret.	(Yoon, 2009)

biology, bioinformatics tools, software and packages are used for the analysis of sequence data, sequence alignment and degree of homology. Sequence analysis may involve using tools such as BLAST, when the new sequence will be aligned to other databases such as GenBank or BOLD in search of a match with sequences that have been identified before (Saeed and Usman, 2019; Hyder et al., 2024). Sequencing data may also be used to predict the phylogenetic tree of the various species using the various databases, as shown in (Table 10.3).

(vi) Alignment and Comparison: Sequences obtained through sequencing methods are aligned to reference sequences to identify similarities and differences. This step ensures accurate comparison and identification. Aligned sequences are compared using bioinformatics tools to assess sequence similarity and divergence. This helps in determining species' identity and detecting any genetic variations. Processes like sequence trimming, error correction, and validation against known sequences are used to develop stability and credibility of the data. This process reduces error and increases the benefits of the barcode library (Johnson et al., 2023; Raclariu-Manolică et al., 2023).

(vii) Annotation: Sequences are annotated with relevant information such as species name, geographic origin, and any known genetic markers. This metadata enhances the utility of sequences for species identification and biodiversity studies (Pearson, 2013).

(viii) Validation and Quality Control: Quality assurance plays a significant role in DNA barcoding where validation and quality control acts as a backbone for the processes. QC's are done at each stage to reduce possible mistakes and to have good levels of data accuracy. Validation means making sure that the methods being used and the results that are obtained are stable and can be replicated.

(ix) Replication: This replicability of an experiment means that it is done several times independently and produces similar data. This entails re-extraction of DNA, PCR, sequencing and data analysis of the same samples over and over again. In all cases, proven samples are run through the whole analytical procedure to check the precision and reliability of the data obtained (Kiran et al., 2010)

(x) Cross-Validation: This is required to avoid the presence of discrepancies in species identification. It can be done using various approaches like using the results of DNA barcoding and conventional morphological identification, or applying different barcode regions.

(xi) **Database development:** The development of powerful database is necessary for storing and assessing DNA barcode data. Different databases like BOLD and ameliorate data supervision enable wide interaction in barcoading research (Hebert et al., 2003; Galimberti et al., 2019).The data of DNA barcoding and other data such as species names, collection process, and topographical information are used to induce accuracy and completeness. Proper organization and explication of data increases its usefulness for different applications (de Boer et al., 2015). Easy interfaces and searching efficacy helps in data recovery, evaluation and utilization by researchers. The public accessibility of the data can induce clarity and interaction. Open access increase the data sharing, research reproducibility, and innovation in barcoading research (Nakazato and Jinbo, 2022; Pennisi, 2019).

10.6 CHECKS FOR QUALITY CONTROL

Quality Control encompasses various checks and procedures throughout the DNA barcoding process. These involve appropriate sample collection to avoid contamination (Mishra et al., 2016), DNA purity check, PCR efficiency, sequencing read length, base calling quality and the number of coverage reads (Lanubile et al., 2024), efficient bioinformatics molecular protocols for sequence data, and comparative examination of results to check, whether they correspond to data available in the open literature, databases, and concerning pertinent knowledge. These standards guarantee that adequate and precise species identification is developed hence boosting the reputation of DNA barcoding in numerous sectors (Droege et al., 2016). Barcode datasets are amended with new techniques and related metadata as additional species are barcoded. Continuous amending can increase the accuracy of the library, reflecting recent advancements in biodiversity (Niemann et al., 2022; Hoban et al., 2022). Quality control audits can supervise the data accuracy, credibility and authenticity of libraries regularly (Gostel and Kress, 2022).

10.7 REGULATORY PERSPECTIVES ON DNA BARCODING

An important aspect to foster the establishment of good structures for the developmental noble legislations and worldwide standardizations of several affairs including the certification of herbal remedies; the following regulatory measures have been deliberated.

Standards and Guidelines: Establishing some guidelines or policy on the handling of data, Polymerase Chain Reaction (PCR), DNA extraction, sequencing, and samples.These criteria have been described by Dormontt et al. (2018) as a way to ensure that the standardization and reliability of results is produced between the laboratories. Specific guidelines tailored for sectors like herbal drug authentication, recommendtarget DNA regions such as rbcL and matK for plants (Huang et al., 2014).

Legal Frameworks and Regulations: Most of the emphasis should be laid on the development of the legal norms governing the conditions for the application of DNA barcoding. This would help in confirming the authenticity of the herbal medicine and also, the recording of food fraud (Fu et al., 2011).

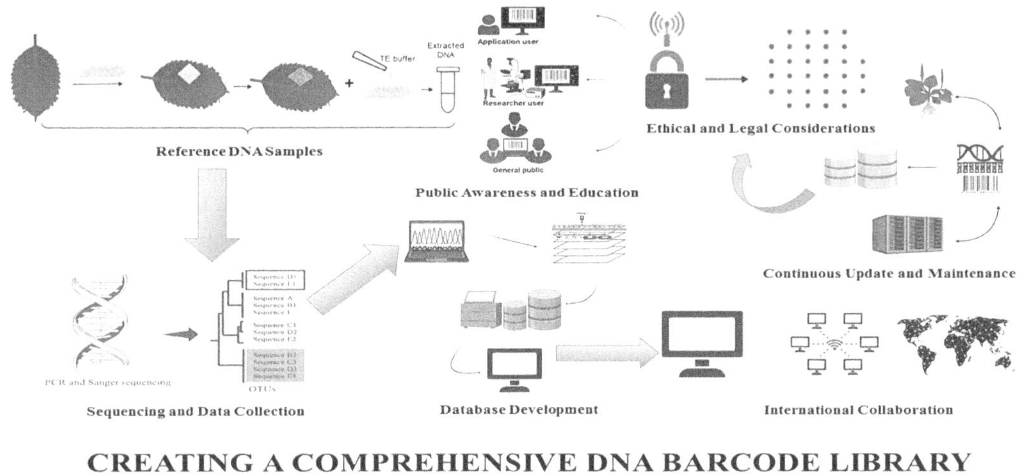

FIGURE 10.3 DNA barcode library for herbal drugs.

Compliance and Validation: The labs are expected to fit regulatory standards to show that they are right, accurate, efficient, and special. It is suggested that labs should take part in proficiency tests and lab directors should not engage in varying from laid down protocols (Mao et al., 2014).

*Global Harmonization:*The functioning of standards and the practices of DNA barcoding is done according to the provisions in the CBD and the cooperation is done through the WHO. An example of the rationale for harmonized regulations is enhanced trade liberalization via the standard DNA barcoding policies that would ensure mutual recognition of tests' results (Huang et al., 2015).

Public Awareness and Education: Besides providing identification data in biochemical and morphological characteristics, emphasis is also on methods that can be used by the populace through the release of the DNA barcode data to sites and servers including Barcode of Life Data Systems (BOLD) (David et al., 2024).

10.8 INTERNATIONAL COLLABORATION

International collaborations are used to standardize DNA barcoding data, assuring uniformity and likeliness of data among regions and studies. Combined techniques can trigger the data merging and support large-scale evaluation (Odah and Odah, 2023). Collaboration of researchers and institutions ameliorates the alteration of the data of DNA barcoding (Figure 10.3). iBOL can increase data sharing and combined techniques can increase the extensiveness of barcode libraries (Dormontt et al., 2018; Droege et al., 2016).

10.9 ETHICAL AND LEGAL CONSIDERATIONS

The collection of genetical data requires IC of individuals during specimen collection. Clear communication and agreement are required to produce ethical practices and respect for individuals right (Deliversky and Deliverska, 2018; Tiwari, 2024). It is also necessary to maintain data privacy, which helps to prevent spreading of sensitive information, like genetical data, which is used to protect the privacy of individuals (Andrews et al., 1994). The increasing of public awareness about DNA barcoding schemes by social news, educational program and other process is important to prevent malpractice, which aims to make people aware about the consequences of DNA barcoding in biodiversity conservation, species recognition, and ecological innovation. Demonstrating the success rates and evidence-based development can help to know the usefulness of DNA barcoding (Septem et al., 2023). Citizen science projects need to be emphasized for DNA barcoding. Schemes like BioBlitz events, where the workers assist to collect specimens for DNA barcoding, enables individuals to actively participate in biodiversity research and innovation. Creating different social media platforms such as webinars and efficient websites can assisting in making the data of DNA barcoding publicly available. This process offers individuals the opportunity to learn about DNA barcoding data and global biodiversity, and to recognize the significance of genetic research data (Castro et al., 2016). The reports from engaged policymakers can also helps to apply the technological knowledge into practical work (Campanaro et al., 2019; De Boer et al., 2017).

10.10 CHALLENGES AND LIMITATIONS OF DNA BARCODING

The development of quality and effective raw herbal drugs still remains a challenge in herbal therapy (Wang et al., 2023; Wachtel-Galor and Benzie, 2011). The production of high-quality and efficacious herbal drugs under precise natural conditions and ecological background is a long-term practice that is helpful to accurately identify what forms the basis for the effective use of potential herbal medicines (Ashraf et al., 2020). In this regards DNA barcoding has become a considerable approach in pharmacognosy to ensure high accuracy and efficiency. However, there are few challenges in herbal drug authentication, which need to be assessed for further growth and development of herbal products in clinical practice (Srivastava and Misra, 2018).

Integrity of Genomic DNA: The integrity of the DNA matrix of herbal substance is the primary factor regulating amplification efficiency and identification. The processing and storage criteria also have potential impact on the integrity (Ngai et al., 2023; Lin et al., 2019). The DNA of herbal products, can damage the integrity of DNA and affects the success rate of molecular marker amplification (Ngai et al., 2022); due to this reason molecular identification is not potentially applied. Additionally, during application of processing methods, such as boiling, steaming, and drying, in the chemical constituents, the DNA integrity is also affected (Lo and Shaw, 2018; Li et al., 2023).

DNA degradation: The molecular authentication can efficiently recognize the original herbs from the counterfeit one. The quality of genomic DNA is a major factor involved in molecular authentication (Ji et al., 2022; Li et al., 2023). Therefore, degradation of DNA matrix during the production processing of herbal medication can make it tough for their molecular authentication. Faller et al. (2019) reported that the fermentation that occurs during the storage condition can also be the reason for the degradation of DNA, leading to expression of shorter fragments which are not suitable for amplification by conventional PCR. Additionally, the inhibitory and interfering property of phenolic components might be responsible for DNA degradation (Li et al., 2015; Lo, Li and Shaw, 2015; Ni and Zhao, 2018). However, Li et al. (2023) proposed a model for genomic DNA degradation of herbal substances and an evaluation process for herbal drugs, which may help to optimize the systemic production process of herbal drugs.

Complex herbal mixtures: The herbal extracts are the complex mixture of a number of constituents, thus it is not properly authenticated during the formulation of drugs (Ansari, Islam, and Sameem, 2012). However, it is necessary to isolate the toxic substances and the narrow therapeutic indexed constituents from the herbal extract during processing of herbal drugs. Additionally, the presence of multiple compounds in the herbal formulation can induce the adverse events of the patients with other complications, and also hamper the stability of the formulation (Bommakanti et al., 2023). Therefore, to improve the safety, effectiveness and formulation quality of herbal drugs isolation of therapeutically active ingredients is necessary (Burapan et al., 2020; Klimek-Szczykutowicz et al., 2020).

Commercial challenges: The limited commercialization of herbal medicine is another major challenge that needs to be overcome. Lack of herbal expertise, mainly for traditional drugs, in the globe has resulted in restricted practice of herbal medicines (Zhang et al., 2012; Ekor, 2013; Kroes, 2014). Thus, to establish worldwide recognition and enhance commercialisation of traditional herbal medicines, thus supporting, nourishing, and strengthening the herbal technologist globally (Sahoo, Manchikanti, and Dey, 2010; Fong, 2002). Furthermore, improving the quality, safety and efficacy of herbal product can improve the marketing of these drugs. The evidence-based trials on human are become increasingly necessary to establish the safety and efficacy of herbal drugs in the domestic and export commercialization; more attention needs to be paid on scientific and technological advancement in this area (Chan, 2003; Im et al., 2023; Raclariu-Manolică and Socaciu, 2023; Mück et al., 2024; Miao et al., 2024).

Intraspecific Variation: It is essential to understand the reasons of evolution by which genetic polymorphisms in different species are altered into genetic divergence between genus and genomic diversity, by assessing the quantity and patterns of nucleotides variation between species (Li et al., 2009; Luikart et al., 2003). The assessing of pathways such as mutation, selection, population structure, and recombination is getting affected by diversities (Korneliussen, Albrechtsen, and Nielsen, 2014). In the DNA barcoding identification of *Chenopodiummurale* similar results are shown. The specimen was assessed as *C. murale* employing BLAST for both *matK* and *rbcL* genes with 100% series matching. Utilizing BOLD, *rbcL* gene shows a high series of analogy to various taxa, like *Chenopodiumambrosioides*, *Chenopodium album*, and *Chenopodiumficifolium*, extending from 96.3% to 100%. The DNA barcode differentiate various species, with no overlap between intraspecific and interspecific disparity (Noor and Bennett, 2009; Tajima, 1989; Ho et al., 2021; Srivastava et al., 2022). Moreover, the efficacy of any DNA barcoding is assessed by the stage of intraspecific and interspecific disparity in a single locus or a stand of loci. The DNA barcoding in few plants is considered an appropriate tool to determine the exact identification of the particular species and/or differentiate the species from other populations of the same family (Hou et al., 2018; Presgraves et al., 2003). For instance, a number of synonym have been found in *Selinum* and the uncertainty in species with different synonyms. *Selinum* species are also differentiated morphologically, because most of the species are identified by different names by different authors, instead of a single name. However, the morphological hallmark alone is inadequate to exactly define the species. In intraspecific variations, anatomical, biochemical and molecular markers are also necessary than gross morphological features (Ellegren, 2014).

Database Completeness: Database completeness is a crucial aspect of herbal drug authentication, yet remains a significant challenge due to the lack of comprehensive databases, incomplete information, variability in formulations, and the emergence of new products in the market (Aruwa and Sabiu, 2024). To address this, stakeholders should collaborate to establish and maintain centralized databases with up-to-date information on herbal drugs. Regular updates, standardization of data collection methods, and increased collaboration can enhance authentication processes and promote consumer safety and confidence in herbal products (Fathifar et al., 2023).

Cost and Accessibility: DNA barcoding is an accurate method that can be utilized in a generic molecular biology, which provides potential background for assessing a variety of genes and thereby improving existing crop differences. However, data obtaining by DNA barcodes require a well-equipped advanced molecular laboratory and is time-consuming, and cost effective. The utilization of short, slandered DNA motif to recognize the species has generally altered, to determine, estimate, and monitor biodiversity within all ecosystems. Initiation of DNA barcoding has markedly contributed in species identification and the constituents of biological communities in the world. Therefore, to obtaining this information through DNA barcodes from specimens, needs many complex steps and thus time and money (Kress and Erickson, 2008). The complex large scale DNA barcoding processes are moving toward activation, but to be able to establish the comprehensive barcode library will ultimately need the acquisition of 100 million barcodes. To fulfil this need, analytical facilities should adopt tools that can assist the rapid, cost-effective assembly of DNA barcodes, in order to identify unique genetic variance. However, the taxonomic recognition accounts for a considerable portion of cost related to bio assessment programs across the US. New analytical tools including DNA barcoding have now promoted as a way to minimize monitoring costs with maximum efficiency, but yet this assumption has not been thoroughly evaluated (Savolainen et al., 2005).

Despite these challenges, DNA barcoding has been effectively used in the authentication of herbal drugs (Table 10.4).

TABLE 10.4
Applications of DNA Bar-coding in Herbal Drug Authentication

Application	Description	Methodology	Advantages	Reference
Accurate Species Identification	DNA barcoding allows precise identification of plant species in herbal drugs by comparing DNA sequences to reference databases, ensuring correct species use.	PCR amplification of specific DNA regions (e.g., ITS, rbcL, matK), sequencing, comparison with reference sequences in DNA databases	Provides rapid and accurate species identification, even in processed or mixed samples.	(Jossan and Singh, 2022)
Detection of Adulteration and Contamination	Detects unwanted species, including contaminants and adulterants, ensuring the safety and quality of herbal products.	DNA extraction, PCR amplification, sequencing, comparison with authentic reference sequences	Helps identify harmful or allergenic substances, ensuring consumer safety.	(Ichim, 2019)
Quality Control and Standardization	Ensures consistency and efficacy of herbal drugs by verifying plant species in each batch, maintaining therapeutic quality.	High-throughput sequencing of multiple samples, comparison with reference sequences, statistical analysis	Enables large-scale screening of herbal products, facilitating batch-to-batch consistency.	(Noviana, Indrayanto, and Rohman, 2022)
Regulatory Compliance	Provides a reliable method for meeting regulatory standards, helping manufacturers avoid legal issues related to mislabeling or adulteration.	DNA extraction, PCR amplification, sequencing, comparison with authenticated reference sequences	Facilitates compliance with regulations, reducing the risk of product recalls and legal penalties.	(Heubl, 2010)
Protection of Endangered Species	Monitors and controls trade of endangered plant species, supporting conservation efforts by verifying species identity in herbal products.	DNA barcoding combined with traditional taxonomy, phylogenetic analysis	Helps prevent exploitation of endangered species, contributing to biodiversity conservation.	(Mishra et al., 2017)

TABLE 10.4 (Continued)
Applications of DNA Bar-coding in Herbal Drug Authentication

Application	Description	Methodology	Advantages	Reference
Consumer Safety and Confidence	Enhances consumer safety and trust by ensuring the authenticity and purity of herbal drugs through accurate species identification.	Metabarcoding, DNA sequencing, bioinformatics analysis	Builds consumer confidence by providing transparent and reliable information about product contents.	(DawanS and Ahn, 2022)
Authentication of Traditional Medicine Ingredients	Verifies the authenticity of plant materials used in traditional medicine, preserving cultural heritage and ensuring the use of correct ingredients.	DNA extraction, PCR amplification, sequencing, comparison with authenticated reference sequences	Helps maintain the integrity and effectiveness of traditional remedies.	(Hyder et al., 2024)
Geographic Origin Verification	Determines the geographic origin of herbal materials, providing valuable information for product authentication, trade regulation, and quality assurance.	DNA stable isotope analysis (DNA-SIA), geographic information systems (GIS)	Helps preserve product integrity and supports geographical indications (GIs) protection.	(Safhi et al., 2023)
Pharmacological and Therapeutic Efficacy Assessment	Supports the correlation of DNA barcode data with pharmacological properties and therapeutic efficacy of herbal drugs, aiding in evidence-based medicine.	DNA barcoding combined with pharmacological assays, clinical trials	Facilitates the development of evidence-based treatments and personalized medicine approaches.	(M. Li et al. 2011)

10.11 FUTURE DIRECTIONS IN DNA BARCODING FOR HERBAL DRUG AUTHENTICATION

Very recently, successful integration of DNA barcoding with gold nanoparticle-based assays for sensitive detection of herbal drug adulterants has been shown to enhance specificity and sensitivity, allowing for the identification of trace amounts of DNA in complex herbal formulations (Lanubile et al., 2024). Advances in single-molecule sequencing technologies, such as Nanopore sequencing for direct, long-read DNA sequencing without PCR amplification, offers more extensive and precise barcode sequences for the authentication of herbal products (Dorey and Howorka, 2024).The eDNA barcoding can also be used to authenticate the medicinal plant from amongst a number of samples (Kress, 2017). The practicability of generating synthetic DNA barcodes to trigger the credibility and specificity of species recognition in DNA barcoding has also been suggested (Zhang et al., 2024). The accommodation of DNA barcoding with block chain techniques for conformation of quality and authenticity throughout the formulation and distribution process has also been reported (Agrawal et al., 2021). The ML techniques are being increasingly used to identify and authenticate herbal drugs (Riza et al., 2023).

10.12 CONCLUSION

The establishment of quality plant-based therapies and their regulatory signalling pathway has become a primary concern for Indian drug producers. Scientists are trying to meet the safety and effectiveness of herbal medicine in order to induce the therapeutic practice of these herbal agents. However, the impaired growth of the herbal medicinal industry due to less efficacious products, lack of research, reduced marketing and finally, the development of herbal industries is hindered by deficiencies in key areas. DNA barcoding methods can identify and authenticate herbal formulations, reducing the adverse events on consumers. The utilization of advanced R&D techniques is also necessary to maintain the quality, effectiveness and novelty of herbal formulations. The marketing management of herbal drugs can induce the awareness of these drugs among individuals. Moreover, developing human resource and educational research can improve the knowledge of industry professionals with advanced techniques for herbal drug developments. All these factors may increase the potentiality of herbal manufacturing plants to produce standard quality products.

medicine, Authentication, Quality control, Adulterants, Regulatory standards

ABBREVIATIONS

SS:	Sanger sequencing
NGS:	Next-Generation Sequencing
PCR:	Polymerase Chain Reaction
ML:	Machine learning
DCA:	Drugs and Cosmetics Act
QC:	Quality control
QA:	Quality assurance
TCM:	Traditional Chinese Medicine

REFERENCES

Abubakar MB, Salleh FM, Omar MSS, Wagiran A. 2017. DNA barcoding and chromatography fingerprints for the authentication of botanicals in herbal medicinal products. *Evidence-Based Complementary and Alternative Medicine*. 2017(1):1352948.

Agrawal, T. K., Kumar, V., Pal, R., Wang, L., Chen, Y. 2021. Blockchain-based framework for supply chain traceability: A case example of textile and clothing industry. *Computers and Industrial Engineering* 154(April).

Al'Khafaji AM, Deatherage D, Brock A. 2018. Control of lineage-specific gene expression by functionalized gRNA barcodes. *ACS Synthetic Biology*. 7(10):2468–74.

Ali Z, Khan IA. 2009. Proceedings of the 8th Annual Oxford International Conference on the Science of Botanicals. April 6–9, 2009, University of Mississippi, University, MS, USA. Edited by Ikhlas Khan (Chairman and Issue Editor), Troy Smillie, and Shabana Khan, (pp. 399–457).

Andrews, L.B., Fullarton, J.E., Holtzman, N.A., Motulsky, A.G. 1994. Assessing Genetic Risks: Implications for Health and Social Policy. In: Institute of Medicine (US) Committee on Assessing Genetic Risks. editors. Washington (DC): National Academies Press (US)

Ansari SH, Islam F, Sameem M. 2012. Influence of nanotechnology on herbal drugs: A review. *Journal of Advanced Pharmaceutical Technology & Research*. 3(3):142.

Aruwa CE, Sabiu S. 2024. Big data application in herbal medicine: The need for a consolidated database. In Izah, S.C., Ogwu, M.C., Akram, M. (Eds.), *Herbal Medicine Phytochemistry: Applications and Trends* (pp. 1713–38). Springer International Publishing.

Ashraf SA, Elkhalifa AEO, Siddiqui AJ, Patel M, Awadelkareem AM, Snoussi M, Ashraf MS, Adnan M, Hadi S. 2020. Cordycepin for health and wellbeing: A potent bioactive metabolite of an entomopathogenic medicinal fungus Cordyceps with its nutraceutical and therapeutic potential. *Molecules*. 25(12): 2735.

Baym M, Shaket L, Anzai IA, Adesina O, Barstow B. 2016. Rapid construction of a whole-genome transposon insertion collection for Shewanella oneidensis by knockout Sudoku. *Nature Communications*. 7(1):13270.

Bommakanti V, Puthenparambil Ajikumar A, Maria S, Prakash G, Mundanat AS, Ahmad F, Haque S, Prieto MA, Rana SS. 2023. An overview of herbal nutraceuticals, their extraction, formulation, therapeutic effects and potential toxicity. *Separations*. 10(3):177.

Burapan S, Kim M, Paisooksantivatana Y, Eser BE, Han J. 2020. Thai curcuma species: Antioxidant and bioactive compounds. *Foods*. 9(9): 1219.

Campanaro A, Tommasi N, Guzzetti L, Galimberti A, Bruni I, Labra M. 2019. DNA barcoding to promote social awareness and identity of neglected, underutilized plant species having valuable nutritional properties. *Food Research International*. 115(January):1–9.

Castro P, Azeiteiro UM, Bacelar-Nicolau P, Leal Filho W, Azul AM. (Eds.). 2016. *Biodiversity and Education for Sustainable Development*. Springer.

Chan K. 2003. Some aspects of toxic contaminants in herbal medicines. *Chemosphere*. 52(9):1361–71.

Chen S, Yao H, Han J, Liu C, Song J, Shi L, Zhu Y, et al. 2010. Validation of the ITS2 region as a novel DNA barcode for identifying medicinal plant species. *PLoS ONE*. 5(1):1–8.

Das K. 2019. Authentic identification and new drug discovery from natural plant-based constituents through DNA bar-coding: A challenging task to the researchers. *Annals of Phytomedicine: An International Journal*. 8(2):19–27.

David A, Arul Priya JD, Gautam A. 2024. DNA sequencing technologies and DNA barcoding. *Methods in Molecular Biology (Clifton, N.J.)*. 2744:139–54.

Dawan S, Jirapat, Ahn J. 2022. Application of DNA barcoding for ensuring food safety and quality. *Food Science and Biotechnology*. 31(11):1355–64.

De Boer HJ, Ghorbani A, Manzanilla V, Raclariu AC, Kreziou A, Ounjai S, Osathanunkul M, Gravendeel B. 2017. DNA metabarcoding of orchid-derived products reveals widespread illegal orchid trade. *Proceedings of the Royal Society B: Biological Sciences*. 284(1863): 20171182.

De Boer HJ, Ichim MC, Newmaster SG. 2015. DNA barcoding and pharmacovigilance of herbal medicines. *Drug Safety*. 38(7):611–20.

Deliversky J, Deliverska M. 2018. Ethical and legal considerations in biometric data usage—Bulgarian perspective. *Frontiers in Public Health*. 6(February):25.

Dorey A, Howorka S. 2024. Nanopore DNA sequencing technologies and their applications towards single-molecule proteomics. *Nature Chemistry*. 16(3):314–34.

Dormontt EE, van Dijk KJ, Bell KL, Biffin E, Breed MF, Byrne M, Caddy-Retalic S, et al. 2018. Advancing DNA barcoding and metabarcoding applications for plants requires systematic analysis of herbarium collections—An Australian perspective. *Frontiers in Ecology and Evolution*. 6(SEP):406014.

Droege G, Barker K, Seberg O, Coddington J, Benson E, Berendsohn WG, Bunk B, et al. 2016. The global genome biodiversity network (GGBN) data standard specification. *Database: The Journal of Biological Databases and Curation*. 2016:1–11.

Edgar RC. 2004. MUSCLE: A multiple sequence alignment method with reduced time and space complexity. *BMC Bioinformatics*. 5(August):113.

Ekor M. 2013. The growing use of herbal medicines: Issues relating to adverse reactions and challenges in monitoring safety. *Frontiers in Pharmacology*. 4: 177.

Ellegren H. 2014. Genome sequencing and population genomics in non-model organisms. *Trends in Ecology & Evolution*. 29(1):51–63.

Elnifro, E. M., Ashshi, A.M., Cooper, R.J., and Klapper, P.E. 2000. Multiplex PCR: Optimization and application in diagnostic virology. *Clinical Microbiology Reviews*. 13(4): 559. https://doi.org/10.1128/CMR.13.4.559-570.2000

Faller AC, Ragupathy S, Shanmughanandhan D, Zhang Y, Lu Z, Chang P, Swanson G, Newmaster SG. 2019. DNA quality and quantity analysis of Camellia sinensis through processing from fresh leaves to a green tea extract. *Journal of AOAC International*. 102(6):1798–807.

Fathifar Z, Kalankesh LR, Ostadrahimi A, Ferdousi R. 2023. New approaches in developing medicinal herbs databases. *Database*. 2023.

Fong HHS. 2002. Integration of herbal medicine into modern medical practices: Issues and prospects. *Integrative Cancer Therapies*. 1(3):287–93.

Fu YM, Jiang WM, Fu CX. 2011. Identification of species within Tetrastigma (Miq.) Planch. (Vitaceae) based on DNA barcoding techniques. *Journal of Systematics and Evolution*. 49(3):237–45.

Galimberti A, Casiraghi M, Bruni I, Guzzetti L, Cortis P, Berterame NM, Labra M. 2019. From DNA barcoding to personalized nutrition: the evolution of food traceability. *Current Opinion in Food Science*. 28:41–8.

Gao T, Yao H, Song J, Liu C, Zhu Y, Ma X, Pang X, Xu H, Chen S. 2010. Identification of medicinal plants in the family Fabaceae using a potential DNA barcode ITS2. *Journal of Ethnopharmacology*. 130(1):116–21.

Gostel MR, Kress WJ. 2022. The expanding role of DNA barcodes: Indispensable tools for ecology, evolution, and conservation. *Diversity*. 14(3):213.

Guo M, Yuan C, Tao L, Cai Y, Zhang W. 2022. Life barcoded by DNA barcodes. *Conservation Genetics Resources*. 14(4):351–65.

Han J, Pang X, Liao B, Yao H, Song J, Chen S. 2016. An authenticity survey of herbal medicines from markets in China using DNA barcoding. *Scientific Reports*. 6: 18723.

Hebert PDN, Cywinska A, Ball SL, DeWaard JR. 2003. Biological identifications through DNA barcodes. *Proceedings of the Royal Society B: Biological Sciences*. 270(1512):313–21.

Heubl G. 2010. New aspects of DNA-based authentication of Chinese medicinal plants by molecular biological techniques. *Planta Medica*. 76(17):1963–74.

Ho VT, Tran TKP, Vu TTT, Widiarsih S. 2021. Comparison of MatK and RbcL DNA barcodes for genetic classification of jewel orchid accessions in Vietnam. *Journal of Genetic Engineering and Biotechnology*. 19(1):93.

Hoban S, Archer FI, Bertola LD, Bragg JG, Breed MF, Bruford MW, Coleman MA, et al. 2022. Global genetic diversity status and trends: Towards a suite of essential biodiversity variables (EBVs) for genetic composition *Biological Reviews*. 97(4):1511–38.

Hollingsworth PM, Graham SW, Little DP. 2011. Choosing and using a plant DNA barcode. *PLoS ONE*. 6(5): e19254.

Hou Z, Wang Z, Ye Z, Du S, Liu S, Zhang J. 2018. Phylogeographic analyses of a widely distributed Populus davidiana: Further evidence for the existence of glacial refugia of cool-temperate deciduous trees in northern East Asia. *Ecology and Evolution*. *8*(24):13014–26.

Howard C, Lockie-Williams C, Slater A. 2020. Applied barcoding: The practicalities of DNA testing for herbals. *Plants*. 9(9):1–25.

Huang Q, Duan Z, Yang J, Ma X, Zhan R, Xu H, Chen W, Leung FCC. 2014. SNP typing for germplasm identification of Amomum villosum Lour. based on DNA barcoding markers. *PLoS ONE*. 9(12): e114940.

Huang WJ, Li FF, Liu YJ, Long CL. 2015. Identification of Crocus sativus and its adulterants from Chinese markets by using DNA barcoding technique. *Iranian Journal of Biotechnology*. 13(1):36.

Huynh OA, Jankowicz-Cieslak J, Saraye B, Hofinger B, Till BJ. 2017. Low-cost methods for DNA extraction and quantification. *Biotechnologies for Plant Mutation Breeding*. 227–39.

Hyder Z, Rizwani GH, Shareef H, Azhar I, Zehra M. 2024. Authentication of important medicinal herbal species through DNA-based molecular characterization. *Saudi Journal of Biological Sciences*. 31(6):103985.

Ichim MC. 2019. The DNA-based authentication of commercial herbal products reveals their globally widespread adulteration. *Frontiers in Pharmacology*. 10:1227.

Im HB, Ghelman R, Portella CFS, Hwang JH, Choi D, Kunwor SK, Moraes SDTA, Han D. 2023. Assessing the safety and use of medicinal herbs during pregnancy: A cross-sectional study in São Paulo, Brazil. *Frontiers in Pharmacology*. 14:1268185.

Ishino S, Ishino Y. 2014. DNA polymerases as useful reagents for biotechnology – the history of developmental research in the field. *Frontiers in Microbiology*. 5:465.

Ji WJ, Zhang YX, Zhao ZL, Ni LH, Li WT, Zhu CX, Chen X, Yang SH. 2022. DNA barcoding of Gentiana crassicaulis in Lijiang and evaluation of root processing methods. *Yaoxue Xuebao*. 57(2):507–13.

Johnson MS, Venkataram S, Kryazhimskiy S. 2023. Best practices in designing, sequencing, and identifying random DNA barcodes. *Journal of Molecular Evolution*. 91(3):263–80.

Jossan JK, Singh R. 2022. Use of DNA barcoding for plant species identification. In Dash, H.R., Shrivastava, P., Lorente, J.A. (Eds), *Handbook of DNA Profiling* (pp. 911–33). Springer Singapore.

Kiran U, Khan S, Mirza KJ, Ram M, Abdin MZ. 2010. SCAR markers: A potential tool for authentication of herbal drugs. *Fitoterapia*. 81(8):969–76.

Klimek-Szczykutowicz M, Dziurka M, Blažević I, Đulović A, Granica S, Korona-Glowniak I, Ekiert H, Szopa A. 2020. Phytochemical and biological activity studies on Nasturtium Officinale (Watercress) microshoot cultures grown in RITA® temporary immersion systems. *Molecules*. 25(22): 5257.

Korbie DJ, Mattick JS. 2008. Touchdown PCR for Increased Specificity and Sensitivity in PCR Amplification. *Nature Protocols.* 3(9):1452–56.

Korneliussen TS, Albrechtsen A, Nielsen R. 2014. ANGSD: Analysis of next generation sequencing data. *BMC Bioinformatics.* 15(1): 356.

Kress WJ, Erickson DL. 2008. DNA barcodes: Genes, genomics, and bioinformatics. *Proceedings of the National Academy of Sciences of the United States of America.* 105(8):2761.

Kress WJ. 2017. Plant DNA barcodes: Applications today and in the future. *Journal of Systematics and Evolution.* 55(4):291–307.

Kroes BH. 2014. The legal framework governing the quality of (Traditional) herbal medicinal products in the European Union. *Journal of Ethnopharmacology.* 158(PART B):449–53.

Lal A, Pant M, Kumar A, Palni LMS, Singh A, Siddiqui OR, Husen A. 2024. DNA barcoding of epiphytic orchids for species-level identification. *Journal of Horticultural Science and Biotechnology*:1–11. DOI: 10.1080/14620316.2024.2371591

Lanubile A, Stagnati L, Marocco A, Busconi M. 2024. DNA-based techniques to check quality and authenticity of food, feed and medicinal products of plant origin: a review. *Trends in Food Science & Technology.* 149(July):104568.

Li H, Handsaker B, Wysoker A, Fennell T, Ruan J, Homer N, Marth G, Abecasis G, Durbin R. 2009. The sequence alignment/map format and SAM tools. *Bioinformatics.* 25(16):2078–79.

Li M, Cao H, But PPH, Shaw PC. 2011. Identification of herbal medicinal materials using DNA barcodes. *Journal of Systematics and Evolution.* 49(3):271–83.

Li W, Yang S, Ni L, Zhao Z, Xu H. 2023. Determination of the genomic DNA degradation rate of the Chinese herb Gentianae crassicaulis Radix during processing and storage. *Pharmacognosy Magazine.* 19(3):520–29.

Li X, Yang Y, Henry RJ, Rossetto M, Wang Y, Chen S. 2015. Plant DNA barcoding: From gene to genome. *Biological Reviews of the Cambridge Philosophical Society.* 90(1):157–66.

Lin C, Cao SM, Chang ET, Liu Z, Cai Y, Zhang Z, Chen G, et al. 2019. Chinese nonmedicinal herbal diet and risk of nasopharyngeal carcinoma: A population-based case-control study. *Cancer.* 125(24):4462–70.

Liu SH, Chuang WC, Lam W, Jiang Z, Cheng YC. 2015. Safety surveillance of traditional Chinese medicine: Current and future. *Drug Safety.* 38(2):117–28.

Lo YT, Li M, Shaw PC. 2015. Identification of constituent herbs in Ginseng decoctions by DNA markers. *Chinese Medicine (United Kingdom).* 10(1).

Lo YT, Shaw PC. 2018. DNA-based techniques for authentication of processed food and food supplements. *Food Chemistry.* 240(February):767–74.

Luikart G, England PR, Tallmon D, Jordan S, Taberlet P. 2003. The power and promise of population genomics: From genotyping to genome typing. *Nature Reviews Genetics.* 4(12):981–94.

Maheaswari R, Kshirsagar JT, Lavanya N. 2016. Polymerase chain reaction: A molecular diagnostic tool in periodontology. *Journal of Indian Society of Periodontology.* 20(2):128.

Mao YR, Zhang YH, Nakamura K, Guan BC, Qiu YX. 2014. Developing DNA barcodes for species identification in Podophylloideae (Berberidaceae). *Journal of Systematics and Evolution.* 52(4):487–99.

Mendonça-Filho RR. 2006. Bioactive phytocompounds: New approaches in the phytosciences. *Modern Phytomedicine: Turning Medicinal Plants into Drugs*:1–24.

Miao P, Li Y, Fan Q, Ni L, Liu S, Li J, Zhang S. 2024. Research on Chinese medicinal materials cultivation: A bibliometric and visual analysis. *Heliyon.* 10(7): e28637.

Mishra P, Kumar A, Nagireddy A, Mani DN, Shukla AK, Tiwari R, Sundaresan V. 2016. DNA barcoding: An efficient tool to overcome authentication challenges in the herbal market. *Plant Biotechnology Journal.* 14(1):8–21.

Mishra P, Kumar A, Sivaraman G, Shukla AK, Kaliamoorthy R, Slater A, Sundaresan V. 2017. Character-based DNA barcoding for authentication and conservation of IUCN Red Listed threatened species of genus Decalepis (Apocynaceae). *Scientific Reports.* 7(1):1–12.

Mück F, Scotti F, Mauvisseau Q, Thorbek BGL, Wangensteen H, de Boer HJ. 2024. Three-tiered authentication of herbal traditional Chinese medicine ingredients used in women's health provides progressive qualitative and quantitative insight. *Frontiers in Pharmacology.* 15:1353434.

Nakazato T, Jinbo U. 2022. Cross-sectional use of barcode of life data system and GenBank as DNA barcoding databases for the advancement of museomics. *Frontiers in Ecology and Evolution.* 10:966605.

Natarajan VP, Zhang X, Morono Y, Inagaki F, Wang F. 2016. A modified SDS-based DNA extraction method for high quality environmental DNA from seafloor environments. *Frontiers in Microbiology*. 7:986.

Ngai HL, Lee HK, Shaw PC. 2023. DNA from herbs can be obtained from air and authenticated by polymerase chain reaction. *Heliyon*. 9(8): e18946.

Ngai HL, Yang X, Chu AJ, Harper R, Jacobsen ABJE, Lau DTW, Yu HY, Lee HK, Shaw PC. 2022. Multi-methodological approach for the quality assessment of Senecionis Scandentis Herba (Qianliguang) in the herbal market. *PLoS ONE*. 17(4): e18946.

Ni LH, Zhao ZL. 2018. A morphometric comparison of three closely related species of Gentiana (Gentianaceae), endemic to the region of the Qinghai-Tibet Plateau. *Botany*. 96(3):209–15.

Niemann HJ, Bezeng BS, Orton RD, Kabongo RM, Pilusa M, van der Bank M. 2022. Using a DNA barcoding approach to facilitate biosecurity: Identifying invasive alien macrophytes traded within the South African aquarium and pond plant industry. *South African Journal of Botany*. 144:364–76.

Noor MA, Bennett SM. 2009. Islands of speciation or mirages in the desert examining the role of restricted recombination in maintaining species. *Heredity*. 103(6):439–44.

Noviana E, Indrayanto G, Rohman A. 2022. Advances in fingerprint analysis for standardization and quality control of herbal medicines. *Frontiers in Pharmacology*. 13(June): 853023.

Odah MAA, Odah MAA. 2023. Unlocking the genetic code: Exploring the potential of DNA barcoding for biodiversity assessment. *AIMS Molecular Science*. 10(4):263–94.

Palhares RM, Drummond MG, Brasil BSAF, Cosenza GP, Brandão MDGL, Oliveira G. 2015. Medicinal plants recommended by the World Health Organization: DNA barcode identification associated with chemical analyses guarantees their quality. *PLoS ONE*. 10(5):1–29.

Pant M, Bist P, Gusain MP, Pant K. 2023. Effect of pH and gelling agents in nutrient medium on in vitro shoot multiplication of *Swertia chirata* Buch.-Ham ex Wall: An endangered medicinal herb. *Materials Today Proceedings*. 73(1):9–12.

Parveen I, Gafner S, Techen N, Murch SJ, Khan IA. 2016. DNA barcoding for the identification of botanicals in herbal medicine and dietary supplements: Strengths and limitations. *Planta Medica*. 82(14):1225–35.

Pavlopoulos GA, Soldatos TG, Barbosa-Silva A, Schneider R. 2010. A reference guide for tree analysis and visualization. *BioData Mining*. 3(1):1.

Pearson WR. 2013. An introduction to sequence similarity ('homology') searching. *Current Protocols in Bioinformatics*. 42(1):1–3.

Pennisi E. 2019. DNA barcodes jump-start search for new species. *Science*. 364(6444):920–21.

Presgraves DC, Balagopalan L, Abmayr SM, Orr HA. 2003. Adaptive evolution drives divergence of a hybrid inviability gene between two species of Drosophila. *Nature*. 423(6941):715–19.

Pretorius E, Van der Bank M, Viljoen AM. 2017. DNA barcoding detects contamination and substitution in herbal products containing Harpagophytum spp. *South African Journal of Botany*. 109:364–65.

Raclariu-Manolică AC, Mauvisseau Q, de Boer HJ. 2023. Horizon scan of DNA-based methods for quality control and monitoring of herbal preparations. *Frontiers in Pharmacology*. 14: 1179099.

Raclariu-Manolică AC, Socaciu C. 2023. In search of authenticity biomarkers in food supplements containing sea buckthorn: A metabolomics approach. *Foods (Basel, Switzerland)*. 12(24): 4493.

Raja K, Balachandran S, Mohanasundaram S, Ramalingam S. 2015. DNA barcoding: A genomic-based tool for authentication of phytomedicinals and its products. *Botanics: Targets and Therapy*. 5:77–84.

Riza LS, Zain MI, Izzuddin A, Prasetyo Y, Hidayat T, Samah KAFA. 2023. Implementation of machine learning in DNA barcoding for determining the plant family taxonomy. *Heliyon*. 9(10):e20161.

Saeed U, Usman Z. 2019. Biological sequence analysis. In Husi, H. and Nat, S.C. (Eds.), *Computational Biology* (pp. 55–69). Exon Publication.

Safhi FA, Alshamrani SM, Bogmaza AFM, El-Moneim DA. 2023. DNA barcoding of wild plants with potential medicinal properties from Faifa mountains in Saudi Arabia. *Genes*. 14(2): 469.

Sahoo N, Manchikanti P, Dey S. 2010. Herbal drugs: Standards and regulation. *Fitoterapia*. 81(6):462–71.

Sahu SK, Thangaraj M, Kathiresan K. 2012. DNA extraction protocol for plants with high levels of secondary metabolites and polysaccharides without using liquid nitrogen and phenol. *International Scholarly Research Notices*. 2012(1):205049.

Satam H, Joshi K, Mangrolia U, Waghoo S, Zaidi G, Rawool S, Thakare RP, et al. 2023. Next-generation sequencing technology: Current trends and advancements. *Biology*. 12(7):997.

Savolainen V, Cowan RS, Vogler AP, Roderick GK, Lane R. 2005. Towards writing the encyclopaedia of life: An introduction to DNA barcoding. *Philosophical Transactions of the Royal Society B: Biological Sciences.* 360(1462):1805–11.

Senapati A, Basak S, Rangan L. 2021. A review on application of DNA barcoding technology for rapid molecular diagnostics of adulterants in herbal medicine. *Drug Safety.* 45(3):193–213.

Septem RL, Zain MI, Prasetyo MIY, Hidayat T, Samah KAFA. 2023. Implementation of machine learning in DNA barcoding for determining the plant family taxonomy. *Heliyon* 9(10). https://doi.org/10.1016/J.HELIYON.2023.E20161.

Singh UA, Kumari M, Iyengar S. 2018. Method for improving the quality of genomic DNA obtained from minute quantities of tissue and blood samples using Chelex 100 Resin. *Biological Procedures Online.* 20(1).

Srivastava RP, Saxena G, Singh L, Singh A, Verma PC, Kaur G. 2022. Interspecific and intraspecific analysis of Selinum spp. collected from Indian Himalayas using DNA barcoding. *Journal of Genetic Engineering & Biotechnology.* 20(1):63.

Srivastava S, Misra A. 2018. Quality control of herbal drugs: Advancements and challenges. In *New Age Herbals: Resource, Quality and Pharmacognosy* (pp. 189–209).

Stranneheim H, Lundeberg J. 2012. Stepping stones in DNA sequencing. *Biotechnology Journal.* 7(9):1063.

Suriya J, Krishnan M, Bharathiraja S, Sekar V, Sachithanandani V. 2020, August. Implications and utility of DNA barcoding. *DNA Barcoding and Molecular Phylogeny*:3–22.

Tajima F. 1989. Statistical method for testing the neutral mutation hypothesis by DNA polymorphism. *Genetics.* 123(3):585–95.

Tan SC, Beow CY. 2009. DNA, RNA, and protein extraction: The past and the present. *Journal of Biomedicine and Biotechnology.*

Techen N, Parveen I, Pan Z, Khan IA. 2014. DNA barcoding of medicinal plant material for identification. *Current Opinion in Biotechnology.* 25:103–10.

Tiwari P. 2024. Legal and ethical considerations in the use of DNA fingerprinting. *Journal of Scientific Research and Reports.* 30(3):236–42.

Tnah LH, Lee SL, Tan AL, Lee CT, Ng KK S, Ng CH, Nurul Farhanah Z. 2019. DNA barcode database of common herbal plants in the tropics: A resource for herbal product authentication. *Food Control.* 95:318–26.

Wachtel-Galor S, Benzie IFF. 2011. Herbal medicine. In *Herbal Medicine: Biomolecular and Clinical Aspects: Second Edition* (pp. 1–10). CRC Press/Taylor & Francis.

Wang H, Chen Y, Wang L, Liu Q, Yang S, Wang C. 2023. Advancing herbal medicine: Enhancing product quality and safety through robust quality control practices. *Frontiers in Pharmacology.* 14:1265178.

Wang Y, Peng C, Ding L, Su Z, Chen X, Wang X, Sun M, Xu J. 2023. An accurate, rapid and cost-effective method for T-Nos detection based on CRISPR/Cas12a. *Foods* 12(3): 615.

Watzinger F, Ebner K, Lion T. 2006. Detection and monitoring of virus infections by real-time PCR. *Molecular Aspects of Medicine.* 27(2): 254. https://doi.org/10.1016/J.MAM.2005.12.001.

Yoon B-J. 2009. Hidden Markov models and their applications in biological sequence analysis. *Current Genomics.* 10(6):402.

Yu J, Wu X, Liu C, Newmaster S, Ragupathy S, Kress WJ. 2021. Progress in the use of DNA barcodes in the identification and classification of medicinal plants. *Ecotoxicology and Environmental Safety.* 208:111691.

Zhang J, Wider B, Shang H, Li X, Ernst E. 2012. Quality of herbal medicines: Challenges and solutions. *Complementary Therapies in Medicine.* 20(1–2):100–6.

Zhang Z, Melzer ME, Arun KM, Sun H, Eriksson CJ, Fabian I, Shaashua S, Kiani K, Oren Y, Goyal Y. 2024. Synthetic DNA barcodes identify singlets in scRNA-seq datasets and evaluate doublet algorithms. *Cell Genomics.* 0(0):100592.

Zhu S, Liu Q, Qiu S, Dai J, Gao X. 2022. DNA barcoding: An efficient technology to authenticate plant species of traditional Chinese medicine and recent advances. *Chinese Medicine.* 17(1):1–17.

11 Biotechnological Approaches to Enhance Pharmaceutically Significant Metabolites in Medicinally Active Plants

Dhurva Prasad Gauchan, Sarita Subedi, Samrat Panta, Alaka Bhantana, Shreya Pathak, Looza Pradhan and Ramakanta Lamicchane*

*Corresponding author

11.1 INTRODUCTION

Biotechnological methods in plant medicine are innovative and futuristic, utilizing biology to enhance plant healing potentials that would otherwise be impossible. For hundreds of years, traditional medicine and modern pharmaceutical industries have relied on medicinally active plants that produce different secondary metabolites be harnessed as medicine to cure various human diseases . Included among these metabolites are alkaloids, terpenoids, flavonoids, and phenolic compounds which exhibit specific pharmacological effects for the treatment of a wide range of diseases (Collin, 2001;Verpoorte et al., 2002; Pant et al., 2023).

The importance of pharmaceutically relevant metabolites cannot be overemphasized because they are the central structures of many therapeutic agents treating anything from infections to cancer (Bortesi and Fischer, 2015). Apart from being used for medicinal purposes, these metabolites possess great economic value (Mahfouz et al., 2014). These natural products have a huge contribution to the market and support the pharmaceutical industry's sustainable practices. Yet, plants' natural synthesis of such metabolites is beset by several challenges including reduced yield, changing environmental conditions, and overexploitation, all of which may threaten plant diversity (Ajikumar et al., 2010). Genetic engineering, metabolic engineering, and plant tissue culture are the key techniques that can regulate the metabolism pathways, thus augmenting more yield in the metabolite production (Rao and Ravishankar, 2002; Zhao et al., 2005). Precise alterations on plant genomes can be made using genetic manipulation techniques, for example, CRISPR/Cas9 and TALENs, for better production of biosynthetic compounds (Raskin et al., 2002; Cushnie and Lamb, 2005). Likewise, metabolic engineering seeks to identify and tweak crucial pathways to enhance output while callus and suspension culture and organogenesis methods adopted in plant tissue culture provide controlled environments for large-scale production of such medicinally important metabolites (Bais et al., 2002).

The secondary metabolites have a wide range of medicinal uses and benefits. Various biological activities, including antioxidant, anti-inflammatory, antibacterial and anti-cancer properties, are found in compounds such as phenolic acids, alkaloids, flavonoids and terpenoids (Cushnie and Lamb, 2005; Zhao et al., 2005). Quercetin, flavonoid and kaempferol are known for their anti-inflammatory and antioxidant effects, which are effective in controlling or preventing diseases such as cancer (Mahfouz et al., 2014). Alkaloids like quinine, with its analgesic and antimalarial properties, or morphine, known for its analgesic capacity, highlight the wide-ranging applications of these metabolites (Ajikumar et al., 2010).

 DOI: 10.1201/9781003482659-11

To be sustainably harvested through conventional means, these metabolites face environmental factors, genetic influences, and demands for large quantities that often exceed what can be harvested. To meet global needs responsibly, modern biotechnological methods such as genome editing paired with advanced cultivation techniques should be employed (Cushnie and Lamb, 2005). These methods also help to save endangered plant species and guarantee the availability of important metabolites (Bortesi and Fischer, 2015). This chapter provides a review of biotechnological interventions for secondary metabolite production and emphasizes the harmonization of omics technologies and environmentally friendly practices to facilitate scale-up approaches required for commercial viability.

11.2 PHARMACEUTICALLY SIGNIFICANT METABOLITES

Pharmacological active ingredients in modern medicine include flavonoids which exhibit antibacterial, antioxidant and anticancer effects, alkaloids with antibacterial/antifungal/ antiviral and anticancer/anti-tumor effects and terpenoids, which are also of antibacterial, antiviral, anti-inflammatory, and anticancer attributes (Pant et al., 2021). Such compounds are of special economic importance, and it has been estimated that they constitute over 50% of the world store of natural products. More importantly, many of these compounds exhibit much stronger therapeutic potential as compared to their synthetic counterparts (Batra et al., 2022) while also overcoming the risk of drug resistance as in case of allopathic medicine (Negi et al., 2024).The current synthesis of these compounds through biotechnological processes has enhanced their economical and sustainable value. However, drawbacks evident in the synthesis of natural metabolite include genetics and environmental factors, which limit production and quality. By using forms of elicitation, genetic improvement, tissue culture, and genetically modifying organisms, it has been made easier to produce secondary metabolites instead of using the wild source, thus saving the unnecessary destruction of threatened plant species. Despite some challenges of uneven metabolic rate, financial constraints, environmental factors, prediction of the reactions that restrict metabolic routes, and regulatory obstacles, significant advancements have been made in biotechnological approaches for plant secondary metabolite production, as discussed below.

11.3 GENETIC ENGINEERING

Genetic engineering methods have completely transformed the secondary metabolite production by enabling precise genome manipulation, enhancing yield and using a wider range of bioactive compounds. Some of the cloning techniques such as gene cloning, expression vector creation, and CRISPR/Cas9 genome editing allow for the precise alteration of plant and microbial genomes to improve the production of secondary metabolites. This allows for the optimization of metabolic pathways, overexpression of key enzymes, and integration of heterologous pathways, enabling valuable therapeutic compounds for various applications (Chen et al., 2012; Zhang et al., 2021; Zhou et al., 2020). Gene editing techniques such as CRISPR/Cas9, TALENs can be used to enhance metabolite production. In CRISPR/Cas9 two main components are i) Cas9, an endonuclease that acts as a scissor and ii) gRNA, which contains a DNA sequence that binds to a DNA sequence at a specified location in the genome. After the DNA is cut, the specific uses of the denomination mentioned above allow the utilization of the cell's biogenesis to heal the break and to place new healthy genetic material at the cut place or delete an existing gene (Figure 11.1).

TALENs employs a FokI nuclease domain and a DNA-binding domain for making precise and accurate changes to DNA sequences (Osakabe and Osakabe, 2015). Both the gene editing tools can be used to increase the secondary metabolite yield by altering the gene promoters, rewiring the metabolic pathways and increasing the precursor availability. Some of TALENs can upregulate

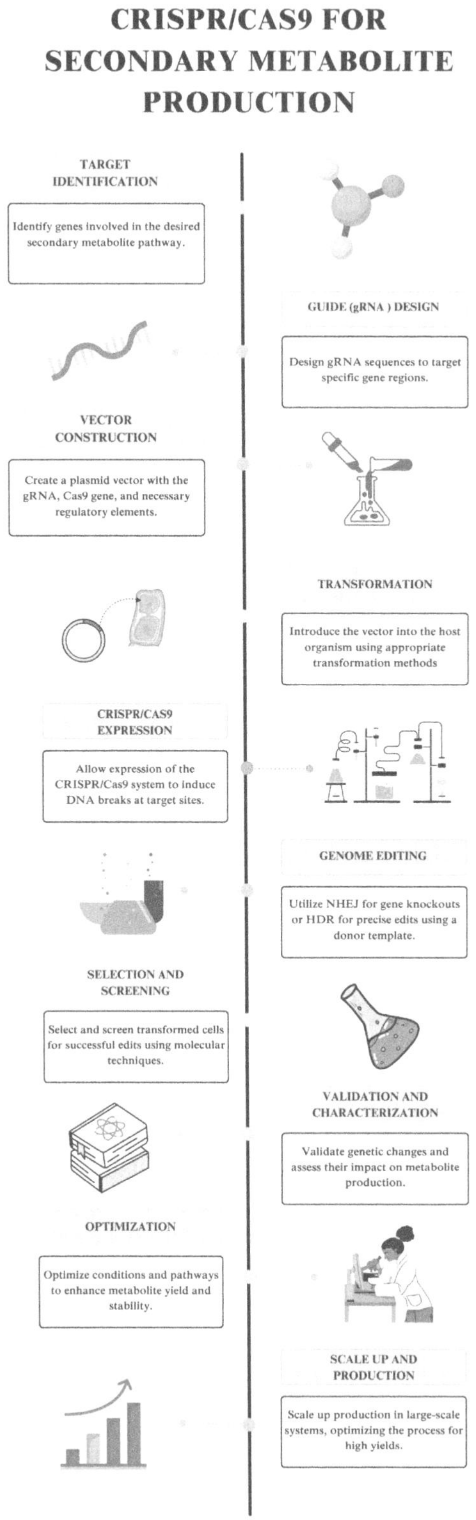

FIGURE 11.1 CRISPR/cas9 for secondary metabolite production.

genes involved in biosynthesis pathway of flavonoids of plants. They also assist in pathway divergence making available precursors for the aroma of their metabolites like phenolics and alkaloids. CRISPR/Cas9 can generate point mutations to knock out the specific genes or enhance the production of the desired metabolite while reducing the production of undesired by-products; an example is the production of low nicotine tobacco. Further, through CRISPR/Cas9, genes can be amplified to improve the levels of precursor molecules to raise production of valuable products like terpenoids.

11.4 METABOLIC ENGINEERING

Metabolic engineering is a branch of biotechnology that works on enzymes and reaction pathways to produce more or less of a substance. Based on the current technologies, complex molecules can be synthesized in plants. Plants generate and secrete different signaling chemicals for interaction with their environment and other organisms, many of which are beneficial to mankind as drugs and nutraceuticals. The metabolism that takes place in plants is divided into primary and secondary metabolism. Primary metabolism synthesis uses primary macromolecules, distal from energy production; while secondary metabolism builds on primary metabolites. At the genetic or enzymatic level there is modulation but there may be external factors which affect it in order to produce a new product (Ashraf et al., 2018). Metabolic engineering has been conducted using various techniques such as:

Transgenic approaches: Production of secondary metabolites biosynthesis has been stated to be likely in tissue culture or by inserting a strange gene with the aid of *Agrobacterium* (Bomzan et al., 2022).

Synthetic biology tools: Synthetic biology involves the creation of novel biological pathways through modular recombination of standardized genetic components like promoter, terminator, and coding sequences. According to previous studies (Zalatan et al., 2015; Li et al., 2017; Zhu et al., 2020) genome editing may be employed in conjunction in numerous strategic ways to enhance the synthesis of target molecules

- (i) Enhancing the expression of genes encoding rate-limiting enzymes in the upstream gene can provide a supply of precursors and metabolites.
- (ii) The expression of key enzymes can be suppressed so that they do not hinder at the branch points or catabolic pathway of the desired product.
- (iii) Simultaneous overexpression of transcription factors that lead to activation of endogenous genes required for the metabolite's biosynthetic pathway.

Pathway Manipulation: Pathway manipulation refers to strategically altering the metabolic pathways within the plants. Some of the approaches applied for pathway manipulation include metabolic engineering, using gene knockout technology, and gene overexpression. Since metabolic engineering involves transferring entire biochemical pathways from other organisms into the host plant it requires balancing the metabolic capacities of the plant to make new compounds that it would not otherwise be able to synthesize (Mora-Vásquez et al., 2022).

Overexpression and Knockout Strategies: Overexpression research on the other hand focuses on enhancing the synthesis of genes that are required in synthesis of target metabolite by using stronger promoters or multiple copies of the gene. By use of potent promoters or multiple copies of the gene, overexpression research aims at increasing the expression of certain genes that are essential for the production of target metabolites. That is why knockout research is used to switch off certain genes that help to analyze their role in certain metabolic pathways and maybe recuse the metabolic flux in the drift of the target metabolite (Lauren et al., 2011; Chen et al., 2013).

Target genes are selected from the transcriptome and proteomic studies, afterwards, these genes are inserted into expression vectors. They are introduced into plants through transformation techniques, and then the expression of these vectors is confirmed by Western blotting and qRT-PCR. Besides, additional precursors can be directed toward the creation of alkaloids by silencing the genes regarding primary metabolism, which can interfere with the formation of the alkaloid biosynthesis pathway (Mora-Vásquez et al., 2022)

11.5 PLANT TISSUE CULTURE TECHNOLOGY

Plant tissue culture technology is employed to produce secondary metabolites to meet its uses like pesticides, drugs, natural insect pest control agents, and diets. Some of the techniques are callus culture, suspension culture, micropropagation, hairy root culture, and biotransformation (Verpoorte et al., 2002). Organogenesis is more appropriate for callus culture for plant regeneration and genetic transformation analysis while, for rapid cell division and production of secondary metabolites, suspension culture is more appropriate (Collin, 2001). This technique is used to synthesize anthraquinones, alkaloids and coumarin compounds which are used in the treatment of tumors, hypersensitivity and sepsis (Bortesi and Fischer, 2015; Mahfouz et al., 2014). Besides, hairy root culture is mostly used for the accumulation of the secondary metabolites produced in the roots of the plant. A quick and preferable technique of large-scale production in in vitro condition is somatic embryogenesis (Hassanein, 2022). The following methods have been effectively used to enhance the metabolite production using plant tissue culture production.

Callus culture: A valuable technique for producing secondary metabolites in medicinal plants which enables the controlled growth of callus tissues by optimizing the bioactive compounds, is callus culture. The yield of the desirable metabolites for commercial products and medicinal uses can be increased by controlling ambient conditions, nutritional availability and hormonal therapies (Babu et al., 2020). The efficiency of the secondary metabolites can be increased through the use of genetic engineering strategies including the gene overexpression and RNA interference, and improvements in callus culture techniques (Goswami et al., 2021). Previously, callus cultures of *Stevia rebaudiana* have yielded the natural sweeteners (Gauchan et al., 2014; Sharma et al., 2015; Kumar et al., 2022), callus cultures of *Hypericum perforatum* have produced secondary metabolites (Siddiqui et al., 2021) in response to chemical or physical stressors, indicating their potential for use in the production of the medicinal plants, and cultures of *Catharanthus roseus* have been employed to produce the alkaloids like ajmalicine and serpentine. If the culture conditions are appropriately maintained, high concentrations of these substances can be produced (Zhao et al., 2001).

Cell suspension culture: Suspension culture is a liquid-based formulation required for the synthesis of secondary metabolites belonging to the medicinal plants. It keeps the plant cells in the culture homogenous which is required for nutrition passage and metabolism. Some of the benefits include the enhancement of scalability, quality environment, and a constant synthesis of bioactive compounds such as alkaloids, terpenoids and flavonoids (Sharma et al., 2015; Ghosh et al., 2019; Bajaj, 2020; Kumar et al., 2022). Improvements in suspension culture methods and biotechnological tools like genetic engineering, have raised the efficiency of production of secondary metabolites by optimizing the metabolic pathways for higher compound yields.

Hairy root culture: Agrobacetrium rhizogenes infection-induced hairy root cultures is a highly efficient technique for extracting the secondary metabolites from medicinal plants (Ghosh et al., 2019). This method is favorable for the high production of secondary metabolites, genetic stability and quick growth conditions. When *A. rhizogenes* transforms plant tissues, Ri plasmids are introduced that result in the hairy roots that induce and accumulate secondary compounds (Zhao et al., 2021). Recent developments have combined methods like genetic engineering and elicitation, demonstrating their adaptability and effectiveness in industrial and pharmaceutical applications to increase the output of particular molecules. The roots of ashwagandha (*Withania somnifera*) have been used to increase hairy roots for withanolide production which has potential medicinal value (Pandey et al., 2017).

Protoplast Culture: Protoplast culture can enhance the production of secondary metabolites because it is possible to culture plant cells without their rigid wall. Secondary metabolites such as phenolics, isoflavonoids, and ligans like podophyllotoxin can be generated from this culture, out of which podophyllotoxin produced from this method is utilized in the synthesis of anticancer medication etoposide (Ramawat et al., 2009; Jain et al., 2020). This method enables the genetic transformation of bioactive chemicals and regeneration of entire plants through the manipulation of metabolic pathways and cellular processes (Xu et al., 2018). The method also aids in the investigation of how elicitors and regulatory variables affect the synthesis of secondary metabolites (Gamborg and Phillips, 1995).

11.6 ELICITATION AND PRECURSOR FEEDING

Elicitation is one of the effective methods for enhancing the production of secondary metabolites in artificial conditions. The addition of biotic and abiotic components to the culture medium (including organ cultures and cell suspensions) aids in metabolite formation significantly increasing the output of plant metabolites (Woerdenbag et al., 1991). Precursor feeding refers to the process of supplying a plant with specific precursor molecules, which are later metabolized by the plants for aiding biosynthetic pathways to produce secondary metabolites. For example, in the case of artemisinin, it is possible to supply the next intermediates closer to the final product in the biosynthetic pathway, by feeding with artemisinic acid or dihydroartemisinic acid. The plant then uses enzymatic processes to metabolize these precursors into actual artemisinin. Mevalonic acid (MVA) or its derivatives which serve as a precursor feeding, enhances the pathway flux and yield of intermediary chemicals which in synthesis form artemisinin (Weathers et al., 2011). Table 11.1 shows a list of biotic and abiotic elicitors that have been used in the production of secondary metabolite artemisinin.

TABLE 11.1
Biotic and abiotic elicitors used for the production of artemisinin

Elicitor	Culture Types	Culture Condition	Yield of Artemisinin	Reference
		Biotic Elicitor		
Cell wall's oligosaccharide from *Colletotrichum* sp. B501	Hairy root culture	MS medium + 20 mg/L elicitor	Increased by 68.29%	Wen et al., 2002
Cerebroside from fungal source	Hairy root culture	MS medium + 10–70 µg/mL cerebroside	Increased by 2.3 folds	Wang et al., 2009
Pencillium oxalium B4	In vitro grown Rooted plantlets	MS medium + 5.0 mg/L BAP + 1.0 mg/L NAA + *P. oxalium* B4 (30 days exposure)	Increased by 43.5%	Zheng et al., 2016
		Abiotic Elicitor		
Ag-SiO2 nanoparticles	Hairy root culture	MS medium + 900 mg/L nano elicitor	Increased by 3.9 folds	Zhang et al., 2013
Methyl jasmonate and mevalonic acid lactone	Cell suspension culture	MS medium + 0.1 mg/L NAA + 0.1 mg/L KN + 50 mg/L mevalonic acid lactone + methyl jasmonate	Increased by 5.93 times	Baldi et al., 2008
Sorbitol and Coronatine	Cell suspension culture	MS medium + 0.1 mg/L NAA + 0.1 mg/L KN + 30 g/L Sorbitol + 0.05 µM Coronatine	Increased by 8 folds	Salehi et al., 2019

11.7 SECONDARY METABOLITE PRODUCTION IN VITRO

The section below discusses secondary metabolite production through biotechnological interventions in some priority medicinal plants.

11.7.1 Artemisinin Production In *Artemisia Annua*

Artemisinin is a sesquiterpene lactone derived from *A. annua* L and is one of the most potent drugs against malaria. It is administered as one of the components of the presently recommended combination treatments of artemisinin against malaria. Due to the expensive production cost and rarity in the traditional method of production, alternative approaches are adopted for addressing this issue. The annual world demand of around 119 metric tons, and the estimated content of artemisinin is 0.01–1.5% (w/w) so, it is insufficient to meet the demand. The ways that this yield of artemisinin can be improved have been explored by various researchers (Nair et al., 2013). Knockdown of SQS gene expression by genome editing through the CRISPR/cas9 method and enhancement of the biosynthesis of artemisinin can be done. Another transgenic approach is to overexpress the native pathways by adding extra copies of those actively participating genes, such as ADS, DBR2, and ALDH1 into the plant genome. There are many other genes that are expressed differently in various parts of the plant that catalyze the artemisinin biosynthesis pathway. Hormones involved in plant defense mechanisms include jasmonic acid and methyl jasmonate that can increase transcription factors and enzymes for the synthesis pathway as well. Specific mutations can be generated by using genetic methods like CRISPR/Cas9. Artemisinin quantity is to be increased in *A. annua* by down-regulating the SQS gene expression. A specific mutation can be introduced to knock-out the function of a specific gene. It is the low off-targets CRISPR/Cas9 that has been changing genome editing methods in recent times. In the genome editing through CRISPR/Cas9, a 20 bps guide RNA sequence guides the Cas9 nuclease to the target region, forming double-strand breaks pursuant to base pairing.

Artemisinin has been examined in callus and suspension cultures of *Artemisia* species. Usually, MS media combined with growth regulators such as BA and NAA are used to induce callus in leaf explants (Keng et al., 2010, Keong et al., 2018). Establishing cell suspension cultures requires careful consideration of callus lines, of which friable callus is favored due to its more rapid and stable growth (Jin & Keng, 2014). When compared to callus cultures, suspension cultures frequently produce more artemisinin, and elicitors such as nitrogen ions can increase the amount of artemisinin (Keng et al., 2010). Valine, cystine, BAP, and NAA have been shown to have favorable impacts on artemisinin synthesis when added to the culture medium along with other amino acids and growth regulators (Zia et al., 2007). Maximizing artemisinin yield in *Artemisia* species callus and suspension cultures requires optimizing culture conditions, including medium composition and growth regulators.

Use of elicitors in artemisinin production has also been studied. Elicitors like jasmonic acid (JA) and methyl jasmonate (MJ), trigger the plant's defense mechanism. In the process of this activation, Jasmonate induces defense response, which leads to secondary metabolite accrual, Artemisinin, as a result of the induction of the perception of attack in a plant. JA medication induces an increase of AaMYC2 transcription factor expression to strengthen artemisinin generation—this by the connection of the AaMYC2 to DBR2 and CYP71AV1 promoters. SiRNA mediated RNA silencing of AaMYC2 has also led to the repression of artemisinin production, while its overexpression has been found to urge the expression of DBR2 and CYP71AV1 involved in artemisinin biosynthesis (Shen et. al., 2016). It was found that exogenously applied MeJA upregulated antimycin expression by the name of ADS, DBR2, CYP71AV1, and ALDH1 associated with the biosynthesis of artemisinin in *A. annua* (Chen et al., 2012). It also induced the transcription factors ERF1 and ORA'- positive regulators of artemisinin biosynthesis. The genes for the biosynthetic pathway of artemisinin, ADS, DBR2, CYP71AV1, and ALDH1, were already expressed before ERF1 and ORAR'. This was expected because MeJA activates the transcription factor, which activates the biosynthesis genes (Xiang et al., 2015). The use of transgenic techniques in the genetic engineering of *A. annua* to increase the plant's capacity in the production of artemisinin has also been of interest among researchers. Critical enzymes in the pathway of artemisinin biosynthesis include farnesyl diphosphate synthase FPS, amorpha-4,11-diene synthase AMS, CYP71AV1, HMGS HMG-CoA synthase, HMGR HMG-CoA reductase, DXR 1-deoxy-D-xylulose 5-phosphate reductase, DXS 1-deoxy-D-xylulose 5-phosphate synthase, ADS amorpha-4, 11-diene synthases, DBR2 artemisinic aldehyde delta-1113-double bond reductase, and ALDH1 alcohol dehydrogenase 1. Overexpression of FPS in *A. annua* increased artemisinin accumulation. According to Maes et al. (2011), *A. annua* leads to an increased accumulation of artemisinin due to the fact that IPP and DMADP are converted into FDP. According to Graham et al. (2010), the genetic map of *A. annua* reveals that DXR is tightly linked with the biosynthesis of artemisinin. Chen et al. (1999) reported an increase in the hairy roots of *A. annua* plants after the infection of altered *A. rhizogenes* with the FPS gene. The content of artemisinin production was better in the transgenic hairy roots, overexpressing FPS, to the extent of three to four times more than in the control (Figure 11.2).

Immobilization techniques are regarded as one of the key strategies for improved secondary metabolite production, followed by adsorption, cross-linking, covalent bonding, entrapment, and encapsulation (Yashmin et al., 2023). The metabolite production was enhanced by maintaining high cell density in the immobilization techniques (Joanna et al., 2016). In this technique, the cells are stabilized and protected from shear stress. It is in this technique that polymer matrices form a controlled microenvironment, modulating parameters like nutrient availability, pH, oxygen supply and other conditions productive of improved secondary metabolites. Figure 11.3 highlights the immobilization of *A. annua* cell lines for metabolite production.

Endophyte assisted metabolite production has been realized as another potential method of secondary metabolite production which utilizes symbiotic bacteria (Ogle et al., 2020). This has been used to increase in the overall yield of artemisinin in *A. annua* (Wani et al., 2021). The conventional method of doing this is the isolation of endophytes from the plant and subsequent screening of these microorganisms for their ability to enhance metabolite production (Hussain et al., 2021). Such identified potential endophytes can be reintroduced into the plant under controlled conditions for testing their effect on artemisinin production. Endophytic fungal species isolated from *A. annua*, could increase the artemisinin content (Zhou et al., 2020) and indicates the potential of the method for secondary metabolite production (Figure 11.4).

Hairy root culture for artemisinin production using *A. rhizogenes* has also been studied (Giri and Narasu, 2000; Patra et al., 2013). Upon infection, the integration of T DNA into the tissue causes root proliferation at the site of infection which results in hairy roots. Unlike the untransformed plant roots, these hairy roots can grow evenly and uniformly in all directions, allowing three-dimensional growth in bioreactors for the production of secondary metabolites (Figure 11.5). Another study by Wang and Tan (2002) showed the impact of nitrogen content on the production of artemisinin. The optimal NO-3/NH4+ ratio was 5:1 which resulted in about a 57% increase compared to production in standard MS medium. Higher concentrations of nitrogen inhibited the growth of hairy roots and promoted callus-like tissue development. The study indicates that media composition needs to be standardized for secondary metabolite production in different plants.

11.7.2 Paclitaxel Production In *Taxus* Spp

Taxol is an anticancer drug that is obtained from *Taxus* spp. Alternative methods include generating Taxol in plant-based medium (B5) supplemented with 2,4-D, from needle explants, and without exploiting the plant's natural vegetation (Bhuju and Gauchan, 2018; Gauchan et al., 2021). Arteca and Wickremesinhe (1993) used stem and needle explants on Gamborg's B5 medium (supplemented with auxins and kinetins) to generate callus from variety of *Taxus* species and taxol concentration

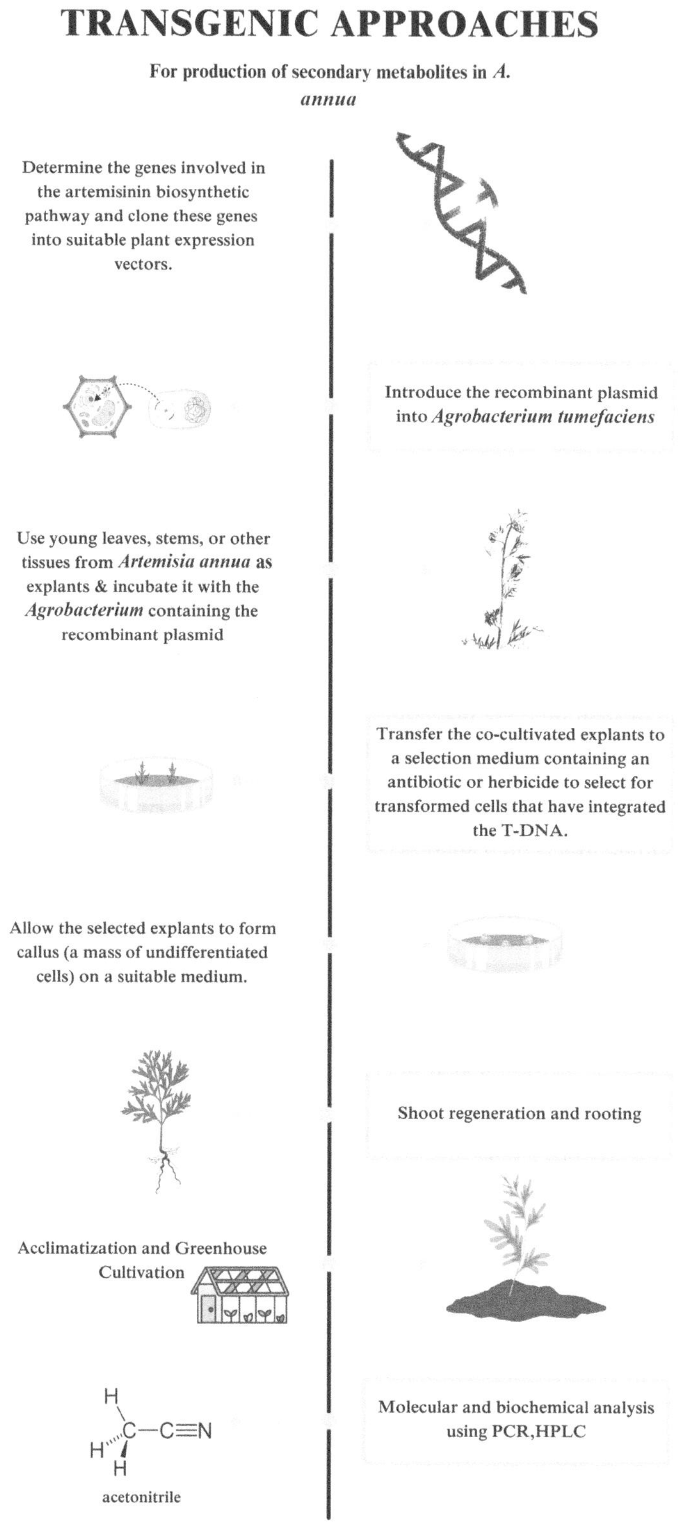

FIGURE 11.2 Transgenic approach for secondary metabolites production.

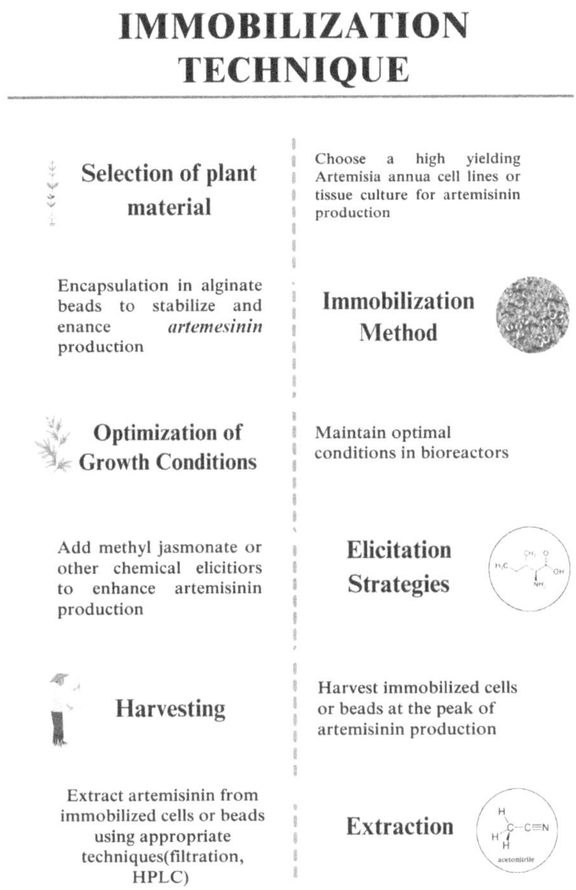

FIGURE 11.3 Immobilization technique for secondary metabolite production.

in the callus ranging from 0.0001 to 0.0131% of dry weight was reported. Researchers worldwide proposed that 2,4-D is a superior source of auxin for a variety of plant species, including wheat (Zheng and Konzak, 1999) and oil palm. Additionally, callus induction of >60% was reported by and that of 70–73% by Saikia et al. (2013) when 2,4-D was combined with kinetin. In another experiment by Ashrafi et al. (2010) to increase the Taxol content, they used needle explants in a 2mg/l of 2-4-D supplemented media for callus induction. During the experiment they observed an increase in the Taxol yield after the second week which then peaked around the conclusion of the third week with maximum observed Taxol content of 0.0035%

In another study on metabolite production from *Taxus* spp. in Nepal, the major phytochemical compounds found were alkaloids, flavonoids, phenol, glycosides, terpenoids, tannins, coumarin, and quinone. Acetone, 50% ethanol, and methanol were found to be the best solvents for screening. Both phenolic and flavonoid content were high in bark samples compared to needle samples and female plants showed higher flavonoid content than male plants (Bhuju et al., 2019)

It is now established that tissue culture, genetic manipulation, and expression can improve paclitaxel production under *in vitro* conditions (Edgar et al., 2023). The amount of Taxol identified in suspension culture samples in the chromatogram was at a retention time close to that of ST-S with a concentration up to 0. 004% (Dmitry et al., 2023). The study on elicitation established that the suspension cultures of *T. baccata* plants alter with regard to elicitation and can be used for the creation of efficient systems for obtaining total phenolic compounds from cell cultures (Katarzyna

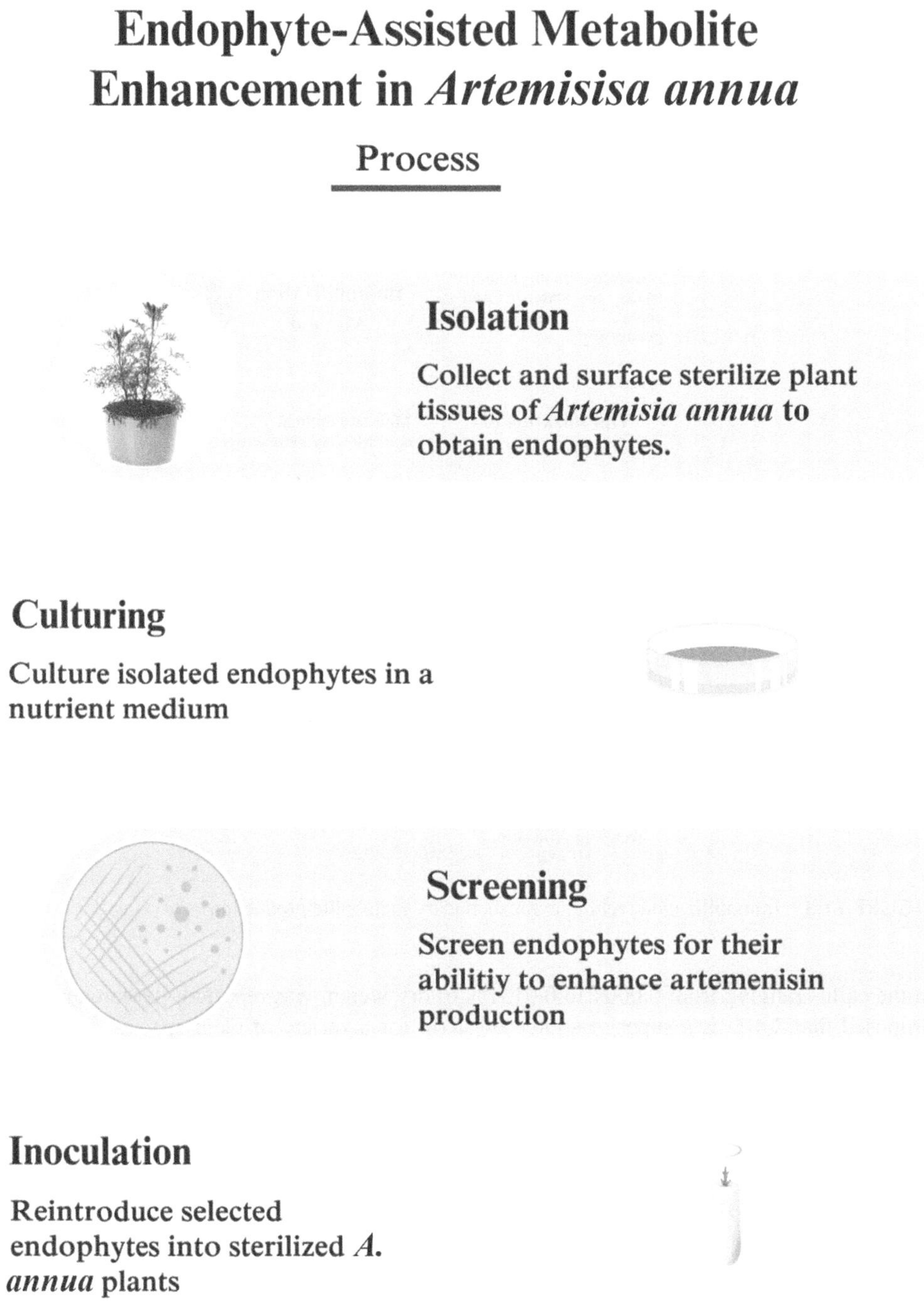

FIGURE 11.4 Process of endophyte assisted metabolite enhancement in *A. annua.*

et al., 2022). *T. wallichiana*, an endangered species in the Himalayan region, is one of the sources of paclitaxel to cure various cancer (Subedi, 2017). However, the intrinsic rate of accumulation is below the threshold, which is necessary for production. These disadvantages have been solved by technologies such as elicitation, genetic engineering, and cell culture technologies. It is advisable

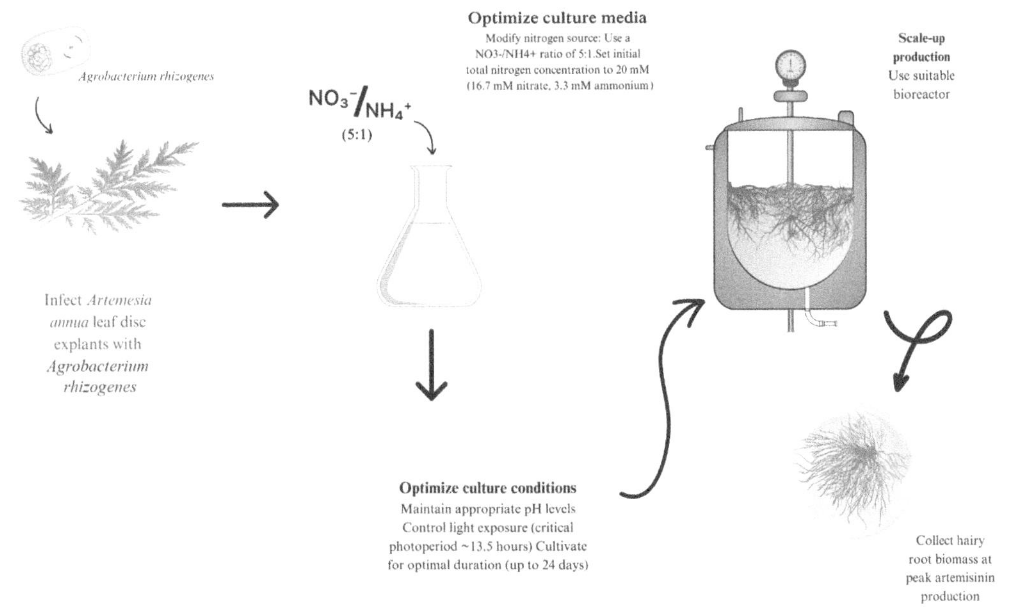

FIGURE 11.5 Biotransformation mechanism in the process of hairy root culture for increasing production of Artemisinin.

to use a combination of the above approaches to improve the production of paclitaxel while satisfying the market demands and the number of patients in need of this drug. Hairy root culture for the production of paclitaxel have been emphasized in the study on *Taxus* spp. (Katarzyna et al., 2022) which involves four steps: bacteria aid, explants surface sterilization, bacteria nourishment, explants hardening and root development (He et al., 2022). Zhao et al. (2023) showed that elicitation could be performed by using suspension cultures of the friable callus line originated from *T. baccata*, followed by sowing on B5 solidified medium for metabolite production. The genetic engineering technique study also assessed the use of hairy root formation using *A. rhizogenes* LBA 9402 in KT hairy root culture, where cultures were maintained in hormone-free modified DCR liquid media for four successive weeks followed by subculture every four weeks. The expression of genes associated with the paclitaxel biosynthesis pathway was determined using qRT-PCR (Katarzyna et al., 2022) (Figure 11.6).

11.7.3 Resveratrol Production In *Vitis Vinifera*

The polyphenolic phytoalexin in grapes known as resveratrol is effective in fighting inflammation, atherosclerosis and hypertension (Das et al., 2011). In variety, it is synthesized in response to fungal pathogen attacks or abiotic stress such as UV radiation (Salehi et al., 2018). The process is tedious and other problems such as infection, low purity or variation in yields could occur. Thus, resveratrol interacts with estrogen receptors in multiple ways, for instance, exposure of trans-resveratrol to light reduces the biological activity of the compound by converting it to cis-resveratrol. Cell culture can be applied to extraction where industries obtain high-quality products. Resveratrol is essential for the prevention and management of several cardiovascular disorders, type 2 diabetes, obesity,

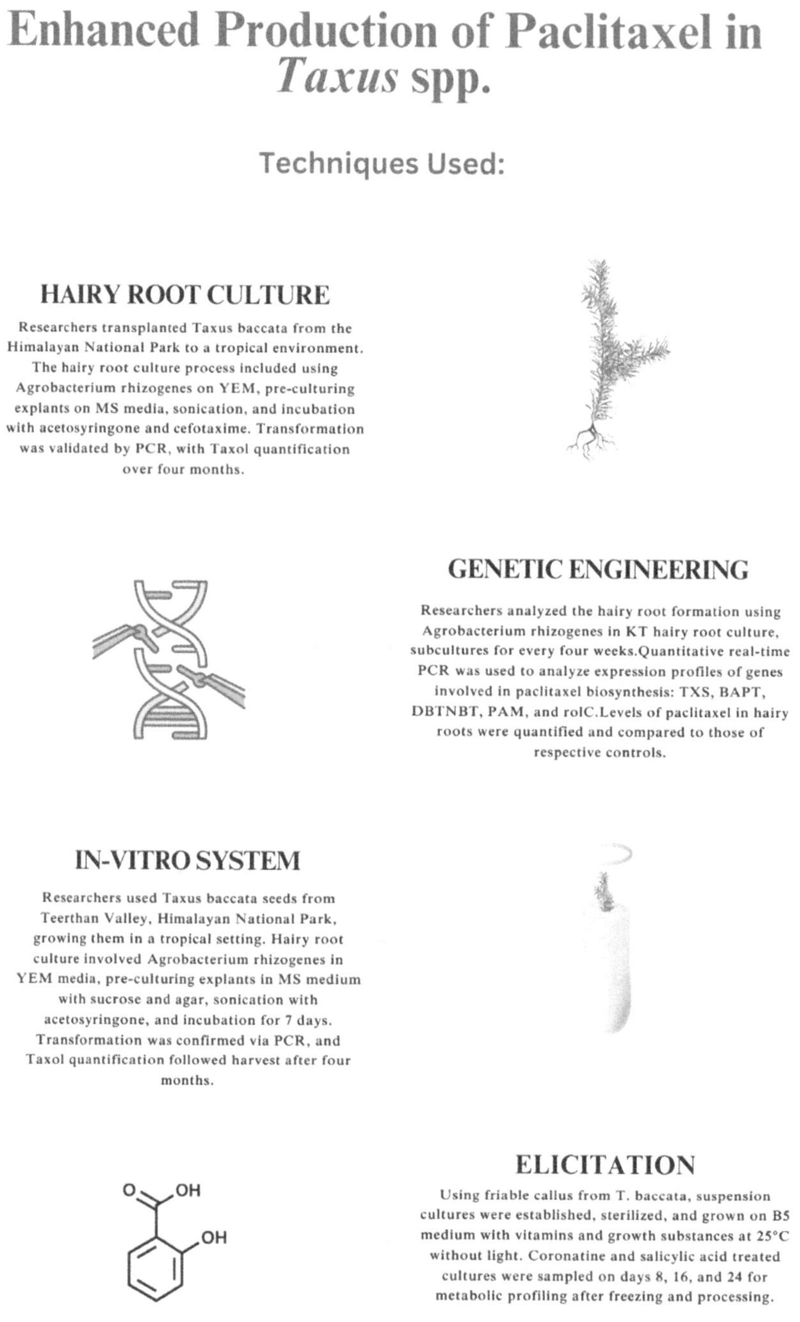

FIGURE 11.6 Techniques for the production of Paclitaxel in *Taxus* spp.

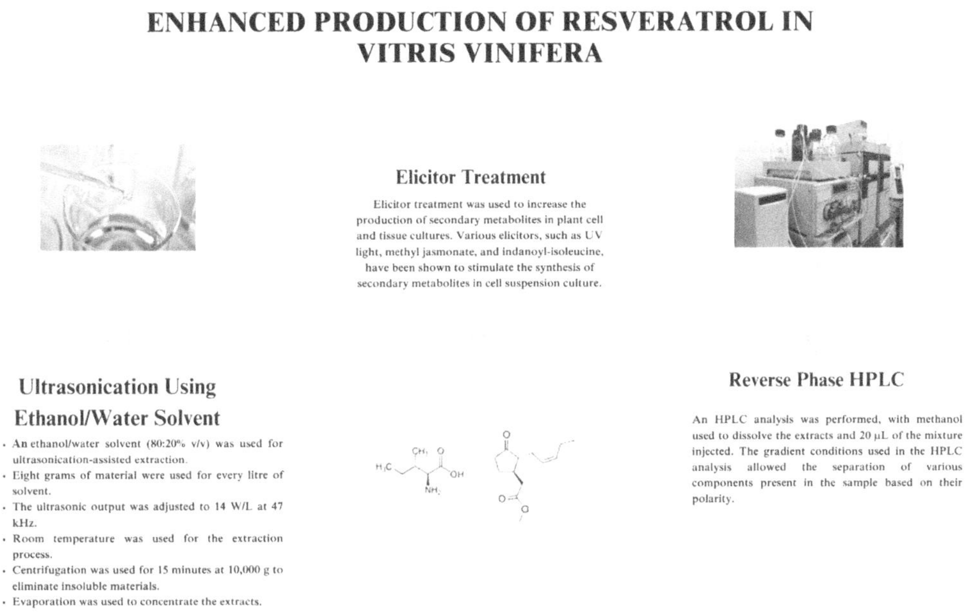

FIGURE 11.7 Techniques involved for enhanced production of Resveratrol in *Vitis vinifera.*

anti-asthma, antiplatelet, anticataract, anti-obesity, anticholinergic, anti-sunburn, wound healing, and antiviral and hypertension (Meng et al., 2021; Sharma et al., 2023).

In recent research, several elicitation techniques have been investigated with the aim of increasing resveratrol production in *Vitis vinifera* cell cultures (Figure 11.7). Among others, methyl jasmonate and cyclodextrin have been observed to boost the production of extracellular trans-resveratrol; ideal outcomes are determined by specific cell densities and concentrations (Belchi-Navaro et al., 2011). Unique elicitors such as insect saliva and synthetic elicitors such as indanoyl-isoleucine and N-linolenoyl-L glutamine have also been shown to promote the production of anthocyanins and resveratrol (Cai et al., 2012). Similarly, trans-resveratrol yields in transgenic cell lines, with modification aimed at overexpression of the stilbene synthase gene, have been enhanced further when used in conjunction with elicitation (Chu et al., 2017). Elicitor treatment has also been employed to induce the production of secondary metabolites in plant cell and tissue cultures with demonstrations revealing cell suspension cultures' responses to various elicitors including methyl jasmonate and indanyl-isoleucine (Vuong et al., 2014).

11.8 SCALING UP TECHNIQUES

For millennia, traditional medicine has relied on plants as a rich supply of medicinal ingredients. Plant cell culture research is being pushed by the increasing need for these "phytochemicals" in order to achieve efficient manufacturing. The most profitable phytochemicals are secondary metabolites, which are in high demand for use in pharmaceutical industries. However, environmental constraints frequently affect wild plant sources. Techniques for growing plant tissue have become an effective means of researching plant growth and the production of secondary metabolites. By integrating diverse cultures and genetic procedures, these in vitro techniques have surmounted obstacles and enhanced the synthesis of metabolites (Hussain et al., 2012). If plants have high shear sensitivity like *C. roseus,* centrifugal impeller bioreactors are used. These are based on the principle of centrifugal pumps. For non-shear-sensitive ones, airlift bioreactors can be used. Bioreactors are used to promote rapid cell division and huge biomass accumulation under controlled conditions of pH, temperature, sugar concentration, dissolved oxygen, etc. One of the methods of producing an excess amount of potentially useful drugs from secondary metabolites is through the hairy root culture method (Srivastava et al., 2007). They simulate their natural counterparts, thereby allowing easy studies of plant regulation of SMs at a molecular level in varied environments. The cells are immobilized in a gel matrix (like alginate or agarose) or solid surface like polyacrylamide) which enhances the cell stability and productivity.

11.9 OPTIMIZATION OF BIOTECHNOLOGICAL PROTOCOLS

For enhanced production of phenolics including epicatechin, luteolin, resveratrol, and curcumin, Mamdouh and Smetanska looked at the best culture conditions for plant cells in liquid suspension culture (Mamdouh et al., 2022). They identified that callus formation was enhanced most efficiently in the presence of both the plant growth hormones, 2,4-D and NAA and a medium with NAA alone used to support more constant cell suspension. Sugar concentrations also influenced the pH, feasibility, and growth of cells in the medium. The active anti-malaria ingredient artemisinin is derived from the *A. annua* plant through breeding and cultivation and other methods of genetic engineering and plant breeding. Metabolic engineering of plants has proven successful in improving yields of artemisinin and guaranteeing the availability of malaria treatment. Different types of the secondary metabolites extracted from various sources have their own optimization techniques (Table 11.2).

TABLE 11.2
Secondary metabolites their uses in medicine and their yield increment process

S. No.	Taxon	Secondary Metabolite	Disease Targeted	Metabolite Yield Increment process	Market demands	References
1	*Galanthus* spp.	Galantamine	Alzheimer's disease, mild cognitive impairment	Solvent extractions and crystalization	High	Corey et al., 2003
2	*Apium graveolens* L., *Cynara cardunculus* L. *var. scolymusdandelion*, and *Chrysanthemum* spp.	Epicatechin	Inflammation, allergies, cancer	HPLC and Solid phase extraction	Increasing	Garcia-Villalba et al., 2009
3	*Theobroma cacao* L. (beans), *Camellia sinensis* L. (leaves)	Luteolin	Heart disease, diabetes, cancer	Enzyme assisted extraction using cellulase, beta-glucosidase and pectinase	Moderate	Ryan et al., 2020
4	*Artemisia annua* L.	Artemisinin	Malaria	Endophyte Assisted Enhancement, Hairy Root Culture (while utilizing elicitors)	Very high	Lapkin et al., 2006
5	*Taxus wallichiana* Zucc.	Paclitaxel	Cancer (breast, ovarian, lung)	Callus culture, suspension culture, Elicitation, In-vitro System	High	Gauchan et al., 2021

TABLE 11.2 (Continued)
Secondary metabolites their uses in medicine and their yield increment process

S. No.	Taxon	Secondary Metabolite	Disease Targeted	Metabolite Yield Increment process	Market demands	References
6	*Vitis vinifera*L. (skins and stems)	Resveratrol	Heart disease, cancer, neurodegenerative diseases	Callus Culture, Ultra Sonication Assisted Extraction	Moderate	Cho et al., 2006; Vuong et al., 2014
7	*Curcuma longa* L.	Curcumin	Inflammation, arthritis, cancer	Pressurized Liquid Extraction	High	Martin et al., 2021
8	*Catharanthus roseus* L.	Vinblastine	Cancer (leukemia, lymphoma)	Callus Culture	Moderate	Yoshiharu et al., 1987
9	*Catharanthus roseus* L.	Vincristine	Cancer (leukemia, lymphoma)	Callus Culture	Moderate	Arora et al., 2010
10	Major Fruits and vegetables	Flavonoids	Anti-inflammatory, Antioxidant	Elicitation	Moderate	Al-Khayri et al., 2022
11	*Catharanthus roseus* L.	Reserpine	Hypertension	Callus Culture	High	Senbagalakshm et al., 2017
12	*Catharanthus roseus* L.	Ajmalicine	Hypertension	Callus Culture	High	Zenk et al., 19 77
14	*Withania somnifera* L.	Withanolide	Osteoarthritis(OA) and RA	Hairy Root Culture	Increasing	Pandey et al., 2017
15	*Solanum lycopersicum* L.	Steroidal Glycoalkaloids	Bacterial infection	Genome Editing using Crispr Cas9	Moderate	Ramirez-Estrada et al., 2017
16	*Salvia. miltiorrhiza* Bunge.	Tanshinones	Heart Disease, Strokes, Neurological Diseases	Shoot and Root Culture	Increasing	Lu et al., 2019
17	*Capsicum chinense* Jacq.	Capsaicin	Pain relief to Arthritis, Psoriasis, or Diabetic Neuropathy	Induced Stress Plant Tissue Culture	Increasing	Kehie et al., 2014
19	*Glycyrrhiza glabra* L.	Glabridin	Anti-inflammatory, Antioxidant and Skin Whitening Agents	Supercritical Fluid Extraction	Low	Hong et al., 2019
20	*Malus domestica* L.	Quercetin	Treat conditions of the heart and blood vessels and to prevent cancer	Methyl Jasmonate Treatment	Increasing	Rudell et al., 2002
21	*Fagopyrum esculentum* Moench.	Rutin	Commonly used for Autism, Aging Skin, Airway Infections caused by exercise	Metabolic Engineering through Hairy Root Culture	Increasing	Joshi et al., 2020
22	*Populus. tremula* L. x *Populus. Alba* L.	Eugenol	treat toothache, gastrointestinal and respiratory complaints	By production of Transgenic Hybrids	Increasing	Lu et al., 2017

(continued)

TABLE 11.2 (Continued)
Secondary metabolites their uses in medicine and their yield increment process

S. No.	Taxon	Secondary Metabolite	Disease Targeted	Metabolite Yield Increment process	Market demands	References
23	*Salvia miltiorrhiza* Bunge.	Salvianolic acid B, Rosmarinic acid	Anti-apoptosis, Anti-inflammatory, Antioxidant and Anti-depressant	Phytohormone elicitation method	Moderate	Xie et al., 2022
24	*Cinchona officinalis* Linn.	Tryptamine	Migrane treatment	Hairy Root Culture	High	Verpoorte et al., 2002
25	*Nicotiana tabacum* L.	Strictosidine	Antimicrobial and Antifeedant	Cell Suspension Culture	Low	Verpoorte et al., 2002
26	*Allium cepa* L.	Genistein	To inhibit breast cancer cells by activating apoptosis	*Agrobacterium* and biolistic-mediated transformation	Moderate	Malla et al., 2020
27	*Artemisia absinthium* L.	Estragole	Antifungal properties	*Agrobacterium* mediated transformation	Moderate	Nin et al., 1997
28	*Panax ginseng* C. A. Meyer	Ginsenoside	Help reduce Inflammation and Oxidative Stress	Shoot Culture/ Callus Induction	Moderate	Luthra et al., (2021)
29	*Artemisia absinthium* L.	Limonene	Treat Stomach and Digestive problems Issues related to Inflammation and Anxiety	*Agrobacterium* mediated transformation	Increasing	Nin et al., 1997
30	*Artemisia absinthium* L.	Borneol	Effectively treats Bronchial Symptoms, Improved Digestion	*Agrobacterium* mediated transformation	Moderate	Nin et al., 1997
31	*Artemisia absinthium* L.	Nerol	Anxiety and Stress Reduction, Pain Relief	*Agrobacterium* mediated transformation	Increasing	Nin et al., 1997

11.10 CONCLUSION

Apart from the health benefits that are offered by bioactive secondary metabolites from medicinal plants, the compounds play a large economic role in the natural products market. It, therefore, becomes imperative to develop efficient metabolite production strategies for priority plants. Optimization strategies such as plant tissue culture techniques, gene stacking, elicitation and precursor feeding enhance metabolite accumulation which is crucial in the generation of high value chemical compounds at industrial level. The use of CRISPR/Cas9 technology and synthetic biology in plant systems have a profound impact for the generation of targeted mutations and the creation of new metabolic pathways (Oksman-Caldentey and Inzé, 2004) thus demonstrating the possibility of using this technology to optimize plant characters and increase the yield of valuable metabolites. Innovative methods in fermentation and microbiome engineering improve these processes through better conditions and positive plant-microbe associations. Moreover, it enhances the solubility of elicitors and precursors indicating better and robust plants into the nanotechnology field. The future for these technologies is defined by the capability to achieve mass production, stabilization and lower prices in a sustainable and environmentally friendly manner.

REFERENCES

Ajikumar, P.K., Xiao, W.H., Tyo, K.E., Wang, Y., Simeon, F., Leonard, E., Mucha, O., Phon, T.H., Pfeifer, B., and Stephanopoulos, G. 2010. Isoprenoid pathway optimization for Taxol precursor overproduction in Escherichia coli. *Science*, *330*(6000), 70–74. www.science.org/doi/10.1126/science.1191652

Al-Khayris, J.M., Sahana, G.R., Nagella, P., Joseph, B.V., Alessa, F.M., and Al-Mssallem, M.Q. 2022. Flavonoids as potential anti-inflammatory molecules: A review. *Molecules*, *27*(9), 2901. www.mdpi.com/1420-3049/27/9/2901

Arora, R., Malhotra, P., Mathur, A.K., Mathur, A., Govil, C.M., and Ahuja, P.S. 2010. Anticancer alkaloids of Catharanthus roseus: Transition from traditional to modern medicine. New Delhi, India: Jaypee Brothers Medical Publishers Pvt. Ltd. www.jaypeedigital.com/book/9788184488418/chapter/ch21

Arteca, R.N., and Wickremesinhe, E.R.M. 1993. Taxus callus cultures: Initiation, growth optimization, characterization and Taxol production. *Plant Cell Tissue and Organ Culture*, *35*, 181–193.

Ashraf, M.A., Iqbal, M., Rasheed, R., Hussain, I., Riaz, M., and Arif, M.S. 2018. Environmental stress and secondary metabolites in plants: An overview. *Plant Metabolites and Regulation Under Environmental Stress*, 153–167. https://doi.org/10.1016/B978-0-12-812689-9.00008-X

Ashrafi, S., Mofid, M.R., Otroshi, M., Ebrahimi, M., and Khosroshahli, M. 2010. Effects of plant growth regulators on the callogenesis and taxol production in cell suspension of *Taxus baccata* L. *Trakia Journal of Science*, *8*(2), 36–43.

Babu, K. S., Bhowmik, D., and Patra, J. K. 2020. Callus culture and secondary metabolite production in medicinal plants: A review. *Journal of Medicinal Plants Research*, *14*(8), 323–336.

Bais, H.P., Weir, T.L., Perry, L.G., Gilroy, S., and Vivanco, J.M. The role of root exudates in rhizosphere interactions with plants and other organisms. *Annual Review of Plant Biology*, *57*(1), 233–266. https://doi.org/10.1146/annurev.arplant.57.032905.105159

Bajaj, Y.P.S. (Ed.) 2020. Plant cell suspension cultures for the production of secondary metabolites. In: *Plant Cell Culture: Essential Methods* (pp. 147–162). Hoboken, NJ: Wiley.

Baldi, A., and Dixit, V.K. 2007. Yield enhancement strategies for artemisinin production by suspension cultures of *Artemisia annua*. *Bioresource Technology*, *99*(11), 4609–4614. https://doi.org/10.1016/j.biortech.2007.06.061

Batra, B., Sudan, S., Pant, M., Pant, K., Lal, A., and Gusain, T. 2022. Molecular docking and TLC analysis of candidate compounds from lesser-used medicinal plants against Diabetes mellitus targets. *Jordan Journal of Biological Sciences*, *15*(2), 339–346.

Belchí-Navarro, S., Almagro, L., Lijavetzky, D.C., Bru, R., and Pedreño, M.Á. 2011. Enhanced extracellular production of trans-resveratrol in Vitis vinifera suspension cultured cells by using cyclodextrins and methyljasmonate. *Plant Cell Reports*, *31*, 81–89. https://api.semanticscholar.org/CorpusID:16618376

Bhuju, S., and Gauchan, D.P. 2018. *Taxus wallichiana* (Zucc.), an endangered anti-cancerous plant: A review. *International Journal of Research*, *5*(21), 10–21. www.researchgate.net

Bhuju, S., Lamichhane, J., Shakya, R., Thapa, N., and Gauchan, D. P. 2019. Phytochemical screening and quantitative screening for total phenolic and flavonoid content in Taxus wallichiana collected from various districts of Nepal. *Journal of Chemical, Biological and Physical Sciences, Section B*, *9*(1), 82–93. https://doi.org/ 10.24214/jcbps.B.9.1.08293

Bomzan, D.P., Shilpashree, H.B., and Nagegowda, D.A. (2022). *Agrobacterium*-mediated in planta transformation in Periwinkle. In Courdavault, V., and Besseau, S. (eds.) *Catharanthus roseus. Methods in Molecular Biology*, vol 2505. New York, NY: Humana. https://doi.org/10.1007/978-1-0716-2349-7_22

Bortesi, L., and Fischer, R. 2015. The CRISPR/Cas9 system for plant genome editing and beyond. *Biotechnology Advances*, *33*(1), 41–52. https://doi.org/10.1016/j.biotechadv.2014.12.006

Cai, Z., Knorr, D., and Smetanska, I. 2012. Enhanced anthocyanins and resveratrol production in Vitis vinifera cell suspension culture by indanoyl-isoleucine, N-linolenoyl-L-glutamine and insect saliva. *Enzyme and Microbial Technology*, *50*(1), 29–34. https://api.semanticscholar.org/CorpusID:5808869

Chen, D.H., Liu, C.J., Ye, H.C., Li, G.F., Liu, B.Y., Meng, Y.L., and Chen, X.Y. 1999. Ri-mediated transformation of Artemisia annua with a recombinant farnesyl diphosphate synthase gene for artemisinin production. *Plant Cell, Tissue and Organ Culture*, *57*, 157–162. https://link.springer.com/article/10.1023/A:1006326818509

Chen, Y., Shen, Q., Wang, Y. *et al*. 2013. The stacked over-expression of *FPS*, *CYP71AV1* and *CPR* genes leads to the increase of artemisinin level in *Artemisia annua* L. *Plant Biotechnology Reports*, *7*, 287–295. https://doi.org/10.1007/s11816-012-0262-z

Cho, Y.J., Hong, J.Y., Chun, H.S., Lee, S.K., and Min, H.Y. 2006. Ultrasonication-assisted extraction of resveratrol from grapes. *Journal of Food Engineering*, *77*(3), 725–730. www.sciencedirect.com/science/article/abs/pii/S0260877405005418?via%3Dihub

Chu, M., Pedreño, M.Á., Alburquerque, N., Faize, L., Burgos, L., and Almagro, L. 2017 . A new strategy to enhance the biosynthesis of trans-resveratrol by overexpressing stilbene synthase gene in elicited Vitis vinifera cell cultures. *Plant Physiology and Biochemistry: PPB*, *113*, 141–148 . https://api.semanticscholar.org/CorpusID:21539883

Corey, J.B. 2003. Galantamine: A review of its use in Alzheimer's disease and vascular dementia. *International Journal of Clinical Practice*, *57*(3), 219–223. https://pubmed.ncbi.nlm.nih.gov/12723727/

Collin, H.A. 2001. Secondary product formation in plant tissue cultures. *Plant Growth Regulation*, *34*, 119–134. https://doi.org/10.1023/a:1013374417961

Cushnie, T.P. T ., and Lamb, A.J. 2005, November. Antimicrobial activity of flavonoids. *International Journal of Antimicrobial Agents*, *26*(5) ,343–356. https://doi.org/10.1016/j.ijantimicag.2005.09.002

Dewick, P.M. 2002. *Medicinal Natural Products: A Biosynthetic Approach*. John Wiley and Son. https://nadjeeb.wordpress.com/wp-content/uploads/2009/10/dewick-natural-prod.pdf

Das, D.K., Mukherjee, S., and Ray, D. Erratum to: Resveratrol and red wine, healthy heart and longevity. *Heart Failure Reviews*, *16*, 425–435. https://doi.org/10.1007/s10741-011-9234-6

Gamborg, O.L., and Phillips, G.C. 1995. Laboratory facilities, operation, and management. In O.L. Gamborg and G.C. Phillips (eds.) *Plant Cell, Tissue and Organ Culture: Fundamental Methods* (pp. 3–20). Berlin Heidelberg: Springer.

Gauchan, D.P., Dhakal, A., Sharma, N., Bhandari, S., Maskey, E., Shrestha, N., Gautam, R., Giri, S., and Gurung, S. 2014. Regenerative callus induction and biochemical analysis of *Stevia rebaudiana* Bertoni. *Journal of Advanced Laboratory Research in Biology* , *45*(3), 41–45.

Gauchan, D.P., Bhuju, S., Lamichhane, J., Shakya, R., and García-Gil, M.R. 2021. Establishment of regenerative callus, cell suspension system, and molecular characterization of Taxus wallichiana Zucc. for the in vitro production of Taxol. *Journal of Applied Pharmaceutical Science*, *11*(6), 022–034. https://japsonline.com/abstract.php?article_id=3374andsts=2

García-Villalba, R., Larrosa, M., Possemiers, S., Tomás-Barberán, F. A., and Espín, J. C. 2009. High-performance liquid chromatography (HPLC) coupled with solid-phase extraction (SPE) for the analysis of epicatechin in plant extracts. *Journal of Agricultural and Food Chemistry*, *57*(14), 6246–6253.

Ghosh, R., Bandyopadhyay, S., and Sen, A. 2019. Hairy root cultures in medicinal plant biotechnology: Progress and prospects. *Biotechnology Advances*, *37*(3), 107345.

Ghosh, R., Singh, R., and Bandyopadhyay, S. 2019. Suspension culture of medicinal plants for secondary metabolite production: Advances and challenges. *Journal of Plant Biotechnology*, *46*(2), 234–245.

Giri A., and Narasu M. L. 2000. Transgenic hairy roots: Recent trends and applications. *Biotechnology Advances*, *18*(1), 1–22. https://doi.org/10.1016/S0734-9750(99)00016-6

Goswami, P., Sahu, P. K., and Singh, M. 2021. Genetic engineering in callus cultures for enhanced secondary metabolite production: Advances and applications. *Plant Cell Reports*, *40*(12), 2087–2100.

Graham, I.A., Besser, K., Blumer, S., Branigan, C.A., Czechowski, T., Elias, L., Guterman, I., Harvey, D., Isaac, P.G., Khan, A.M., and Larson, T.R. 2010. The genetic map of Artemisia annua L. identifies loci affecting yield of the antimalarial drug artemisinin. *Science*, *327*, 5963, 328–331. www.science.org/doi/10.1126/science.1182612

Hassanein, A.M. 2022. *Application of Tissue Culture Techniques to Improve the Productivity of Medicinal Secondary Products from Medicinal Plants*. IntechOpen. https://doi.org/10.5772/intechopen.105193

He, J., Plácido, J.P., Pateraki, I., Kampranis, S., Favero, B.T., and Lütken, H. 2022. Hairy root induction of Taxus baccata L. by natural transformation with Rhizobium rhizogenes. *Horticulturae*, *9*(1), 4. www.mdpi.com/2311-7524/9/1/4

Hong, J.H., Jung, I.I., Cho, Y.K., Haam, S., Lee, S.Y., Lim, G., and Ryu, J.H. 2019. Preparation of high-quality glabridin extract from *Glycyrrhiza glabra*. *Biotechnology and Bioprocess Engineering*, *24*, 666–674. https://link.springer.com/article/10.1007/s12257-019-0121-7

Hussain, M.S., Fareed, S., Ansari, S., Rahman, M.A., Ahmad, I.Z., and Saeed, M. 2012. Current approaches toward production of secondary plant metabolites. *Journal of Pharmacy and Bioallied Sciences*, *4*(1), 10–20. https://journals.lww.com/jpbs/fulltext/2012/04010/current_approaches_toward_production_of_secondary.3.aspx

Hussain, T., Javed, M., Shaikh, S., Tabasum, B., Hussain, K., Ansari, M.S., Khan, A., and Khan, A.A. 2021. Enhancement of plant secondary metabolites using fungal endophytes. In *Biotechnological Approaches to Enhance Plant Secondary Metabolites* (pp. 61–70). CRC Press. https://doi.org/10.1201/9781003034957

Jain, S. K., Sharma, P., and Bhatnagar, P. 2020. Protoplast culture of *Withania somnifera* for the enhanced production of withanolides. *Plant Cell Reports*, *39*(10), 1391–1404.

Jin, C.S., and Keng, C.L. 2014. Factors affecting the selection of callus cell lines and the preparation of the cell suspension culture of Artemisia annua L. *Plant Tissue Culture and Biotechnology*, *23*, 157–163. https://api.semanticscholar.org/CorpusID:86371062

Joanna, Z., Wojcieszyńska, D., and Guzik, U. 2016. Metabolic responses of bacterial cells to immobilization. *Molecules*, *21*(7), 958. www.mdpi.com/1420-3049/21/7/958

Joshi, D.C., Zhang, K., Wang, C., Chandora, R., Khurshid, M., Li, J., He, M., Georgiev, M.I., and Zhou, M. 2020. Strategic enhancement of genetic gain for nutraceutical development in buckwheat: A genomics-driven perspective. *Biotechnology Advances*, *39*, 107479. www.sciencedirect.com/science/article/abs/pii/S073497501930179X?via%3Dihub

Kehie, M., Kumaria, S., and Tandon, P. 2014. Manipulation of culture strategies to enhance capsaicin biosynthesis in suspension and immobilized cell cultures of *Capsicum chinense* Jacq. cv. Naga King Chili. *Bioprocess and Biosystems Engineering*, *37*, 1055–1063. https://link.springer.com/article/10.1007/s00449-013-1076-2

Keng, C.L., Singaram, N., and Lim, B.P. 2010. Production of artemisinin from cell suspension culture of Artemisia annua L. In *Proceedings Asia Pacific Conference on Plant Tissue and Agribiotechnology (APaCPA)*, vol. 17, p. 21. https://api.semanticscholar.org/CorpusID:85680676

Keong, C.C, Nordin, F.A., Bhatt, A., and Keng, C.L. 2018. Culture medium and varietal selection for establishment of callus culture of *Artemisia annum*. *EDUCATUM Journal of Science, Mathematics and Technology*, *5*(2), 47–56. https://doi.org/10.37134/ejsmt.vol5.2.6.2018

Kochkin, D.V., Demidova, E.V. Globa, E.B., and Nosov, A.M. 2023. Profiling of taxoid compounds in plant cell cultures of different species of yew (Taxus spp.). *Molecules*, *28*(5), 2178. https://doi.org/10.3390/molecules28052178

Kumar, S., Raj, N., and Singh, R. 2022. Production of steviosides in callus cultures of *Stevia rebaudiana*: Optimization and potential applications. *Journal of Plant Biochemistry and Biotechnology*, *31*(1), 35–45.

Kumar, V., Bhardwaj, V., and Verma, P. 2022. Elicitation strategies for secondary metabolite production in suspension cultures of *Catharanthus roseus*. *Frontiers in Plant Science*, *13*, 834761.

Lapkin, A.A., Plucinski, P.K., and Cutler, M. 2006. Comparative assessment of technologies for extraction of artemisinin. *Journal of Natural Products*, *69*(11), 1653–1664. https://pubs.acs.org/doi/10.1021/np060375j

Li, Y., Kong, D., Fu, Y., Sussman, M.R., and Wu, H. 2020. The effect of developmental and environmental factors on secondary metabolites in medicinal plants. *Plant Physiology and Biochemistry*, *148*, 80–89. https://doi.org/10.1016/j.plaphy.2020.01.006

Li, Z., Zhang, D., Xiong, X., Yan, B., Xie, W., Sheen, J., and Li, J.F. 2017. A potent Cas9-derived gene activator for plant and mammalian cells. *Nature Plants*, *3*(12), 930–936. https://doi.org/10.1038/s41477-017-0046-0

Lu, S. 2019. Salvia miltiorrhiza: An economically and academically important medicinal plant. *The Salvia miltiorrhiza Genome*, 1–15. https://link.springer.com/chapter/10.1007/

Lu, D., Yuan, X., Kim, S.J., Marques, J.V., Chakravarthy, P.P., Moinuddin, S.G., Luchterhand, R., Herman, B., Davin, L.B., and Lewis, N.G. 2017. Eugenol specialty chemical production in transgenic poplar (*Populus tremula* × *P. alba*) field trials. *Plant Biotechnology Journal*, *15*(8), 970–981. https://onlinelibrary.wiley.com/doi/10.1111/pbi.12692

Luthra, R., Roy, A., Pandit, S., and Prasad, R. 2021. Biotechnological methods for the production of ginsenosides. *South African Journal of Botany*, *141*, 25–36. www.sciencedirect.com/science/article/pii/S025462992100154X?via%3Dihub

Maes, L., Van Nieuwerburgh, F.C., Zhang, Y., Reed, D.W., Pollier, J., Vande Casteele, S.R., Inzé, D., Covello, P.S., Deforce, D.L., and Goossens, A. 2011. Dissection of the phytohormonal regulation of trichome formation and biosynthesis of the antimalarial compound artemisinin in Artemisia annua plants. *New Phytologist*, *189*(1), 176–189. https://nph.onlinelibrary.wiley.com/doi/10.1111/j.1469-8137.2010.03466.x

Mahfouz, M.M., Piatek, A., and Stewart, Jr C.N. 2014. Genome engineering via TALENs and CRISPR/Cas9 systems: challenges and perspectives. *Plant Biotechnology Journal*, *12*(8), 1006–1014. https://doi.org/10.1111/pbi.12256

Maghraby, Y.R., El-Shabasy, R.M., Ibrahim, A.H., and El-Said Azzazy, H.M. 2023. Enzyme immobilization technologies and industrial applications. *ACS Omega*, *8*(6), 5184–5196. https://pubs.acs.org/doi/10.1021/acsomega.2c07560

Malla, A., Shanmugaraj, B., Srinivasan, B., Sharma, A., and Ramalingam, S. 2020. Metabolic engineering of isoflavonoid biosynthesis by expressing Glycine max isoflavone synthase in Allium cepa L. for genistein production. *Plants*, *10*(1), 52. www.mdpi.com/2223-7747/10/1/52

Mamdouh, D., and Smetanska, I. 2022. Optimization of callus and cell suspension cultures of Lycium schweinfurthii for improved production of phenolics, flavonoids, and antioxidant activity. *Horticulturae*, *8*(5), 394. www.mdpi.com/2311-7524/8/5/394

Martin, A., Chopra, R., and Nicassio, P.M. 2021. Nonpharmacologic pain management in inflammatory arthritis. *Rheumatic Disease Clinics*, *47*(2), 277–295. www.sciencedirect.com/science/article/abs/pii/S0889857X20301447?via%3Dihub

Meng, T., Xiao, D., Muhammed, A., Deng, J., Chen, L., and He, J. 2021. Anti-inflammatory action and mechanisms of resveratrol. *Molecules*, *26*(1), 229. www.mdpi.com/1420-3049/26/1/229

Mora-Vásquez, S., Wells-Abascal, G.G., Espinosa-Leal, C., Cardineau, G.A., and García-Lara, S. 2022. Application of metabolic engineering to enhance the content of alkaloids in medicinal plants. *Metabolic Engineering Communications*, *14*, e00194. https://doi.org/10.1016%2Fj.mec.2022.e00194

Nair, P., Misra, A., Singh, A., Shukla, A.K., Gupta, M.M., Gupta, A.K., Gupta, V., Khanuja, S.P.S., and Shasany, A.K. 2013. Differentially expressed genes during contrasting growth stages of Artemisia annua for artemisinin content. *PLoS One*, *8*(4), e60375. www.ncbi.nlm.nih.gov/pmc/articles/PMC3616052/

Negi, S., Syed, N., and Pant, M. 2024 . Toxic compounds associated with some antimalarial plant species. In: Husen, A. (ed.) *Antimalarial Medicinal Plants* .CRC Press, Taylor and Francis Group . https://doi.org/10.1201/9781003378396

Nin, S., Bennici, A., Roselli, G., Mariotti, D., Schiff, S., and Magherini, R. 1997. Agrobacterium-mediated transformation of Artemisia absinthium L. (wormwood) and production of secondary metabolites. *Plant Cell Reports, 16* , 725–730. https://link.springer.com/article/10.1007/s002990050310

Ogbe, A.A., Finnie, J.F., and Van Staden, J. 2020. The role of endophytes in secondary metabolites accumulation in medicinal plants under abiotic stress. *South African Journal of Botany*, *134*, 126–134. https://doi.org/10.1016/j.sajb.2020.06.023

Oksman-Caldentey, K.M., and D. Inzé. 2004. Plant cell factories in the post-genomic era: New ways to produce designer secondary metabolites. *Trends in Plant Science,9*(9), 433–440. www.cell.com/trends/plant-science/abstract/S1360-1385(04)00177-3?_returnURL=https%3A%2F%2Flinkinghub.elsevier.com%2Fretrieve%2Fpii%2FS1360138504001773%3Fshowall%3Dtrue

Osakabe, Y., and Osakabe, K. 2015. Genome editing with engineered nucleases in plants. *Plant and Cell Physiology*, *56*(3), 389–400.

Pant, M., Bist, P., Gusain, M.P., and Pant, K. 2023. Effect of pH and gelling agents in nutrient medium on in vitro shoot multiplication of *Swertia chirata* Buch.-Ham ex Wall: An endangered medicinal herb. *Materials Today Proceedings*, *732*(1), 9–12. https://doi.org/10.1016/j.matpr.2022.08.502

Pant, P., Pandey, S., and Dall'Acqua, S. 2021. The influence of environmental conditions on secondary metabolites in medicinal plants: A literature review. *Chemistry & Biodiversity*, *18*(11) e2100345. https://doi.org/10.1002/cbdv.202100345

Pandey, V., Ansari, W.A., Misra, P., and Atri, N. 2017. Withania somnifera: Advances and implementation of molecular and tissue culture techniques to enhance its application. *Frontiers in Plant Science*, *8*, 1390. www.frontiersin.org/journals/plant-science/articles/10.3389/fpls.2017.01390/full

Patra, N., Srivastava, A.K., and Sharma, S. 2013. Study of various factors for enhancement of artemisinin in Artemisia annua hairy roots. *International Journal of Chemical Engineering and Applications*, *4*(3), 157. www.ijcea.org/papers/284-I20008.

Perez-Matas, E., Hidalgo-Martinez, D., Escrich, A., Alcalde, M.A., Moyano, E., Bonfill, M., and Palazon, J. 2023. Genetic approaches in improving biotechnological production of taxanes: An update. *Frontiers in Plant Science*, *14*, 1100228. www.ncbi.nlm.nih.gov/pmc/articles/PMC9909606/

Pickens, L.B., Tang, Y., and Chooi, Y.-H. 2011. Metabolic engineering for the production of natural products. *Annual Review of Chemical and Biomolecular Engineering*, *2*(1), 211236. https://doi.org/10.1146%2Fannurev-chembioeng-061010-114209

Ramawat, K.G., Goyal, S., Kumar, N., and Sonie, K.C. 2009. Natural products of plant origin with therapeutic potentiality and their biotechnological production. *Plant Tissue Culture and Molecular Markers*, *92*, 247. https://books.google.com.np/books?id=iEGmRvdlV38C

Ramirez-Estrada, K., Castillo, N., Lara, J.A., Arro, M., Boronat, A., Ferrer, A., and Altabella, T. 2017. Tomato UDP-2glucose sterol glycosyltransferases: a family of developmental and stress regulated genes that encode cytosolic and membrane-associated forms of the enzyme. *Frontiers in Plant Science*, *8*, 984. https://europepmc.org/article/med/28649260

Rao, S.R., and Ravishankar, G.A. 2002. Plant cell cultures: Chemical factories of secondary metabolites. *Biotechnology Advances*, *20*(2), 101–153.

Raskin, I., Ribnicky, D.M., Komarnytsky, S., Ilic, N., Poulev, A., Borisjuk, N., Brinker, A., Moreno, D.A., Ripoll, C., Yakoby, N., and O'Neal, J.M. 2002. Plants and human health in the twenty-first century. *Trends in Biotechnology*, *20*(12), 522–531.

Rudell, D.R., Mattheis, J.P., Fan, X., and Fellman, J.K. 2002. Methyl jasmonate enhances anthocyanin accumulation and modifies production of Phenolics and Pigments in 'Fuji' Apples. *Journal of the American Society for Horticultural Science, 127*(3), 435–441. https://journals.ashs.org/jashs/view/journals/jashs/127/3/article-p435.xml

Ryan, C.M. 2020. Anti-diabetic and anti-obesity activities of cocoa (*Theobroma cacao*) via physiological enzyme inhibition. Doctoral dissertation, Virginia Tech. www.semanticscholar.org/paper/Anti-Diabetic-and-Anti-Obesity-Activities-of-Cocoa-Ryan/9462d87ec97da9ec7367c693c0c686d36a495cd2

Saikia, M., Shrivastava, K., and Singh, S. S. 2013. Effect of culture media and growth hormones on callus induction in Aquilaria malaccensis Lam., a medicinally and commercially important tree species of North East India. *Asian Journal of Biological Science*, *6*(2), 96–105.

Salehi, B., Mishra, A.P., Nigam, M., Sener, B., Kilic, M., Sharifi-Rad, M., Fokou, P.V.T., Martins, N. and Sharifi-Rad, J. 2018. Resveratrol: A double-edged sword in health benefits. *Biomedicines*, *6*(3), 91. www.mdpi.com/2227-9059/6/3/91

Salehi, M., Karimzadeh, G, and Naghavi, M.R. 2019. Synergistic effect of coronatine and sorbitol on artemisinin production in cell suspension culture of Artemisia annua L. cv. Anamed. *Plant Cell, Tissue and Organ Culture (PCTOC), 137*, 587597. https://link.springer.com/article/10.1007/s11240-019-01593-8

Sharma, N., Gauchan, D.P., Dhakal, A., Luitel, A., Shakya, S., and Shakya, R. 2015. Establishment of regenerative callus, cell suspension system and molecular characterization of Stevia rebaudiana Bertoni for the production of stevioside in in vitro. *International Journal for Research in Applied Science and Engineering Technology*, *3*(3), 133–148. www.researchgate.net

Sharma, G., Poudel, P., Lamichhane, S., Lamichhane, G., and Bastola, T. 2023. Vitis spp.(Vitis elongata (Roxb.) Wall. ex MA Lawson, Vitis flexuosa Thunb, Vitis heyneana Roem. & Schult, Vitis himalayana [Royle] Brandis, Vitis repanda (Vahl) Wight & Arn, Vitis vinifera L.). In *Himalayan Fruits and Berries* (pp. 465–479). Academic Press. www.sciencedirect.com/science/article/abs/pii/B9780323855914000374?via%3Dihub

Shen, Q., Lu, X., Yan, T., Fu, X., Lv, Z., Zhang, F., Pan, Q., Wang, G., Sun, X., and Tang, K. 2016. The jasmonate-responsive Aa MYC 2 transcription factor positively regulates artemisinin biosynthesis in Artemisia annua. *New Phytologist*, *210*(4), 1269–1281. https://nph.onlinelibrary.wiley.com/doi/10.1111/nph.17453

Siddiqui, M. H., Ali, S., and Khan, M. N. 2021. Elicitation of secondary metabolites in callus cultures of *Hypericum perforatum*: A comprehensive review. *Plant Cell Tissue and Organ Culture*, *145*(1), 13–29.

Srivastava, S., and Srivastava, A.K. 2007. Hairy root culture for mass-production of high-value secondary metabolites. *Critical Reviews in Biotechnology*, *27*(1), 29–43. www.tandfonline.com/doi/full/10.1080/07388550601173918

Sykłowska-Baranek, K., Sygitowicz, G., Maciejak-Jastrzębska, A., Pietrosiuk, A., and Szakiel, A. 2022. Application of priming strategy for enhanced paclitaxel biosynthesis in Taxus×media hairy root cultures. *Cells,11*(13), 2062.

Subedi, T. Phytochemical studies of Taxus species and their uses in cancer treatment. *Journal of Interdisciplinary Studies, 6*, 160–171. www.nepjol.info/index.php/JJIS/article/view/19317/15808#.

Verpoorte, R., Contin, A., and Memelink, J. 2002. Biotechnology for the production of plant secondary metabolites. *Phytochemistry Reviews, 1*, 13–25. https://doi.org/10.1023/A:1015871916833

Vuong, T.V., Franco, C.M., and Zhang, W. 2014. Treatment strategies for high resveratrol induction in Vitis vinifera L. cell suspension culture. *Biotechnology Reports, 1–2*, 15–21. https://doi.org/10.1016/j.btre.2014.04.002

Wang, J.W., and Tan, R.X. 2022. Artemisinin production in *Artemisia annua* hairy root cultures with improved growth by altering the nitrogen source in the medium. *Biotechnology Letters , 24* , 1153–1156. https://link.springer.com

Wang, J.W., Zheng, L.P., Zhang, B., and Zou, T. 2009. Stimulation of artemisinin synthesis by combined cerebroside and nitric oxide elicitation in Artemisia annua hairy roots. *Applied Microbiology and Biotechnology, 85*, 285–292. https://doi.org/10.1007/s00253-009-2090-9

Wani, K.I., Choudhary, S., Zehra, A., Naeem, M., Weathers, P., and Aftab, T. 2021. Enhancing artemisinin content in and delivery from Artemisia annua: A review of alternative, classical, and transgenic approaches. *Planta,254* , 1–15.

Weathers, P.J., Arsenault, P.R., Covello, P.S., McMickle, A., Teoh, K.H., and Reed, D.W. 2011. Artemisinin production in Artemisia annua: studies in planta and results of a novel delivery method for treating malaria and other neglected diseases. *Phytochemistry Reviews, 10*, 173–183. https://link.springer.com/article/10.1007/s11101-010-9166-0

Wen, W.J., Xia, Z.H., and Tan, R. X. 2022. Elicitation on artemisinin biosynthesis in Artemisia annua hairy roots by the oligosaccharide extract from the endophytic Colletotrichum sp. *Journal of Integrative Plant Biology,44* (10), 1233.

Wink, M. 2018. Introduction: Biochemistry, role and biotechnology of secondary metabolites (From APR Volume 2). *Annual Plant Reviews,* 1–15. https://doi.org/10.1002/9781119312994.apr0012

Xiang, Y., Song, X., Qiao, J., Zang, Y., Li, Y., Liu, Y., and Liu, C. 2015. An ultrahigh-performance liquid chromatography method with electrospray ionization tandem mass spectrometry for simultaneous quantification of five phytohormones in medicinal plant Glycyrrhiza uralensis under abscisic acid stress. *Journal of Natural Medicines, 69*, 278–286. http://dx.doi.org/10.1055/s-2006-960369

Xie, Y., Meiling, D., Xuecui, Y., Guanfeng, W., Zhang, B., Lingxiang, C., Pengda,M., and Juane, D. 2022. MAPKK2/4/5/7-MAPK3-JAZs modulate phenolic acid biosynthesis in *Salvia miltiorrhiza. Phytochemistry, 199* , 113177.

Xu, H., Zhang, H., and Li, J. 2018. Protoplast culture and its application in secondary metabolite production. *Frontiers in Plant Science, 9*, 1484.

Yoshiharu, M.I.U.R.A., and Kazumasa, H.I.R.A.T.A. 1987. Isolation of vinblastine in callus culture with differentiated roots of Catharanthus roseus (L). G. Don. *Agricultural and Biological Chemistry, 51*(2), 611–614. www.jstage.jst.go.jp/article/bbb1961/51/2/51_2_611/_article

Zalatan, J.G., Lee, M.E., Almeida, R., Gilbert, L.A., Whitehead, E.H., La Russa, M., Tsai, J.C., Weissman, J.S., Dueber, J.E., Qi, L.S., and Lim, W.A. 2015. Engineering complex synthetic transcriptional programs with CRISPR RNA scaffolds. *Cell, 160*(1), 339–350. https://doi.org/10.1016/j.cell.2014.11.052

Zenk, M.H., El-Shagi, H., Arens, H., Stöckigt, J., Weiler, E.W., and Deus, B. 1977. Formation of the indole alkaloids serpentine and ajmalicine in cell suspension cultures of Catharanthus roseus. In *Plant Tissue Culture and Its Bio-technological Application: Proceedings of the First International Congress on Medicinal Plant Research, Section B, held at the University of Munich, Germany September 6–10, 1976* (pp. 27–43). Berlin, Heidelberg: Springer. https://link.springer.com/chapter/10.1007/978-3-642-66646-9_3

Zhang, X., Xu, G., Cheng, C., Lei, L., Sun, J., et al. (2021). Establishment of an Agrobacterium-mediated genetic transformation and CRISPR/Cas9-mediated targeted mutagenesis in Hemp (Cannabis Sativa L.). *Plant Biotechnology Journal, 19*, 1979–1987. https://doi.org/10.1111/pbi.13611

Zhang, B., Zheng, L.P, Yi Li, W., and Wen Wang, J. 2013. Stimulation of artemisinin production in Artemisia annua hairy roots by Ag-SiO2 core-shell nanoparticles. *Current Nanoscience*, *9*(3), 363–370. www.ingentaconnect.com/content/ben/cnano/2013/00000009/00000003/art00012

Zhao, J., Hu, Q., Guo, Y.Q., and Zhu, W.H. 2001. Effects of stress factors, bioregulators, and synthetic precursors on indole alkaloid production in compact callus clusters cultures of Catharanthus roseus. *Applied Microbiology and Biotechnology*, *55*, 693–698. https://doi.org/10.1007/s002530000568

Zhao, J., Davis, L.C., and Verpoorte, R. 2005, June. Elicitor signal transduction leading to production of plant secondary metabolites. *Biotechnology Advances*, *23* (4), 283–333.

Zhao, L., Wang, X., and Zhang, J. 2021. Hairy root culture: An efficient system for the production of secondary metabolites. *Frontiers in Plant Science*, *12*, 678908.

Zhao, Z., Zhang, Y., Li, W., Tang, Y., and Wang, S. 2023. Transcriptomics and physiological analyses reveal changes in Paclitaxel production and physiological properties in Taxus cuspidata suspension cells in response to elicitors. *Plants*, *12*(22), 3817. www.mdpi.com/2223-7747/12/22/3817

Zheng, L.P., Tian, H., Yuan, Y.F., and Wang, J.W. 2016. The influence of endophytic Penicillium oxalicum B4 on growth and artemisinin biosynthesis of in vitro propagated plantlets of Artemisia annua L. *Plant Growth Regulation*, *80*, 93–102. https://doi.org/10.1007/s10725-016-0162-2

Zheng, M.Y., and Konzak, C.F. 1999. Effect of 2,4-dichlorophenoxyacetic acid on callus induction and plant regeneration in another culture of wheat (Triticum aestivumL.). *Plant Cell Rep ort*, *19*(1), 69–73.

Zhou, Z., Tan, H., Li, Q., Li, Q., Wang, Y., Bu, Q., Li, Y., Wu, Y., Chen, W., and Zhang, L. 2020. Trichome and artemisinin regulator 2 positively regulates trichome development and artemisinin biosynthesis in *Artemisia annua*. *New Phytologist*, *228*(3), 932–945. https://nph.onlinelibrary.wiley.com/doi/10.1111/nph.16777

Zhu, Q., Wang, B., Tan, J., Liu, T., Li, L., and Liu, Y.G. 2020. Plant synthetic metabolic engineering for enhancing crop nutritional quality. *Plant Communications*, *1*(1). https://doi.org/10.1016/j.xplc.2019.100017

Zia, M., Mannan, A., and Chaudhary, M.F. 2007. Effect of growth regulators and amino acids on artemisinin production in the callus of Artemisia absinthium. *Pakistan Journal of Botany (Pakistan)*, 39(3). https://api.semanticscholar.org/CorpusID:10082233

12 Genetic Control of Stress Hormone Production in Medicinal Plants

Mukesh Saran, Arshi Khanam, Rishabh Shrivastava Ronin*, Anupriya Chatterjee and Ashima Bagaria*

*Corresponding authors

12.1 INTRODUCTION

Plants constantly face biotic stressors like pathogens and herbivores, and abiotic stressors such as drought and extreme temperatures. Being immobile, they have evolved complex mechanisms to detect and respond to these challenges. Central to these responses are plant hormones, which regulate the perception and transmission of environmental signals. In the post-genomic era, research into plant stress tolerance has grown significantly. The regulation of stress-responsive genes is crucial for a plant's ability to endure stressful conditions and advances in functional genomics have significantly expanded our knowledge of gene expression and its link to stress tolerance. As stationary organisms, plants must adapt to changing conditions, which leads to the accumulation of various intracellular substances such as nucleic acids, proteins, carbohydrates, and amino acids. These substances act as signaling molecules, moving between cells through plasmodesmata.

Over the past two decades, intensive research has shed light on how these hormones orchestrate defense responses, revealing the sophisticated nature of plant adaptation and resilience. Understanding these pathways is crucial for enhancing plant growth and survival under stress conditions (Ku et al., 2018). Various signaling pathways help transduce external stimuli into cellular responses, working either independently or together to activate stress-responsive genes. These pathways create a cascade of events essential for the plant's stress response. Stress signaling in plants involves intricate processes such as receptor-coupled phosphorelay, phosphoinositol-induced calcium changes, and mitogen-activated protein kinase cascades, all of which lead to the activation of stress-responsive genes.

Stress-inducible genes fall into two main categories: those that produce proteins directly protecting cells from stress, and those involved in signal transduction and gene expression regulation. The first category includes proteins like chaperones, antifreeze proteins, and enzymes for osmolyte biosynthesis. The second category involves protein kinases, phosphatases, transcription factors, and enzymes of phospholipid metabolism.

To date, nine major classes of plant hormones have been identified that play roles in responding to environmental signals. These include auxin, cytokinin, gibberellin (GA), abscisic acid (ABA), ethylene (ET), brassinosteroid (BR), jasmonic acid (JA), salicylic acid (SA), and strigolactone. Among these, abscisic acid and jasmonic acid are particularly widespread in higher plants and have acquired significant attention in the study of plant stress responses and defense mechanisms (Choudhary et al., 2021). The expression of different hormones can be influenced by factors like chromatin structure and DNA methylation. The levels of chromatin acetylation and DNA methylation directly impact gene expression related to hormone production and release. Furthermore, RNA

DOI: 10.1201/9781003482659-12

polymerase-II, which attaches to the promoter region at the TATA box, needs various regulatory proteins to start transcription. Transcription factors (TFs) are crucial regulatory proteins that control genes involved in stress responses, affecting metabolism, physiological balance, cell development, and environmental interactions.

The increasing demand for pharmaceutical ingredients calls for the development of new medicinal plant varieties that can withstand harsh environmental conditions. To achieve this, it is crucial to understand how plants respond to these stresses as these hormones play a vital role in developing plants that can tolerate stress.

12.2 GENETIC REGULATION OF STRESS HORMONES IN PLANT ADAPTATION

Stress hormones play a crucial role in the adaptation of plants to environmental stresses, both abiotic and biotic (Bolaji et al., 2022; Wang et al., 2021). These stresses, along with climate change, cause significant crop losses and yield reduction globally. Understanding how stress hormones are genetically regulated is crucial for enhancing the survival capabilities of plants.

To combat these threats, plants have evolved adaptive mechanisms that detect stress signals and generate cellular responses through phytohormones like jasmonates, salicylates, and abscisic acid (ABA). These hormones help plants adapt by forming a defense network against environmental stresses, enhancing survival, and influencing the biosynthesis of medicinal compounds like alkaloids and flavonoids with therapeutic properties (Ghassemi et al., 2021; Rahman et al., 2023). The interaction between stress hormones and other factors such as auxins, gibberellins (GAs), and cytokinins is essential for regulating plant defenses against both abiotic and biotic stresses. For example, ABA is vital for abiotic stress defense, while salicylic acid (SA), jasmonic acid (JA), and ethylene (ET) respond to biotic stresses. Recent research highlights the interplay among these hormones, which collectively regulate plant defenses(Verma et al., 2016).

Plants adapt to stresses with numerous factors contributing to their resistance, including signaling molecules, DNA-binding proteins, and stress-responsive genes, which coordinate tolerance at the molecular level. Stress-related molecules like ABA, JA, reactive oxygen species (ROS), calcium ions, various transcription factors (TFs), and other regulatory proteins interact in a complex manner to facilitate adaptation, referred to as "cross-talk"(Peleg and Blumwald, 2011;Salvi et al., 2021).

Understanding stress signaling pathways reveals a sequence involving signal perception, transduction, gene expression, and the activation of physiological and metabolic responses. Drought and salinity trigger signals perceived by ion channels, receptors, kinases, and other proteins at the membrane level. This perception initiates intracellular signals, including Ca^{2+}, ROS, and phytohormones, inducing stress-responsive genes that encode products enhancing tolerance, such as secondary metabolites and regulatory molecules.

Stress-inducible genes are categorized into those directly enhancing tolerance [e.g., heat shock proteins, LEA (Late embryogenesis abundant) proteins, detoxification enzymes] and those regulating downstream genes and signaling cascades [e.g., phytohormones like ABA, JA, SA, ethylene; signaling molecules like MAPKs (mitogen-activated protein kinase), CDPKs (Calcium-dependent protein kinases); stress-responsive TFs], illustrating how plants coordinate gene expression to adapt and survive in challenging environments (Qari and Tarbiyyah, 2021; Shinozaki and Yamaguci-Shinozaki, 2022).

Functional genomics studies gene functions through expression analysis. By examining gene expression, researchers discern gene functions under different conditions. Techniques such as Amplified Fragment Length Polymorphism (AFLP), Serial Analysis of Gene Expression (SAGE), and massively parallel signature sequencing allow for the simultaneous assessment of thousands of genes without prior genomic or transcriptomic knowledge. Advanced methods like microarrays and real-time PCR require sequence knowledge and are used to monitor gene activity during biological processes. Microarrays identify genes with differing expression levels between experimental and control conditions, providing insights into transcriptional changes and functional genomics. It does so by identifying genes with significant expression changes under varying conditions and clusters genes with similar patterns, revealing functional or regulatory similarities. Grouping genes by expression profiles helps identify those involved in the same biological processes.

12.3 THE ROLE OF CALCIUM IONS AND REACTIVE OXYGEN SPECIES IN PLANT STRESS RESPONSE

When plants are exposed to drought or salinity, their genetic regulation mechanisms activate complex signaling pathways. Various intracellular signaling molecules interact intricately to help plants adapt to these stresses. Key secondary messengers like reactive oxygen species (ROS) and calcium ions (Ca^{2+}) play significant roles in the initial response to these abiotic factors, orchestrating the plant's protective measures (Feno et al., 2019).

Calcium ions (Ca^{2+}) are essential secondary messengers in plants, particularly in response to salinity and drought, which cause ionic and osmotic stress. To maintain cellular homeostasis, plants use Salt Overly Sensitive (SOS) genes to manage ionic balance. The SOS3 gene encodes a protein that acts as a receptor for cytosolic Ca^{2+}, triggering a cascade that activates protective genes against salt stress. For example, the SOS1 gene encodes a plasma membrane Na^+/H^+ antiporter, which helps maintain osmotic balance by compartmentalizing or expelling excess Na^+. This process is mediated by SOS2, which encodes a CDPK. Ca^{2+} also interacts with other signaling molecules, including CDPKs, which play crucial roles in plant stress responses. Calcium-dependent protein kinases are crucial for plants' cellular and molecular responses to abiotic stress, enhancing resistance and adaptation. They modulate metabolic pathways and are vital in synthesizing important organic molecules through phosphorylation. CDPKs can phosphorylate phenylalanine ammonia-lyase (PAL), a key enzyme in stress defense, linking it to secondary metabolite biosynthesis and stress resistance (Demidchik and Shabala, 2018).

During abiotic stresses like salinity and drought, excess ROS disrupts the cellular redox state, leading to oxidative stress. However, at low levels, ROS serve as crucial secondary messengers in stress response and secondary metabolism due to their spatiotemporal flexibility (Liu and He, 2016).

There is a synergistic interaction between Ca^{2+} and ROS signaling in plant defense (Figure 12.1). The RBOH gene family encodes NADPH oxidase, a key enzyme in producing extracellular ROS. This enzyme is activated by phosphorylation through CDPKs, requiring increased Ca^{2+} levels. Conversely, elevated ROS levels raise Ca^{2+} concentrations by interacting with Ca^{2+}-permeable cation channels like VICCs and HACCs, leading to a transient increase in cytosolic Ca^{2+}, which is crucial for various physiological and regulatory activities.

Like Ca^{2+}, ROS interact with signalling molecules such as MAPKs, initiating cascades that transduce stress signals. ROS enhances the activity of the ANP1 kinase in Arabidopsis and its homolog NPK1 in tobacco, which triggers downstream MAPK cascades. ANP1 activates AtMPK3 and AtMPK6, which phosphorylate defense-related transcription factors like WRKY and ERF. Numerous studies confirm the link between ROS and MAPKs in various plants, highlighting their role in abiotic stress responses (Singh et al., 2017).

These biosynthesis pathways are controlled by a complex network of genes and transcription factors responding to internal and external cues. Understanding these genetic controls allows researchers to manipulate stress hormone levels through breeding or genetic engineering, resulting in plants with enhanced stress resistance and improved medicinal properties. This offers sustainable solutions for agriculture and the pharmaceutical industry (Kajla et al., 2023).

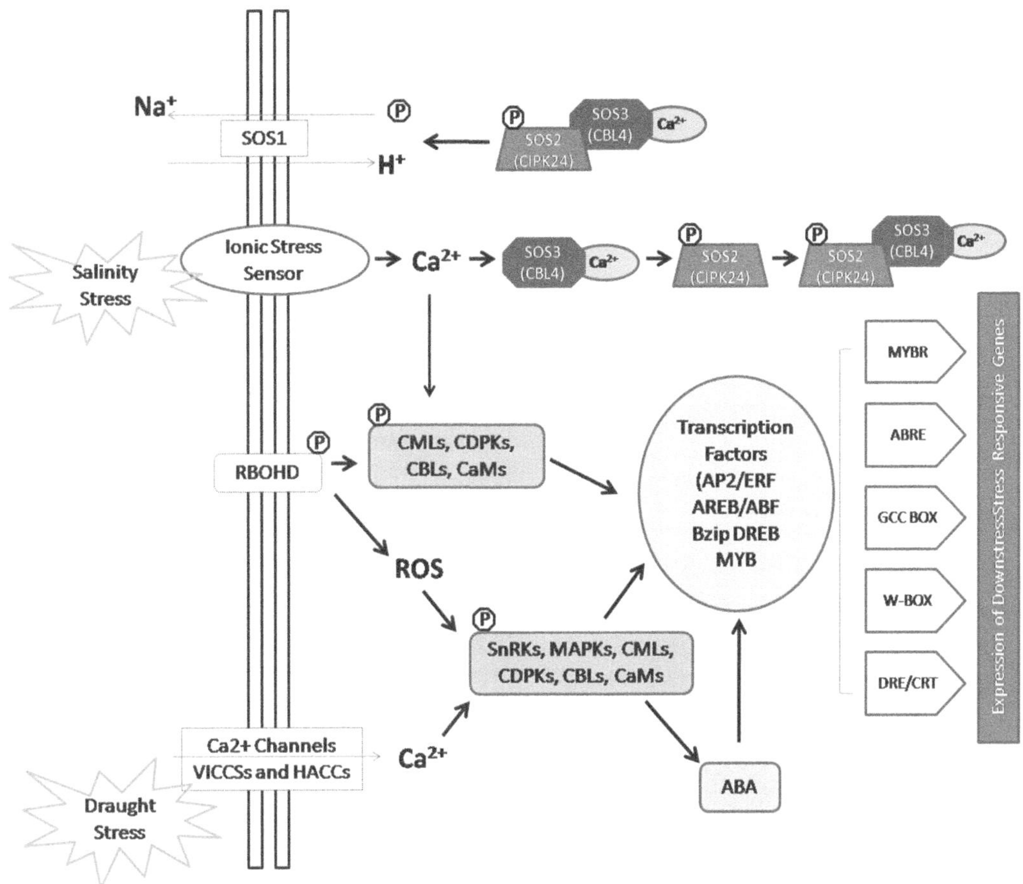

FIGURE 12.1 Synergistic interaction between Ca^{2+} and ROS signaling in plant defense. (Modified from Qari and Tarbiyyah, 2021.)

12.4 BIOSYNTHESIS PATHWAYS OF STRESS HORMONES IN PLANTS

Metabolic engineering is crucial for improving the production of beneficial secondary metabolites and enhancing plant resistance. There is a strong correlation between stress-induced genes and secondary metabolite production. Some secondary metabolite-encoding genes are considered stress-induced genes since they play protective roles and are induced during exposure to stress (Pandey et al., 2023).

Plants host diverse phytohormone pathways, and significant advancements have been made in the study of phytohormone biology and synthesis over the past decade. New tools and methods have been developed, leading to the discovery of phytohormone substrates, intermediates, and final products. The biosynthetic pathways of plant hormones have been elucidated, and extensive research has identified the genes in the plant genome encoding the enzymes that catalyze various stages of phytohormone synthesis. The various chemical structures of phytohormones are pivotal for their diverse biological functions and biosynthesis (Qari and Tarbiyyah, 2021).

12.4.1 Abscisic Acid (ABA) Biosynthesispathway

Abscisic acid is essential for plant physiological processes such as seed dormancy, germination, and abiotic stress responses, as well as cellular processes like seed development and vegetative growth.

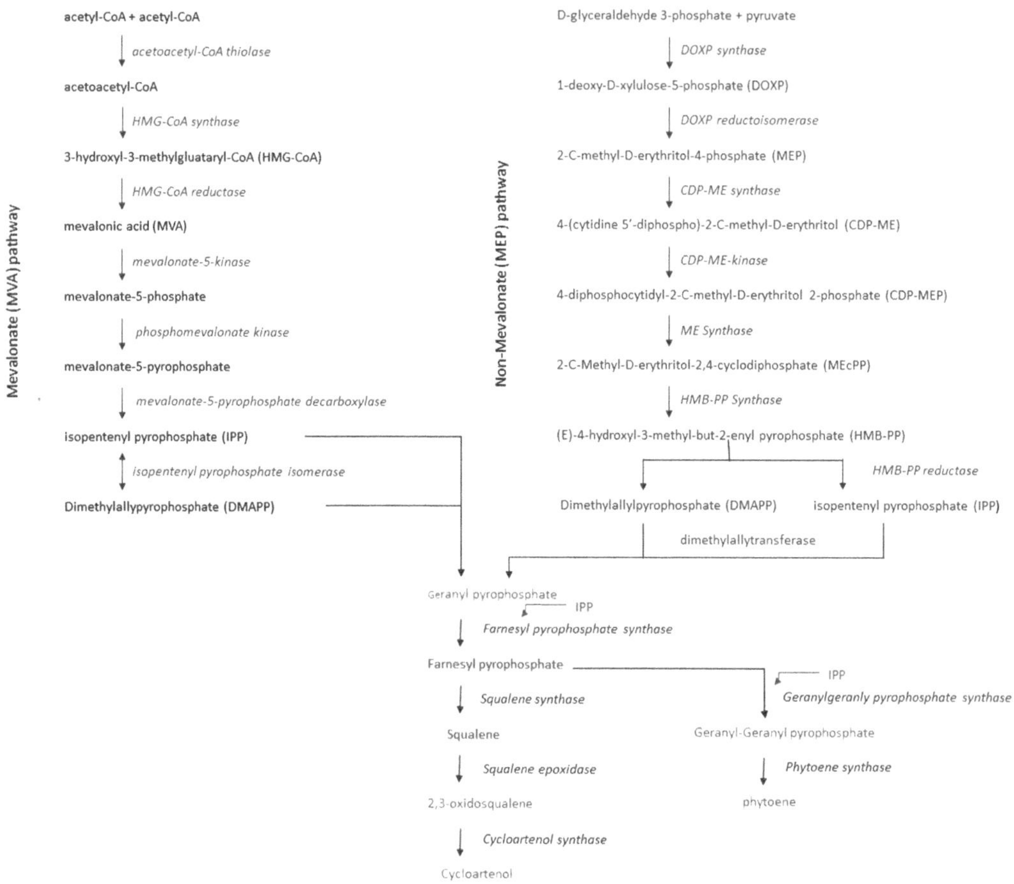

FIGURE 12.2 Biosynthesis pathways of isoprenoids through the non-mevalonate (MEP) pathway and the mevalonate (MVA) pathway. (Modified from Bajguz and Piotrowska-Niczyporuk, 2023.)

High levels of ABA are typically found in young tissues, and it is primarily synthesized in vascular tissues and transported to target tissues via the xylem and phloem, allowing for bidirectional transport between roots and shoots. ABA is found in both lower and higher plants. The naturally occurring form of ABA is S-(+)-ABA, which is biologically active, whereas trans-ABA is inactive. Figure 12.2 shows the biosynthesis pathways of isoprenoids through the non-mevalonate (MEP) pathway (- common in plastids of plants and many bacteria) and the mevalonate (MVA) pathway (-typical in eukaryotic cytosol and some archaea) (Seo and Koshiba, 2002).

The MEP pathway starts with D-glyceraldehyde 3-phosphate and pyruvate, progressing through a series of enzyme-catalyzed steps to produce 1-deoxy-D-xylulose-5-phosphate (DOXP), which is then converted to 2-C-methyl-D-erythritol-4-phosphate (MEP). Subsequent steps lead to the formation of 4-diphosphocytidyl-2-C-methyl-D-erythritol-2-phosphate (CDP-MEP), then to (E)-4-hydroxy-3-methyl-but-2-enyl pyrophosphate (HMB-PP), and finally to isopentenyl pyrophosphate (IPP) and dimethylallyl pyrophosphate (DMAPP).

The MVA pathway begins with acetyl-CoA, which undergoes several transformations, including conversion to 3-hydroxy-3-methylglutaryl-CoA (HMG-CoA), mevalonic acid (MVA), and then to mevalonate-5-pyrophosphate, eventually yielding IPP and DMAPP. Both pathways converge at IPP and DMAPP, which are precursors for geranyl pyrophosphate and subsequently farnesyl pyrophosphate, squalene, and 2,3-oxidosqualene. These intermediates lead to the synthesis of vital isoprenoids like cycloartenol and phytoene, essential for sterols, carotenoids, and other terpenoids.

The biosynthesis pathway of abscisic acid (ABA), begins with phytoene and progresses through several carotenoid intermediates and enzymatic reactions. Phytoene is converted to phytofluene, then to ζ-carotene, neurosporene, and lycopene. Lycopene is cyclized to all-trans-β-carotene, which is hydroxylated to zeaxanthin. Zeaxanthin undergoes epoxidation to form antheraxanthin and then all-trans-violaxanthin. All-trans-violaxanthin is converted to all-trans-neoxanthin and 9-cis-violaxanthin, which are further transformed into xanthoxin. Xanthoxin is oxidized toabscisic aldehyde, which is finally oxidized to abscisic acid. Abscisic aldehyde can also be converted to abscisic alcohol, and abscisic acid can be transformed into abscisic acid-trans-diol.

12.4.2 Jasmonic Acid (JA) Biosynthesis Pathway

Jasmonates, including JA and its derivatives such as methyl jasmonate (MeJA), are crucial phytohormones involved in plant development and stress responses such as those against herbivores and pathogens. The biosynthesis pathways of jasmonates have been extensively explored in dicotyledonous plants like *Arabidopsis thaliana*, tobacco, and tomato. JA synthesis involves multiple branches of the LOX pathway, where LOX enzymes catalyze the production of fatty acid hydroperoxides from polyunsaturated substrates. In Arabidopsis, 13-LOX enzymes such as AtLOX2, 3, 4, and 6 are localized in the chloroplast, while 9-LOX enzymes like AtLOX1 and AtLOX5 are likely found in the cytosol.

The biosynthesis of JA initiates with the oxygenation of α-linolenic acid, a fatty acid found in chloroplast membranes, catalyzed by lipoxygenase (LOX) enzymes. This process leads to the formation of 12-oxo-phytodienoic acid (OPDA), which is further converted into JA through enzymes like allene oxide synthase (AOS) and allene oxide cyclase (AOC). Figure 12.3 outlines

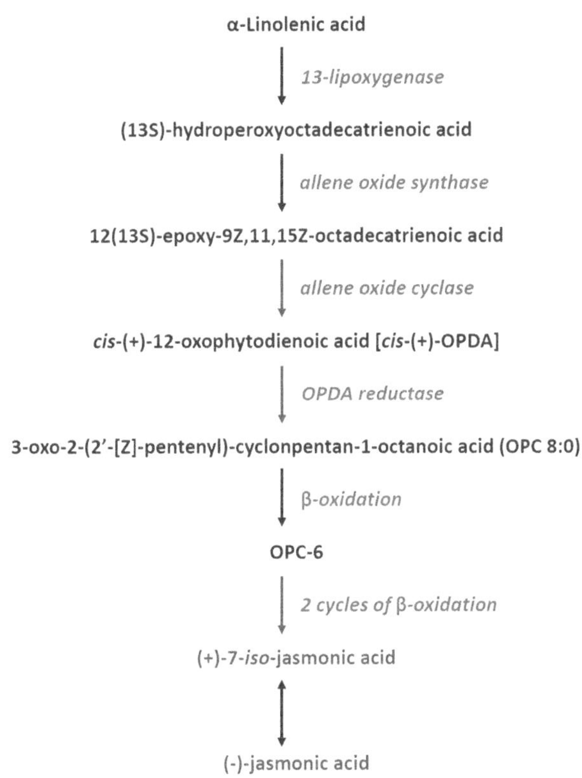

FIGURE 12.3 Biosynthesis of jasmonic acid. (Modified from Bajguz and Piotrowska-Niczyporuk, 2023.)

the biosynthesis pathway of jasmonic acid from α-linolenic acid. The process begins with the conversion of α-linolenic acid to (13S)-hydroperoxyoctadecatrienoic acid by 13-lipoxygenase. This is then transformed into 12(13S)-epoxy-9Z,11,15Z-octadecatrienoic acid by allene oxide synthase. The compound is subsequently converted to cis-(+)-12-oxophytodienoic acid (cis-(+)-OPDA) by allene oxide cyclase. Cis-(+)-OPDA is reduced to 3-oxo-2-(2'-[Z]-pentenyl)-cyclopentan-1-octanoic acid (OPC 8:0) by OPDA reductase, which then undergoes β-oxidation to form OPC-6. After two cycles of β-oxidation, OPC-6 is converted to (+)-7-iso-jasmonic acid, which is finally converted to (-)-jasmonic acid. Recent studies suggest an alternative pathway where JA originates from dinner-OPDA (dn-OPDA), indicating the diversity within the jasmonate family (Solomon and Muruganantham, 2016).

The production of JA is tightly regulated by a network of regulatory genes and transcription factors. Environmental cues such as wounding or pathogen attack induce the expression of these genes, leading to elevated JA levels and activation of plant defense mechanisms. Understanding these genetic controls enables scientists to develop strategies for enhancing JA production in medicinal plants, thereby augmenting their capacity to produce bioactive compounds with therapeutic potential.

12.5 GENETIC REGULATION OF STRESS HORMONE PRODUCTION

Genetic regulation of stress hormone production is crucial for plants to adapt and thrive under challenging conditions. Key stress hormones like abscisic acid (ABA), salicylic acid (SA), jasmonic acid (JA), and ethylene are controlled by intricate genetic networks. These networks include signaling pathways where specific genes respond to stress signals. For example, drought triggers the ABA biosynthesis pathway, activating genes that conserve water and regulate stomatal closure. Similarly, high salinity disrupts cellular ion balance, prompting ABA production to maintain osmotic equilibrium. Cold stress also induces ABA synthesis, stabilizing cellular structures and activating genes that protect against freezing temperatures. Transcription factors such as DREB (dehydration-responsive element-binding protein) and MYC/MYB are pivotal in these responses by binding to promoters of stress-responsive genes. Important genes in the ABA pathway, such as zeaxanthin epoxidase (ABA1) and 9-cis-epoxy carotenoid dioxygenase (NCED), catalyze essential steps in ABA production. NCED is a critical enzyme in the ABA biosynthesis pathway, with its gene expression finely tuned by environmental signals. Transcription factors like ABA-responsive element-binding proteins (AREB/ABF) and MYC/MYB play significant roles in ABA signaling by binding to ABA-responsive elements (ABREs) in stress-responsive gene promoters. ABI5, another crucial transcription factor in ABA signaling, oversees the expression of genes linked to stress responses. Moreover, epigenetic changes like DNA methylation and histone modifications further fine-tune the expression of stress-related genes (Esmaeili et al., 2022).

Similarly, mechanical damage from herbivores or detection of pathogens initiates rapid JA biosynthesis through regulatory genes (Figure 12.4). JA orchestrates defense responses by inducing toxic metabolites that deter herbivores and inhibit pathogen growth. It also regulates defense-related gene expression, bolstering the plant's resistance to biotic stress. Genes like AOS and AOC are pivotal in converting precursor molecules into active JA, in the JA biosynthesis pathway. MYC2, a central transcription factor in JA signaling, regulates the expression of multiple JA-responsive genes to coordinate defense responses and stress adaptation. Jasmonate ZIM-domain (JAZ) proteins act as repressors in the JA pathway, degraded upon JA exposure to release MYC2 and activate target genes. Understanding the promoters and transcription factors that govern these genes is essential for adjusting JA levels in plants (Parmar et al., 2017).

Both ABA and JA pathways exemplify intricate regulatory networks involving multiple genes and transcription factors that fine-tune plant responses to environmental challenges. These networks not only enhance our grasp of plant physiology but also provide potential avenues for enhancing

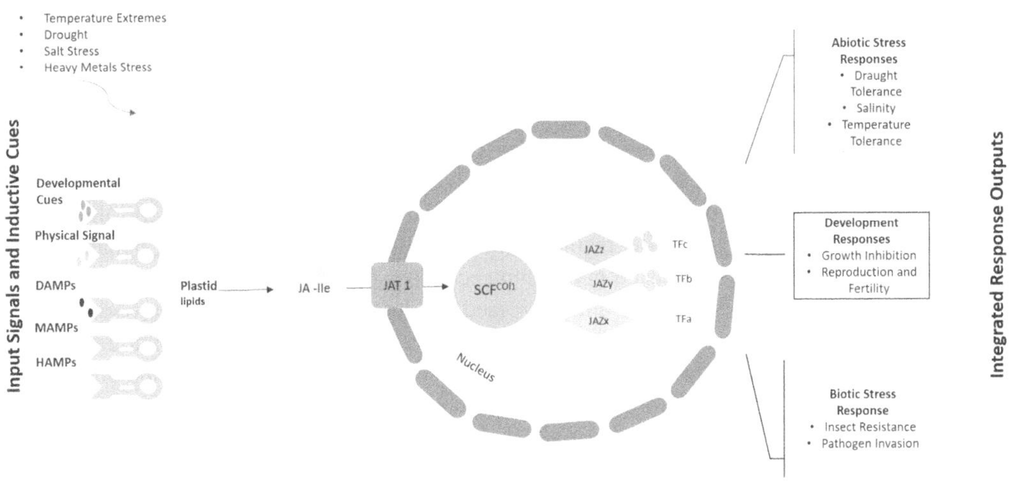

FIGURE 12.4 Core components of Jasmonoyl-L-Isoleucine (JA-Ile) signaling pathway. (Modified from Wang et al., 2021.)

TABLE 12.1
Regulatory genes associated with plant hormone

Plant Hormone	Regulatory Genes	Transcription Factors
Salicylic Acid (SA)	PAL (phenylalanine ammonia-lyase)	TGA (TGACG-binding factor)
Salicylic Acid (SA)	ICS1 (iso-chorismate synthase)	NPR1 (non-expression of pathogenesis-related genes 1)
Jasmonic Acid (JA)	AOS (allene oxide synthase)	MYC2
Abscisic Acid (ABA)	ABA1 (ZEP zeaxanthin epoxidase)	MYC/MYB
Jasmonic Acid (JA)	AOC (allene oxide cyclase)	JAZ (jasmonatezimdomain)
Gibberellins (GAs)	GA3ox (GA 3oxidase)	GAMYB
Ethylene	ACO (ACC Oxidase)	ERF (ethylene response factors)
Ethylene	ACS (ACC Synthase)	EIN3/EIL (ETHYLENEINSENSITIVE3/EIN3LIKE)
Gibberellins (GAs)	GA20ox (GA 20Oxidase)	DELLA Proteins
Auxins	YUC (YUCCA)	AUX/IAA
Cytokinins	IPT (Isopentenyltransferase)	ARR (arabidopsis response regulators)
Auxins	TAA1 (tryptophan aminotransferase of arabidopsis)	ARF (Auxin Response Factors)
Abscisic Acid (ABA)	NCED (9-cis-epoxy carotenoid dioxygenase)	AREB/ABF (ABA-responsiveelementbinding proteins)
Cytokinins	CKX (cytokinin oxidase/dehydrogenase)	-

crop resilience through genetic engineering (Basu et al., 2023). For instance, manipulating the activity of transcription factors like ABI5 in the ABA pathway and MYC2 in the JA pathway can significantly impact stress tolerance and promote hormone production. Table 12.1 shows transcription factors related to regulatory genes associated with Plant Hormones(Singh et al., 2023).

Moreover, stress-induced genes and secondary metabolite production are closely linked. Many genes for secondary metabolites activate in response to stress, producing compounds that protect against pathogens, herbivores, and abiotic stressors like drought and salinity. For instance, drought increases flavonoid and phenolic compound production, which have antioxidant properties. Alkaloids and terpenoids deter herbivores and pathogens (Abdulraheem et al., 2024). Understanding this link helps develop crops with enhanced stress tolerance and beneficial secondary metabolites through genetic and metabolic engineering.

Genetic and metabolic engineering approaches to improve plant stress resilience stress-induced genes and secondary metabolite production are strongly correlated in plants. When plants encounter stress, specific stress-induced genes are upregulated, increasing the synthesis of secondary metabolites which in turn helps in stress tolerance, resilience, and survival. By understanding and manipulating the correlation between stress-induced genes and secondary metabolites, researchers can develop crops with enhanced stress tolerance and increased production of beneficial compounds. Gene mining and transcriptome analysis provide comprehensive insights into the molecular mechanisms governing biosynthetic pathways. "Gene mining" involves the systematic exploration and identification of genes related to specific traits or biochemical pathways within an organism's genome and "Transcriptome analysis" involves studying all RNA molecules produced in a cell or organism, providing insights into which genes are active and to what extent under different conditions. "Metabolic engineering" refers to the manipulation of metabolic pathways in organisms, such as plants, to optimize the production of desired compounds or enhance specific traits, using information from Gene mining and transcriptome analysis.

Enhancing stress hormone production involves sophisticated strategies integrating both breeding and genetic engineering with metabolic pathway engineering.

In breeding, traditional methods focus on selecting and crossing plants with naturally high-stress hormone levels to consolidate desirable traits across generations, leveraging natural genetic variation to enhance stress resilience and hormone production. Whereas, in medicinal plants, genetic engineering tools such as CRISPR/Cas9, RNA interference (RNAi), and overexpression of regulatory genes can be used to modify stress hormone biosynthesis pathways. Where CRISPR/Cas9 can knock out genes that suppress hormone biosynthesis, RNAi silences competing genes, and overexpression of regulatory genes boosts hormone production (Das et al., 2024). By targeting genes involved in biosynthesis and regulation, such as those for ABA and JA, and/or by overexpressing enzymes in ABA biosynthesis or manipulatingJA pathways we can improve stress tolerance as these approaches optimize stress hormone levels, enhancing medicinal properties and resilience to environmental stressors (Esmaeili et al., 2022).

Metabolic pathway engineering, on the other hand, plays a pivotal role in redesigning biochemical pathways within plants. By identifying and modifying key enzymes and regulatory factors, researchers can enhance the efficiency and yield of stress hormone synthesis. This approach often involves integrating insights from fields like transcriptomics, proteomics, and metabolomics to comprehensively understand and manipulate metabolic networks (Tan et al., 2022). Transcriptomics analyzes RNA transcripts to identify stress-induced genes and their roles, while proteomics studies protein structures and functions to determine their involvement in stress responses and secondary metabolite production. Metabolomics examines metabolites to link them to biosynthetic genes and pathways. Bioinformatics integrates data from these fields to identify key regulatory genes and networks involved in secondary metabolite biosynthesis.

Together, these three strategies synergistically advance our ability to tailor plants for heightened stress tolerance and improved hormone production. They hold promise, not only for enhancing agricultural productivity and sustainability, but also for developing new therapeutic applications in medicinal plants.

12.6 APPLICATIONS OF TRANSCRIPTOME ANALYSIS

Transcriptome analysis describes and quantifies all RNA in a cell, offering insights into genetic regulation. It identifies actively transcribed genes and their expression variations under different conditions, revealing patterns of gene expression, alternative splicing events, non-coding RNA functions, and gene-expression dynamics. This is crucial for constructing gene regulatory networks and identifying transcription factors regulating biosynthetic pathways, facilitating advancements in genomics, plant breeding, and metabolic engineering (Lowe et al., 2017).

Technological approaches for transcriptome analysis encompass hybridization-based and sequencing-based methods (Liu et al., 2007). Hybridization-based methods utilize DNA microarrays to detect and quantify RNA transcripts, offering a broad overview of gene activity whereas Sequencing-Based Methods, such as Expressed Sequence Tag (EST) analysis, Serial Analysis of Gene Expression (SAGE), and Massively Parallel Signature Sequencing (MPSS), provide detailed and comprehensive insights into gene expression patterns.

Next-generation sequencing (NGS), particularly RNA-seq, revolutionizes transcriptome analysis by rapidly collecting vast amounts of data. RNA-seq provides comprehensive gene expression profiles, detects known and novel transcripts, uncovers alternative splicing events, and measures expression levels. It does not require prior genome knowledge, making it useful for studying non-model organisms and discovering new genes. RNA-seq explores complex regulatory networks, gene functions, and molecular processes, offering unparalleled resolution and depth for research in genomics, biology, and biotechnology (Satam et al., 2023).

For example, Transcriptome analysis of *C. roseus* reveals essential genes and pathways in terpenoid indole alkaloid (TIA) manufacture, including vinblastine and vincristine. It shows enzymes that convert precursor compounds as well as regulatory factors that influence gene expression, resulting in the discovery of novel elements in TIA biosynthesis. This reveals gene expression patterns under a variety of situations and informs techniques for increasing TIA production, which is critical for cancer treatment and biotechnology. RNA-seq data demonstrates that gene expression in TIAs varies between root and leaf tissues. These patterns are confirmed by quantitative RT-PCR, which also shows the functions of transcription factors, that identify diverse regulatory mechanisms and metabolic processes in several organs, with roots serving as the principal site for alkaloid synthesis which in turn paves the path for research into regulatory networks and metabolic engineering to increase alkaloid production for medications (Zhou and Chen, 2022).

12.7 FUTURE PROSPECTS

The intricate link between plant hormones and environmental stress is becoming more apparent, emphasizing their critical roles in plant adaptation and resilience. The future of plant hormones and stress management is extremely promising for sustainable agriculture, new medications, and a better knowledge of plant-environment interactions. Where advances in genomic technologies, like CRISPR-Cas9, have enabled researchers to precisely target specific genes involved in hormone biosynthesis and signaling, systems biology techniques and computational modeling are providing insights into the dynamic interconnections of hormone networks, allowing for more effective genetic interventions (Abdelnour et al., 2021). Enhancing plants' natural stress responses is especially crucial in the face of climate change and rising food security demands, as it promotes more sustainable and environmentally friendly agricultural operations. Harnessing genetic regulation of stress hormones can result in crops that are more resistant to environmental challenges, leading to consistent yields, less reliance on pesticides and fertilizers, and enhanced general crop health. Similarly, modulating stress hormone pathways in therapeutic plants has considerable promise. Plants that increase specific stress hormones can produce more secondary metabolites with therapeutic characteristics, perhaps leading to more effective herbal medications. Elevated JA levels, for example, may increase anti-inflammatory or anticancer-associated phytochemical synthesis.

12.8 CONCLUSION

Genetic engineering to fine-tune ABA and JA biosynthesis can increase the production of important chemicals, opening up new therapeutic opportunities. The journey is fraught with challenges. Ethical and regulatory issues in genetic engineering must be thoroughly addressed, as public perception and acceptance of genetically modified organisms vary widely. Addressing safety and environmental concerns is crucial and for that, collaboration among scientists, policymakers, and stakeholders is essential for developing robust regulatory frameworks and maintaining transparency in biotechnological advancements. Embracing the challenges and opportunities in the genetic control of stress hormone production can lead to a more resilient agricultural system and new therapeutic possibilities in medicinal plants. Understanding the genetic mechanisms underlying stress hormone production opens new avenues and continued research and application of these genetic pathways will drive significant advancements for improving crop resilience, boosting agricultural productivity, and harnessing the pharmacological potential of medicinal plants.

REFERENCES

Abdelnour, Sameh A., Long, Xie, Abdallah A. Hassanin, Erwei, Zuo, and Yangqing, Lu. 2021. The Potential of CRISPR/Cas9 Gene Editing as a Treatment Strategy for Inherited Diseases. *Frontiers in Cell and Developmental Biology* 9:699597.

Abdulraheem, Mukhtar Iderawumi, Yani, Xiong, Abiodun Yusuff Moshood, Gregorio Cadenas-Pliego, Hao Zhang, and Jiandong Hu. 2024. Mechanisms of Plant Epigenetic Regulation in Response to Plant Stress: Recent Discoveries and Implications. *Plants* 13(2):163.

Bajguz, Andrzej, and Alicja, Piotrowska-Niczyporuk. 2023. Biosynthetic Pathways of Hormones in Plants. *Metabolites* 13(8):884.

Basu, Umer, Syed Riaz Ahmed, Basharat Ahmad Bhat, Zunaira Anwar, Ahmad Ali, Aqsa Ijaz, Addafar Gulzar, Amir Bibi, Anshika Tyagi, Suresh M. Nebapure, Chengeshpur Anjali Goud, Shafat Ahmad, Ahanger, Sajad Ali, and Muntazir Mushtaq. 2023. A CRISPR Way for Accelerating Cereal Crop Improvement: Progress and Challenges. *Frontiers in Genetics* 13:866976.

Bolaji Umar, Olayinka, Lawal Amudalat Ranti, Abdulbaki Shehu Abdulbaki, Abdulra'uf Lukman Bola, Abdulkareem Khadijat Abdulhamid, Murtadha Ramat Biola, and Kayode Oluwagbenga Victor. 2022.Stresses in Plants: Biotic and Abiotic. in *Current Trends in Wheat Research*:1–8.

Choudhary, Sadaf, Andleeb Zehra, Mohammad Mukarram, M. Naeem, M. Masroor A. Khan, Khalid Rehman Hakeem, and Tariq Aftab. 2021. An Insight into the Role of Plant Growth Regulators in Stimulating Abiotic Stress Tolerance in Some Medicinally Important Plants. *Plant Growth Regulators*:75–100.

Das, Swati, Moonhyuk Kwon, and Jae-Yean Kim. 2024. Enhancement of Specialized Metabolites Using CRISPR/Cas Gene Editing Technology in Medicinal Plants. *Frontiers in Plant Science* 15:1279738.

Demidchik, Vadim, and Sergey Shabala. 2018. Mechanisms of Cytosolic Calcium Elevation in Plants: The Role of Ion Channels, Calcium Extrusion Systems and NADPH Oxidase-Mediated 'ROS-Ca2+ Hub.' *Functional Plant Biology* 45(2):9.

Esmaeili, Nardana, Guoxin Shen, and Hong Zhang. 2022. Genetic Manipulation for Abiotic Stress Resistance Traits in Crops. *Frontiers in Plant Science* 13:1011985.

Feno, Simona, Gaia Butera, Denis Vecellio Reane, Rosario Rizzuto, and Anna Raffaello. 2019. Crosstalk between Calcium and ROS in Pathophysiological Conditions. *Oxidative Medicine and Cellular Longevity* 1:9324018.

Ghassemi, Saeid, Nasser Delangiz, Behnam Asgari Lajayer, Davood Saghafi, and Filippo Maggi. 2021. Review and Future Prospects on the Mechanisms Related to Cold Stress Resistance and Tolerance in Medicinal Plants. *Acta Ecologica Sinica* 41(2):120–29.

Kajla, Mohini, Amit Roy, Indrakant K. Singh, and Archana Singh. 2023. Regulation of the Regulators: Transcription Factors Controlling Biosynthesis of Plant Secondary Metabolites during Biotic Stresses and Their Regulation by miRNAs. *Frontiers in Plant Science* 14:1126567.

Ku, Yee-Shan, Mariz Sintaha, Ming-Yan Cheung, and Hon-Ming Lam. 2018. Plant Hormone Signaling Crosstalks between Biotic and Abiotic Stress Responses. *International Journal of Molecular Sciences* 19(10):3206.

Liu, Fang, Tor-Kristian Jenssen, Jeff Trimarchi, Claudio Punzo, Connie L. Cepko, Lucila Ohno-Machado, Eivind Hovig, and Winston Patrick Kuo. 2007. Comparison of Hybridization-Based and Sequencing-Based Gene Expression Technologies on Biological Replicates. *BMC Genomics* 8(1): 153.

Liu, Yukun, and Chengzhong He. 2016. Regulation of Plant Reactive Oxygen Species (ROS) in Stress Responses: Learning from AtRBOHD. *Plant Cell Reports* 35(5):995–1007.

Lowe, Rohan, Neil Shirley, Mark Bleackley, Stephen Dolan, and Thomas Shafee. 2017. Transcriptomics Technologies. *PLOS Computational Biology* 13(5):e1005457.

Pandey, Praveen, Arpita Tripathi, Shweta Dwivedi, Kanhaiya Lal, and Tripta Jhang. 2023. Deciphering the Mechanisms, Hormonal Signaling, and Potential Applications of Endophytic Microbes to Mediate Stress Tolerance in Medicinal Plants. *Frontiers in Plant Science* 14:1250020.

Parmar, Nehanjali, Kunwar Harendra Singh, Deepika Sharma, Lal Singh, Pankaj Kumar, J. Nanjundan, Yasin Jeshima Khan, Devendra Kumar Chauhan, and Ajay Kumar Thakur. 2017. Genetic Engineering Strategies for Biotic and Abiotic Stress Tolerance and Quality Enhancement in Horticultural Crops: A Comprehensive Review. *3 Biotech* 7:1–35.

Peleg, Zvi, and Eduardo Blumwald. 2011. Hormone Balance and Abiotic Stress Tolerance in Crop Plants. *Current Opinion in Plant Biology* 14(3):290–95.

Qari, Sameer Hasan, and Ibrahim Tarbiyyah. 2021. The Genetic Regulation of Secondary Metabolic Pathways in Response to Salinity and Drought as Abiotic Stresses. *Applied Sciences* 11(15):6668.

Rahman, Shakeelur, Muhammad Iqbal, and Azamal Husen. 2023. Medicinal Plants and Abiotic Stress: An Overview. *Medicinal Plants*1–34.

Salvi, Prafull, Mrinalini Manna, Harmeet Kaur, Tanika Thakur, Nishu Gandass, Deepesh Bhatt, and Mehanathan Muthamilarasan. 2021. Phytohormone Signaling and Crosstalk in Regulating Drought Stress Response in Plants. *Plant Cell Reports* 40(8):1305–29.

Satam, Heena, Kandarp Joshi, Upasana Mangrolia, Sanober Waghoo, Gulnaz Zaidi, Shravani Rawool, Ritesh P. Thakare, Shahid Banday, Alok K. Mishra, Gautam Das, and Sunil K. Malonia. 2023. Next-Generation Sequencing Technology: Current Trends and Advancements. *Biology* 12(7):997.

Seo, Mitsunori, and Tomokazu Koshiba. 2002. Complex Regulation of ABA Biosynthesis in Plants. *Trends in Plant Science* 7(1):41–48.

Shinozaki, Kazuo, and Kazuko Yamaguchi-Shinozaki. 2022. Functional Genomics in Plant Abiotic Stress Responses and Tolerance: From Gene Discovery to Complex Regulatory Networks and Their Application in Breeding. *Proceedings of the Japan Academy, Series B* 98(8):470–92.

Singh, Chandrashekhar, Rajesh Saini, Richa Upadhyay, and Kavindra Nath Tiwari. 2023. Revealing the Epigenetic Mechanisms Underlying the Stress Response in Medicinal Plants. *Stress-responsive Factors and Molecular Farming in Medicinal Plants*:107–22.

Singh, Rachana, Parul Parihar, Samiksha Singh, Rohit Kumar Mishra, Vijay Pratap Singh, and Sheo Mohan Prasad. 2017. Reactive Oxygen Species Signaling and Stomatal Movement: Current Updates and Future Perspectives. *Redox Biology* 11:213–18.

Solomon, S., N. Muruganantham, and M. M.Senthamilselvi. 2016. Anticancer activity of abelmoschus esculentus (Flowers) against human liver cancer. *International Journal of Pharmacy and Biological Sciences* 6(3):154–57.

Tan, Yung Cheng, Asqwin Uthaya Kumar, Ying Pei Wong, and Anna Pick Kiong Ling. 2022. Bioinformatics Approaches and Applications in Plant Biotechnology. *Journal of Genetic Engineering and Biotechnology* 20(1):106.

Verma, Vivek, Pratibha Ravindran, and Prakash P. Kumar. 2016. Plant Hormone-Mediated Regulation of Stress Responses. *BMC Plant Biology* 16:1–10.

Wang, Yun, Salma Mostafa, Wen Zeng, and Biao Jin. 2021. Function and Mechanism of Jasmonic Acid in Plant Responses to Abiotic and Biotic Stresses. *International Journal of Molecular Sciences* 22(16):8568.

Zhou, Pengfei, and Mingxiang Chen. 2022. Exploration of the Mechanisms of Differential Indole Alkaloid Biosynthesis in Dedifferentiated and Cambial Meristematic Cells of Catharanthus Roseus Using Transcriptome Sequencing. *Frontiers in Genetics* 13:867064.

13 Genetic Engineering for Abiotic Stress Tolerance in Medicinal and Aromatic Plants

Shiwani Latwal, Anju Rao and Prabhleen Kaur*

*Corresponding author

13.1 INTRODUCTION

From time immemorable medicinal and aromatic plants (MAPs) have been an integral part of human health and customs. Secondary metabolism produces plant bioactive chemicals such as tannins, alkaloids, flavonoids, terpenoids, steroids, phenolics, saponins, sesquiterpenes, diterpenes, polyketides, and fatty acid derivatives, that are employed in medicine and therapy (Muhammad et al., 2019; Erb and Kliebenstein, 2020; Pant et al., 2022). Like primary metabolites, these compounds are not an absolute requirement for plants but are integral to evolutionary and ecological interactions with the surroundings including biotic and abiotic stresses. These metabolites are a result of plant resistance against extreme environmental conditions as they can serve as an energy reserve when facing a hostile climate. Nowadays, phytochemical compounds used in healthcare are derived from plants having medicinal fortune e.g. Taxol, a widely used anti-cancerous drug is obtained from *Taxus brevifolia* (Lai et al., 2013). Curcumin having anti-oxidant and anti-inflammatory properties is obtained from *Curcuma longa* (Hewlings and Kalman, 2017). Wogonine from *Scutellaria baicalensis* is an immunogenic cell death inducer (Li et al., 2018) and Genistein extracted from *Genista tinctoria* is anti-cancerous and effective against Alzheimer's disease (Tuli et al., 2019). Artemisinin and Quinine have anti-malarial activity procured from *Artemesia annua* and *Cinchona officinalis* respectively (Brown et al., 2003; Mohammadi et al., 2020). Similar to this, aromatic plants store volatile essential oils with characteristic order in their buds, stems, leaves, wood, roots, and other plant parts. The sweet and earthy aroma of these plants is due to the essential oil they contain. The plant's ability to communicate through essential oil synthesis can also help to protect itself from pathogenic attacks. These essential oils rich in terpenoids, phenols, and aromatic aldehydes are examined to have therapeutic effects on human health (Christaki et al., 2012; Zhang and Piao, 2023). Apart from this, fragrant essential oils hold a unique position in the cosmetic and flavoring industries (Sharma et al., 2023). A summary of various MAPs along with their bioactive compounds is given in Table 13.1. Studies conducted by various scientists suggested that physiological and environmental elements are driving factors for the synthesis of these metabolites. Many MAPs build and accumulate secondary metabolites in response to various abiotic stressors like temperature, drought, flood, heavy metals and salinity. However, prolonged exposure to such hostile conditions may disrupt biochemical pathways and genes that produce these metabolites (Tuteja, 2007; Pant et al., 2023). The free radicles generated react with other molecules, altering their metabolic activity and ultimately affecting total plant production (Zehra et al., 2020). Also, overexposure to such a hostile climate may cause serious cell damage and membrane disintegration ultimately leading to plant death.

DOI: 10.1201/9781003482659-13

TABLE 13.1
Therapeutic application of secondary metabolites in some common MAPs

Botanical Source	Family	2° Metabolites	Therapeutic Application	References
Artemesia annua	Asteraceae	Artemisinin, artemether, artemetin,artemin, artesin, rosmarinic acid, casticin, cirsilineol, chrysoplenetin	Antimalarial, anti-inflammatory, anti-cancerous, immune-modulatory properties	Shinyuy et al., 2023
Withania somnifera	Solanaceae	Withaferin, withanolides withanone, withasomniferols	Cytokine modulation, asthma, bronchitis, edema, rheumatism, carcinoma and arthritis	Sangwan and Sangwan, 2013
Eleutherococci radix	Araliaceae	Syringin (Eleutheroside B)	Increased phagocytosis of human leukocytes and granulocytes, increased level of interleukins	Asano et al., 1986
Panax ginseng	Araliaceae	Panaxyne, ginsenoynes and ginsenosides	Enhance memory and improve blood circulation, boost the immune system	Todorova et al., 2021 Truong and Jeong, 2022
Cannabis sativa	Cannabaceae	Cannabidiolic acid, cannabidiol, cannabigerol, cannabichromene, cannabinolic acid, cannabinoid and cannabinol.	Oxidative stress and inflammation	Appendino et al., 2011 Kopustinskiene et al., 2022
Berberis vulgaris	Berberidaceae	Berberine, ceptisine and palmatine	Antidiabetic, antibacterial, anti-inflammatory	Singh et al., 2021
Moringa oleifera	Moringaceace	Pterygospermin, moringine and niazirin	Anti-epileptic, anti-diabetic, anti-fertility, anti-asthmatic, anti-inflammatory	Paikra and Gidwani, 2017
Catharanthus roseus	Apocynaceae	Vingramine, vinblastine, catharanthamine, vincristine, cavincine, roseamine, neoleurocristine, vindolicine, vindolidine.	Anti-fungal, anti-mitotic, anti-inflammatory, cardiotonic, anti-viral, cytotoxic	Aslam et al., 2010
Taxus brevifolia	Taxaceae	Paclitaxel (taxol), taxuyunnanin C, taxifolin, taxiresinol, taxine B	Antileukemic, anti-cancerous and anti-tumor	Devanesan, 2023 Tomilova et al., 2023
Thymus algeriensis	Lamiaceae	Thymol, carvacrol, borneol, linalool, γ-terpinene	Cytotoxic activity, anti-microbial, dyspepsia, bronchitis, laryngitis, tonsillitis	Nikolić et al., 2014
Eucalyptus globus	Myrtaceae	Limonene, pinene, cymene, eucalyptin, linalool and pinocarveol.	Treatment of infections, anti-histamic, bronchitis, pneumonia, aching and stiffness	Shala and Gurrani, 2021
Mentha piperita	Lamiaceae	Eriocitrin and rosmarinic acid	Antiviral, antioxidant, antispasmodic, anti-microbial, anti-allergic	Trevisan et al., 2017
Santalum album	Santalaceae	Santalol, santalol, santene, santenol, teresantalol	Coolant, sedative, astringent activity, perfume industry, anti-poison and blood purifier	Sindhu et al., 2010
Cymbopogon winterianus	Poaceae	Citronellol, citronellal,geraniol	Antiseptic and antispasmodic	Simic et al., 2008
Salvia officinalis	Lamiaceae	Borneol, camphor, cineole, humulene, camphene, thujone and inflammatory limonene	Treatment of seizures, rheumatism, inflammation, dizziness, anticancer	Ghorbani and Esmaeilizadeh, 2017
Swertia chirata	Gentianaceae	Amarogentin, swerchirin, xanthones, iridoid glycosides, mangiferin, C-glucoflavones	Treatment of chronic fever, stomach and liver infections, anemia, liver and skin diseases	Pant et al., 2023

It has been evaluated that approximately 4,000 new secondary metabolites from different plant species are being discovered each year (Verpoorte et al., 2000; Gomez-Galera et al., 2007). According to data from the World Health Organisation (WHO), 80% of people in underdeveloped countries still utilise plants as a source of medicine. There has been an abrupt surge in demand for plant-derived drugs in recent years. Such an increment in the requirement for MAPs makes it unavoidable to use genomics tools to create genetically engineered plants that are much more resistant to unfavorable abiotic factors. Such novel genetic engineering techniques not only improve plant's stress tolerance but are also beneficial for its physiological growth and development ultimately boosting commercial yield.

13.2 CLASSIFICATION AND STATUS OF SECONDARY METABOLITES UNDER ABIOTIC STRESS

13.2.1 Terpenes

Terpenes constitute the largest group of secondary metabolites. The structure is composed of repeating five-carbon isoprene units, classifying them into: monoterpenes (C10), sesquiterpenes (C15), diterpenes (C20), and triterpenes (C30) and tetraterpenes (C40). So far, 30,000 terpenes have been discovered (Teoh, 2016). Several medicinally important compounds comprise terpenoids, for example, menthol obtained from mint is a monoterpenoid having a pleasant aroma and cooling effect, taxol is a diterpenoid having anti-cancerous properties and artemisinin a sesquiterpenoid exhibiting an anti-malarial effect. Sesquiterpenoids and triterpenoids are synthesized via cytosolic MVA pathway (Mevalonic acid pathway), while the plastidial MEP (Methylerythritol-4-phosphate pathway) pathway is a route in charge of the production of monoterpenoids, diterpenoids, and tetraterpenoids (Boncon et al., 2020). Under stressful circumstances, certain terpenoids may serve as defense mechanisms to combat both biotic and abiotic stress. It has recently been noted that salinity stress changed the volatile organic chemicals in sweet basil plants, causing them to release certain airborne signals that encouraged kin recipients to flower earlier and thus increase their chances of successful reproduction (Verdeguer et al., 2020). Also, temperature stress brought on by 4-10 °C drop in temperature encourages the buildup of terpenoids and flavonoids (He et al., 2023). Similarly, the accumulation of cineole, thujone and camphor (monoterpenoids) elevated under drought stress conditions in *Salvia officinalis* (Radwan et al., 2017). Essential oils concentration varies depending on the type of plant species and the degree of stress experienced.

13.2.2 Phenolics

Phenolics are essential for MAPs to survive under unfavorable surroundings. Increased phenolic synthesis ensures that the plant will withstand abiotic stress and remain resilient and competitive throughout unfavorable conditions (Kumar et al., 2020). Plant phenolics can be characterized by 'phenol', constituting a phenyl (C_6H_5) ring with attached hydroxyl substituents. The shikimic acid pathway is accountable for the emergence of phenolic compounds in plants (Mandal et al., 2010). The most prevalent compounds among phenolics are flavonoids. It comprises seven subgroups: flavonols, flavones, flavanones, isoflavonoids, anthocyanins and catechins. Phenolics improve plant metabolism, development and growth, biomass accumulation and seed germination (Raskin, 1992; Yalpani et al., 1994; Senaratna et al., 2000). Furthermore, stress-induced flavonoids can scavenge damaging reactive oxygen species. The major defensive strategy to stabilize deleterious outcomes of drought stress by plants is boosting the accumulation of phenolic substances. These compounds, for example polyphenols, shield plants from oxidative damage by assisting ascorbate-dependent antioxidant defense systems during stress (Sakihama et al., 2002). As part of an additional defensive mechanism against ROS, lipid alkoxyl radicals, a potent oxidant that participates in lipid peroxidation, is reduced by flavonoids and other polyphenols. (Zhu et al., 2009).

13.2.3 Nitrogen-Containing Secondary Compounds

Secondary metabolites with nitrogen in their structure include: Cyanogenic glycosides, alkaloids, glucosinolates and non-protein amino acids. The best-known function of alkaloids is their protective function as herbivores and insect deterrents (Saunders et al., 1992). Alkaloids are known for their numerous pharmacological effects. Examples of some of the commonly known alkaloids in plants are: morphine obtained from *Papaver somniferum*, quinine from *Cinchona officinalis*, atropine from *Atropa belladonna*, nicotine from *Nicotiana tabacum* and colchicine from *Colchicum autumnale* (Matsuura et al., 2015). However, their biosynthesis and accumulation are remarkably under the check of environmental factors. Da Silva Magedans et al. (2017), observed heat shock and cold stress conditions in *Psychotria brachyceras* leads to the accumulation of alkaloid brachycerine, which helps mitigate the ROS. *Catharanthus roseus* treated under short-term heat shock results in the rise of vinblastine, vindoline and catharanthine in the leaves of the seedlings. He finally concluded high temperatures could encourage the biosynthesis of different alkaloids (Guo et al., 2007). The roots and leaves of *Diplotaxis tenuifolia*, indicated a reduction in glucosinolate accumulation due to salinity and heat stress (Cocetta et al., 2018).

13.3 BIOCHEMICAL AND PHYSIOLOGICAL MECHANISMS OF ABIOTIC STRESS

Abiotic stress is accountable for a variety of physiochemical alterations in medicinal plants, including an excess of ROS (H_2O_2, $O2^-$, OH) leading to cellular impairment. To overcome such situations plants have evolved endogenous non-enzymatic and enzymatic ROS scavenging systems. An antioxidant defense system's enzymatic components include superoxide dismutase (SOD), ascorbic acid, catalase (CAT), glutathione peroxidase (GPX), and peroxiredoxins. The non-enzyme antioxidants such as glutathione, vitamin E and carotenoids also take part in the cellular control of toxic ROS. Furthermore, the stress tolerance process also implies reconciliation of phytohormones, overexpression of defense-related genes, synthesis of stress proteins, production of metabolic compounds, cellular membrane restructuration, etc (Figure 13.1) (Khan et al., 2020). It has been observed that, at lower concentrations, the hydrogen peroxide signaling pathway (H_2O_2,) holds a pivotal role in mitigating stress, as it activates the plant's defense system to attenuate the detrimental effect of abiotic conditions. However, greater concentrations negatively impact the plant's

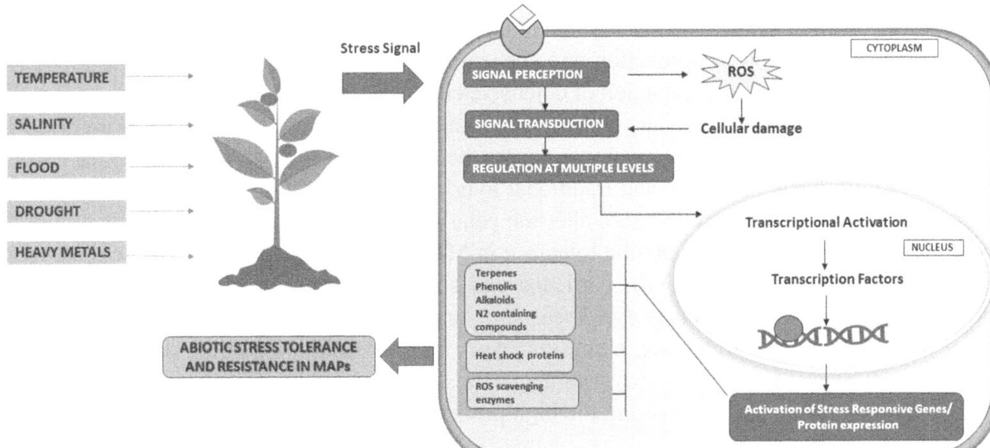

FIGURE 13.1 Diagrammatic representation of abiotic stress signalling and tolerance in plants.

TABLE 13.2
Abiotic stress responses in some medicinally important plants

Botanical Name	Family	Type of Abiotic Stress	Stress Response	Reference
Glycyrrhiza glabra	Fabaceae	Drought	Drought enhanced the synthesis of glycyrrhizin	Hosseini et al., 2018
Carthamus tinctorius	Asteraceae	Salinity	Enhanced phenolics and flavonoids content	Golkar and Taghizadeh, 2018
Camellia sinensis	Theaceae	Drought	Caffeine level amplified	Munivenkatappa et al., 2018
Ocimum tenuiflorum	Lamiaceae	Heavy metals	Enhanced biosynthesis of eugenol	Rai et al., 2004
Daucus carota	Apiaceae	Heat	Decrease in biosynthesis of α-terpinolene content	Semmar, 2024
Withania somnifera	Solanaceae	Heavy metal	Increase in phenolics	Khatun et al., 2008

defense mechanism (Xia et al., 2009). PP2C (protein phosphatase type 2C) implicated in abscisic acid (ABA) biosynthesis and defense-related genes help to reduce the harmful outcomes of drought stress. Similarly, for salt tolerance and solute homeostasis, the activation of a stress signaling mechanism SOS (Salt Overly Sensitive) was observed (Gupta and Huang, 2014). Ethylene-responsive transcription factors (ERF-VIIs) have been found to contribute to plants' ability to withstand submersion and waterlogging (Wei et al., 2019., Yu et al., 2019). Heat stress proteins Hsf and Hsp genes are responsible for thermotolerance and play a pivotal role in attenuating heat stress in plants (Qu et al., 2013). Apart from this, a wide range of osmolytes having less molecular mass, including proline, sugar alcohols, ammonium compounds, and sulphonium compounds, can accumulate in certain plant species to combat stress (Wahid et al., 2007). Some abiotic stress and their effects on plants are enumerated in Table 13.2.

13.4 MITIGATING ABIOTIC STRESS IN MAPS USING GENETIC ENGINEERING TOOLS

Transgenic MAPs using gene target approach: MAPs growing under different environmental stress conditions also express differences in concentration of chemical constituents. Many plant species have shown a surge in secondary metabolites under controlled stress. The two mechanisms that account for variation in production and concentration of chemical constituents are a passive shift and an active shift. The production of defensive chemicals uses plant energy, which causes the passive shift (Selmar and Kleinwachter, 2013) and the up-regulation of the enzymes involved in the secondary metabolite synthesis pathway causes the active shift (Radwan et al., 2017). As stress tolerance is multigenic in nature, and traditional hybridization practices took a long time to show results, genetic advancement utilized to alter the gene can lessen the detrimental effects of abiotic pressures on plants. To improve abiotic tolerance in MAPs in a short period, transgenic plant technology can be employed to target the particular genes and pathways responsible for heat, drought and salt stress using biological (via *Agrobacterium*), and direct gene transfer approaches (physical, chemical and electrical). Moreover, significant work has been done regarding the production of stress-tolerant plants due to the discoveries about stress promoters. The strong promoters are used for creating transgenic plants, ensuring the high productivity of secondary metabolites throughout the year.

As more than one gene at the quantitative trait locus (QTL) controls the complex abiotic stress mechanism, it is foremost to recognize the prospective candidate genes linked with multiple abiotic tolerance. Overexpression of genes responsible for encoding proteins, transcription factors and enzymes provides insights into the function of these genes in biological systems. Transgenic grapevine has been shown to develop tolerance against drought and cold when ABA-mediated DREB1b gene is overexpressed (Jin et al., 2009). Hence, by lowering the gene expression of ABI1, PP2C and ABI2, and increasing the expression level of RAB18, ABF4, DREB, SnRK2-2, MYB2 and PYL2 genes we may be able to create transgenic medicinal and aromatic plants.

Regulation of channel proteins: The amount of water content inside the plant cell is critical for its survival during stress. Aquaporins (AQP) are water and ions transporting channel proteins found in almost all living organisms including plants. Plant aquaporins are confined in the vacuoles, endoplasmic reticulum, cell membrane and plastids (Kruse et al., 2006). The function of AQPs in drought tolerance and resilience of genetically altered AQPs to environmental stress are currently being evaluated. A total of 36 AQPs have been identified in *Ziziphus jujuba*, a medicinal plant with excellent capacity to survive under water-deficient conditions (Kumar et al., 2020). Therefore, characterization of AQPs may guide us in understanding plant's ability to withstand extreme stress.

Modifications in HSPs and LEA proteins: Plants own stress response proteins include heat-shock proteins (HSP), chaperones (CSPs), and late embryogenesis abundant (LEA) proteins. In the medicinal plant *Clematis florida,* increased level of water-soluble protein prolines including expression of two HSPs (HSP 70 and HSP 18) was observed in heat-resistant variety. However, RNA chaperones CspA and CspB from *Bacillus. subtilus* and *Escherichia. coli* when inserted via recombinant technology enhanced abiotic stress tolerance in rice, maize, and *Arabidopsis* as well (Castiglioni et al., 2008). This strategy can also be utilized for medicinal and aromatic plants. While LEA proteins, CSPs, and HSPs have been frequently proven to be associated with stress response, only a few studies have exploited these proteins to develop abiotic stress tolerance in MAPs.

ROS detoxification using *Agrobacterium*-mediated transformation: ROS like $^1O_2$, H_2O_2, and HO are hazardous chemicals that induce oxidative damage to functional macromolecules. High antioxidant activity and more effective ROS removal contribute to improved tolerance when genes associated with antioxidant activity are used. Furthermore, greater antioxidant activity has been linked to better tolerance response to a variety of stimuli. Detoxification of ROS using root locus gene (rol) in *Agrobacterium rhizogenes*, is an encouraging tool in the augmentation of secondary metabolite. Recently, *Artemisia tilesii*, a medicinal herb when transformed using stable gene integration of the rolB and rolC transgenes demonstrated an increase in total flavonoid content along with antioxidant activities (Matvieieva et al., 2020). Similarly, *Agrobacterium*-mediated rol B and rol C transgenics showed high antioxidant activity in *Artemisia carvifolia* (Dilshad et al., 2016). *Agrobacterium*-mediated gene transformation optimized in *Trachyspermum ammi* (Ajowan) to pass on the BADH gene results in an increase in profile of essential oil thymol along with resistance to salinity and drought (Niazian et al., 2019)

Modulating Plant Hormone expression: Phytohormones and plant growth regulators are low-concentration molecules regulating growth and development during different phases of plant life. It has been discovered that the sense of abiotic stress triggers phytohormones, initiating the signaling cascade and ultimately affecting cellular dynamics (Kohli et al., 2013). Abscisic acid (ABA), a stress hormone helps plants adapt when exposed towards environmental challenges by modulating signaling pathways and gene expression levels (Sreenivasulu et al., 2010; O'Brien and Benková, 2013). In-vitro transformation methods using *Agrobacterium tumefaciens* technology have been used for decades for stable functional integration of foreign genes in the plant genome. Recently some non-*Agrobacterium* species (e.g. *Mesorhizobium loti* and *Sinorhizobium meliloti*) found to be effective for the successful genetic alteration of plants. These bacterial strains can be tested for

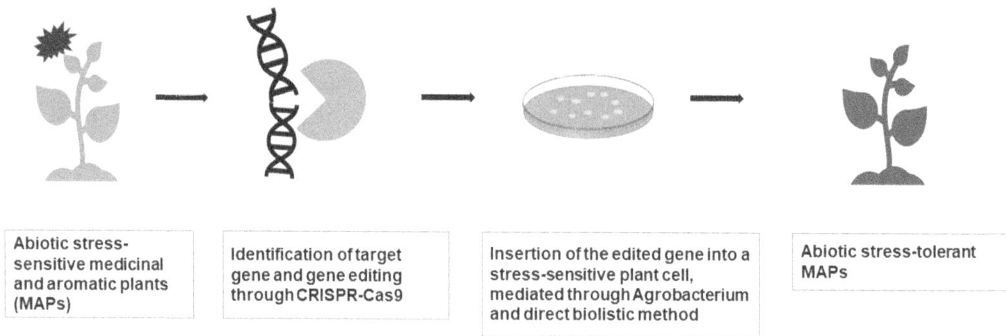

FIGURE 13.2 CRISPR/Cas-9 technology mediated production of transgenic plants.

genetic transformation of commercially important MAPs as well (Broothaerts et al., 2005). In 2016, Zhang et al., studied MsZEP, a zeaxanthin epoxidase gene from *Medicago sativa* essential for producing ABA, confers tolerance against drought and salinity in transgenic tobacco. Similarly, drought resistance in Petunia is increased by overexpressing 9-cis-epoxy carotenoid dioxygenase (NCED) gene using a stress-inducible promoter (Estrada-Melo et al., 2015). Also, an isopentenyltransferase gene (IPT) enhances transgenic tobacco's resistance to salt stress by escalating cytokinin biosynthesis (Qiu et al., 2012).

CRISPR/CAS Approach: CRISPR/Cas9 (clustered regularly interspaced short palindromic) technique to target abiotic stress in MAPs is quite a new concept and only few research has been conducted regarding this concept. However, the results have shown that it has a promising role in molecular breeding programmes in the coming years. The CRISPR-Cas9 approach is based on simple RNA/DNA hybrids having sequence specificity that can alter any particular sequence in the entire genome to reveal its function. This system can be used for gene replacement, gene duplication, gene deletion and gene insertion resulting in loss-off function and activation of mutants, leading to the generation of abiotic stress-tolerant plants (Figure 13.2). This novel technique has made it possible to mutate specific sites in various plants. SmCPS1, a dedicated diterpene synthase gene involved in tanshinone production, was precisely knocked out using Cas9 technology in the Chinese medicinal herb *Salvia miltiorrhiza.* In Papaya, CRISPR-mediated overexpression of the CpDreb2 gene is accountable for mediating signals under water stress. Similarly, in Cassava, CRISPR-generated analysis of the MeKUPs gene helps in maintaining osmotic balance during stress. In tomatoes, the CRISPR-Cas9 technique is exploited to check on mitogen-activated tomato protein kinase 3 (slmapk3) to understand the regulatory cascade underpinning SlMAPK3-mediated drought tolerance. Furthermore, in *Citrullus lanatus* editing technology was used to edit ALS gene to improve herbicide resistance. Hence, various experiments on commercially important MAPs open up new opportunities for CRISPR/Cas9-mediated genome engineering to amplify plant development during stress conditions (Sami et al., 2021)

Regulating the Role of MicroRNA: MicroRNA are 21-24 nucleotide, non-coding, small RNA molecules that help in development by regulating gene expression. The target gene is negatively regulated when the miRNA binds to the target messenger RNA (mRNA) and forms a miRNA-induced silencing complex (MIRISC), which blocks translation. miRNAs are involved directly or indirectly in many kinds of abiotic and biotic stress. It was found that miR398, OsmiR399 and miR395 participate in the plant's mechanism of responding to nutritional stress (Sun et al., 2019). Numerous studies have demonstrated that miRNAs associated with salinity tolerance have a significant impact on salt stress response in sorghum, tomatoes and rice (Mondal and Ganie, 2014;

Bouzroud et al., 2018; Nagaraju et al., 2018). Furthermore, PTO-miR156a plays a significant regulatory role in the stress response of Poplar to low nitrogen content. Hence, miRNA can modulate a wide range of plant physiochemical activities. By employing miRNAs to modulate stress-related genes in MAPs, we can achieve several good plant phenotypes. This allows us to cultivate improved medicinal plants which are more tolerant of changing environmental conditions.

13.5 CONCLUSION

New technologies are making it possible to create transgenic plants that can maintain their growth and productivity in harsh and changing environmental conditions. Many genes implicated in mechanisms of abiotic stress have been discovered and used to create stress-resistant plants. However, most of the investigations are carried out in crop plants in laboratories under artificial stress circumstances with proper growth conditions. Further, research needs to be conducted on economically important medicinal plants in actual environmental conditions in open fields to reveal the intricacy of the stress-linked signal-transduction pathway. Also, promising genetic engineering technologies to alter stress-responsive genes, promoters, hormones, and regulatory proteins can facilitate novel transgenic plants exhibiting a high degree of tolerance and high productivity of secondary metabolites. This will be the key approach to releasing plant-based medicine security but there is still a long way to go before the technology is exploited to its full potential.

REFERENCES

Appendino, G., Chianese, G., and Taglialatela-Scafati, O. 2011. Cannabinoids: Occurrence and medicinal chemistry. *Current Medicinal Chemistry* 18(7): 1085–1099.

Asano, K., Takahashi, T., Miyashita, M., Matsuzaka, A., Muramatsu, S., Kuboyama, M., and Imai, J. 1986. Effect of *Eleutherococcus senticosus* extract on human physical working capacity. *Planta Medica* 52(3): 175–177.

Aslam, J., Khan, S. H., Siddiqui, Z. H., Fatima, Z., Maqsood, M., Bhat, M. A., and Sharma, M. P. 2010. *Catharanthus roseus* (L.) G. Don. An important drug: It's applications and production. *Pharmacie Globale (IJCP)* 4(12): 1–16.

Boncan, D. A. T., Tsang, S. S., Li, C., Lee, I. H., Lam, H. M., Chan, T. F., and Hui, J. H. 2020. Terpenes and terpenoids in plants: Interactions with environment and insects. *International Journal of Molecular Sciences* 21(19): 7382.

Bouzroud, S., Gouiaa, S., Hu, N., Bernadac, A., Mila, I., Bendaou, N., Smouni, A., Bouzayen, M., and Zouine, M. 2018. Auxin response factors (ARFs) are potential mediators of auxin action in tomato response to biotic and abiotic stress (Solanum lycopersicum). *PloS One* 13(2): e0193517.

Broothaerts, W., Mitchell, H. J., Weir, B., Kaines, S., Smith, L. M., Yang, W., Mayer, J. E., Roa-Rodriguez, C., and Jefferson, R. A. 2005. Gene transfer to plants by diverse species of bacteria. *Nature* 433(7026): 629–633.

Brown, G. D., Liang, G. Y., and Sy, L. K. 2003. Terpenoids from the seeds of *Artemisia annua*. *Phytochemistry* 64(1): 303–323.

Castiglioni, P., Warner, D., Bensen, R. J., Anstrom, D. C., Harrison, J., Stoecker, M., and Heard, J. E. 2008. Bacterial RNA chaperones confer abiotic stress tolerance in plants and improved grain yield in maize under water-limited conditions. *Plant Physiology* 147(2): 446–455.

Christaki, E., Bonos, E., Giannenas, I., and Florou-Paneri, P. 2012. Aromatic plants as a source of bioactive compounds. *Agriculture* 2(3): 228–243.

Cocetta, G., Mishra, S., Raffaelli, A., and Ferrante, A. 2018. Effect of heat root stress and high salinity on glucosinolates metabolism in wild rocket. *Journal of Plant Physiology* 231: 261–270.

Da Silva Magedans, Y. V., Matsuura, H. N., Tasca, R. A. J. C., Wairich, A., de Oliveira Junkes, C. F., de Costa, F., and Fett-Neto, A. G. 2017. Accumulation of the antioxidant alkaloid brachycerine from *Psychotria brachyceras* Müll. Arg. is increased by heat and contributes to oxidative stress mitigation. *Environmental and Experimental Botany* 143: 185–193.

Devanesan, A. 2023. *Taxus baccata* and *Taxus brevifolia*: The source of paclitaxel, docetaxel and cabazitaxel. In: *Modern Medicines from Plants*: Taylor and Francis.

Dilshad, E., Ismail, H., Haq, I. U., Cusido, R. M., Palazon, J., Ramirez-Estrada, K., and Mirza, B. 2016. Rol genes enhance the biosynthesis of antioxidants in *Artemisia carvifolia* Buch. *BMC Plant Biology* 16: 1–9.

Erb, M., and Kliebenstein, D. J. 2020. Plant secondary metabolites as defences, regulators, and primary metabolites: The blurred functional trichotomy. *Plant Physiology* 184(1): 39–52.

Estrada-Melo, A. C., Chao, C., Reid, M. S., and Jiang, C. Z. 2015. Overexpression of an ABA biosynthesis gene using a stress-inducible promoter enhances drought resistance in petunia. *Horticulture Research* 2: 15013.

Ghorbani, A., and Esmaeilizadeh, M. 2017. Pharmacological properties of *Salvia officinalis* and its components. *Journal of Traditional and Complementary Medicine* 7(4): 433–440.

Golkar, P., and Taghizadeh, M. 2018. In vitro evaluation of phenolic and osmolite compounds ionic content, and antioxidant activity in safflower (*Carthamus tinctorius* L.) under salinity stress. *Plant Cell, Tissue and Organ Culture (PCTOC)* 134: 357–368.

Gomez-Galera S., Pelacho, A. M., Gene, A., Capell, T., and Christou, P. 2007. The genetic manipulation of medicinal and aromatic plants. *Plant Cell Reports* 26: 1689–1715.

Guo, X. R., Yang, L., Yu, J. H., Tang, Z. H., and Zu, Y. G. 2007. Alkaloid variations in *Catharanthus roseus* seedlings treated by different temperatures in short term and long term. *Journal of Forestry Research* 18(4): 313–315.

Gupta, B., and Huang, B. 2014. Mechanism of salinity tolerance in plants: Physiological, biochemical, and molecular characterization. *International Journal of Genomics* 2014(1): 701596.

He, J., Yao, L., Pecoraro, L., Liu, C., Wang, J., Huang, L., and Gao, W. 2023. Cold stress regulates accumulation of flavonoids and terpenoids in plants by phytohormone, transcription process, functional enzyme, and epigenetics. *Critical Reviews in Biotechnology* 43(5): 680–697.

Hewlings, S. J., and Kalman, D. S. 2017. Curcumin: A review of its effects on human health. *Foods* 6(10): 92.

Hosseini, M. S., Samsampour, D., Ebrahimi, M., Abadía, J., and Khanahmadi, M. 2018. Effect of drought stress on growth parameters, osmolyte contents, antioxidant enzymes and glycyrrhizin synthesis in licorice (*Glycyrrhiza glabra L.)* grown in the field. *Phytochemistry* 156: 124–134.

Jin, W., Dong, J., Hu, Y., Lin, Z., Xu, X., and Han, Z. 2009. Improved cold-resistant performance in transgenic grape (*Vitis vinifera* L.) overexpressing cold-inducible transcription factors AtDREB1b. *HortScience* 44(1): 35–39.

Khan, M., Jannat, A., Munir, F., Fatima, N., and Amir, R. 2020. Biochemical and molecular mechanisms of abiotic stress tolerance. In: Hasanuzzaman, M. (eds) *Plant Ecophysiology and Adaptation under Climate Change: Mechanisms and Perspectives II*: Springer.

Khatun, S., Ali, M. B., Hahn, E. J., and Paek, K. Y. 2008. Copper toxicity in *Withania somnifera*: Growth and antioxidant enzymes responses of in vitro grown plants. *Environmental and Experimental Botany* 64(3): 279–285.

Kohli, A., Sreenivasulu, N., Lakshmanan, P., and Kumar, P. P. 2013. The phytohormone crosstalk paradigm takes center stage in understanding how plants respond to abiotic stresses. *Plant Cell Reports* 32: 945–957.

Kopustinskiene, D. M., Masteikova, R., Lazauskas, R., and Bernatoniene, J. 2022. *Cannabis sativa* L. Bioactive compounds and their protective role in oxidative stress and inflammation. *Antioxidants* 11(4): 660.

Kruse, E., Uehlein, N., and Kaldenhoff, R. 2006. The aquaporins. *Genome Biology* 7: 1–6.

Kumar, M., Tak, Y., Potkule, J., Choyal, P., Tomar, M., Meena, N. L., and Kaur, C. 2020. Phenolics as plant protective companion against abiotic stress. *Plant Phenolics in Sustainable Agriculture* 1: 277–308.

Kumar, N., Kumawat, S., Khatri, P., Singla, P., Tandon, G., Bhatt, V., and Deshmukh, R. 2020. Understanding aquaporin transport system in highly stress-tolerant and medicinal plant species *Jujube* (*Ziziphus jujuba* Mill.). *Journal of Biotechnology* 324: 103–111.

Lai, H. C., Singh, N. P., and Sasaki, T. 2013. Development of artemisinin compounds for cancer treatment. *Investigational New Drugs* 31: 230–246.

Li, X., Dong, W., Nalin, A. P., Wang, Y., Pan, P., Xu, B., and Yu, J. 2018. The natural product chitosan enhances the anti-tumor activity of natural killer cells by activating dendritic cells. *Oncoimmunology* 7(6): e1431085.

Mandal, S. M., Chakraborty, D., and Dey, S. 2010. Phenolic acids act as signalling molecules in plant- microbe symbioses. *Plant Signaling and Behaviour* 5(4): 359–368.

Matvieieva, N. A., Morgun, B. V., Lakhneko, O. R., Duplij, V. P., Shakhovsky, A. M., Ratushnyak, Y. I., and Yevtushenko, D. P. 2020. *Agrobacterium rhizogenes*-mediated transformation enhances the antioxidant potential of *Artemisia tilesii* Ledeb. *Plant Physiology and Biochemistry* 152: 177–183.

Matsuura., H. N., and Fett-Neto, A. G. 2015. Plant alkaloids: Main features, toxicity, and mechanisms of action. *Plant Toxins* 2(7): 1–15.

Mohammadi, S., Jafari, B., Asgharian, P., Martorell, M., and Sharifi-Rad, J. 2020. Medicinal plants used in the treatment of Malaria: A key emphasis to *Artemisia, Cinchona, Cryptolepis*, and *Tabebuia* genera. *Phytotherapy Research* 34(7): 1556–1569.

Mondal, T. K., and Ganie, S. A. 2014. Identification and characterization of salt responsive miRNA- SSR markers in rice (*Oryza sativa*). *Gene* 535(2): 204–209.

Muhammad, K. K., Farid, C., Sattar, S., and Imran, M. 2019. Extraction of bioactive molecules: Conventional versus novel methods. In: Ansar, H. Suleria, R. (Eds.), *Bioactive Compounds from Plant Origin*: Apple Academic Press.

Munivenkatappa, N., Sarikonda, S., Rajagopal, R., Balakrishnan, R., and Nagarathana, C. K. 2018. Variations in quality constituents of green tea leaves in response to drought stress under south Indian condition. *Scientia Horticulturae* 233: 359–369.

Nagaraju, M., Reddy, P. S., Kumar, S.A, Kumara, A. S., Kumar, A., Suravajhala, P., Ali, A., and Rao, D. M. 2018. Genome-wide in silico, analysis of dehydrins in *Sorghum bicolor, Setaria italica*, and *Zea mays*, and quantitative analysis of dehydrin gene expressions under abiotic stresses in *Sorghum bicolor*. *Plant Gene* 13: 64–75.

Niazian, M., Sadat-Noori, S. A., Tohidfar, M., Galuszka, P., and Mortazavian, S. M. M. 2019. *Agrobacterium*-mediated genetic transformation of ajowan (*Trachyspermum ammi* (L.) Sprague): An important industrial medicinal plant. *Industrial Crops and Products* 132: 29–40.

Nikolić, M., Glamočlija, J., Ferreira, I. C., Calhelha, R. C., Fernandes, A., Marković, T., and Soković, M. 2014. Chemical composition, antimicrobial, antioxidant and antitumor activity of *Thymus serpyllum* L., *Thymus algeriensis* Boiss. and Reut and *Thymus vulgaris* L. essential oils. *Industrial Crops and Products* 52: 183–190.

O'Brien, J. A., and Benková, E. 2013. Cytokinin cross-talking during biotic and abiotic stress responses. *Frontiers in Plant Science* 4: 451.

Paikra, B. K., and Gidwani, B. 2017. Phytochemistry and pharmacology of *Moringa oleifera* Lam. *Journal of Pharmacopuncture* 20(3): 194.

Pant, M., Dolma, S., Gahlot, M., Sharma, A., and Mundepi, S. 2023. Phytoremediation of Heavy Metals. In: Singh, R. P., Singh, P., and Srivastava, A. (eds) *Heavy Metal Toxicity: Environmental Concerns, Remediation and Opportunities*. Springer. https://doi.org/10.1007/978-981-99-0397-9_15

Pant, M., Bist, P., Gusain, M. P., and Pant, K. 2023. Effect of pH and gelling agents in nutrient medium on in vitro shoot multiplication of *Swertia chirata* Buch.-Ham ex Wall: An endangered medicinal herb. *Materials Today Proceedings* 73(1): 9–12.

Pant, M., Chaudhary, S., Mishra, A., and Pant, K. 2022. Plant bioactives can delay aging and senescence through sirtuin activation: A review. *Research Journal of Biotechnology* 17(11): 155–163.

Qiu, W., Liu, M., Qiao, G., Jiang, J., Xie, L., and Zhuo, R. 2012. An isopentyl transferase gene driven by the stress-inducible rd29A promoter improves salinity stress tolerance in transgenic tobacco. *Plant Molecular Biology Reporter* 30: 519–528.

Qu, A. L., Ding, Y. F., Jiang, Q., and Zhu, C. 2013. Molecular mechanisms of the plant heat stress response. *Biochemical and Biophysical Research Communications* 432(2): 203–207.

Radwan, A., Kleinwächter, M., and Selmar, D. 2017. Impact of drought stress on specialised metabolism: Biosynthesis and the expression of monoterpene synthases in sage (*Salvia officinalis*). *Phytochemistry* 141: 20–26.

Rai, V., Vajpayee, P., Singh, S. N., and Mehrotra, S. 2004. Effect of chromium accumulation on photosynthetic pigments, oxidative stress defense system, nitrate reduction, proline level and eugenol content of *Ocimum tenuiflorum* L. *Plant Science* 167(5): 1159–1169.

Raskin, I. 1992. Role of salicylic acid in plants. *Annual Review of Plant Biology* 43(1): 439–463.

Sakihama, Y., Cohen, M. F., Grace, S. C., and Yamasaki, H. 2002. Plant phenolic antioxidant and prooxidant activities: Phenolics-induced oxidative damage mediated by metals in plants. *Toxicology* 177(1): 67–80.

Sami, A., Xue, Z., Tazein, S., Arshad, A., He Zhu, Z., Ping Chen, Y., and Jin Zhou, K. 2021. CRISPR–Cas9-based genetic engineering for crop improvement under drought stress. *Bioengineered* 12(1): 5814–5829.

Sangwan, N. S., and Sangwan, R. S. 2013. Secondary metabolites of traditional medical plants: A case study of Ashwagandha (*Withania somnifera*). In Nick, P. and Opatrny, Z. (Eds.), *Applied Plant Cell*

Biology: Cellular Tools and Approaches for Plant Biotechnology (pp. 325–367). Springer Berlin Heidelberg.

Saunders, J. A., O'neill, N. R., and Romeo, J. T. 1992. Alkaloid chemistry and feeding specificity of insect herbivores. *Alkaloids: Chemical and Biological Perspectives* 8:151–196.

Selmar, D., and Kleinwächter, M. 2013. Stress enhances the synthesis of secondary plant products: The impact of stress-related over-reduction on the accumulation of natural products. *Plant and Cell Physiology* 54(6): 817–826.

Semmar, N. 2024. Temperature-linked constraints and plant protection responses. In: Baluška, F. (Ed.), *In Secondary Metabolites in Plant Stress Adaptation: Analytic Space of Secondary Metabolites*: Springer International Publishing.

Senaratna, T., Touchell, D., Bunn, E., and Dixon, K. 2000. Acetyl salicylic acid (Aspirin) and salicylic acid induce multiple stress tolerance in bean and tomato plants. *Plant Growth Regulation* 30(2): 157–161.

Shala, A. Y., and Gururani, M. A. 2021. Phytochemical properties and diverse beneficial roles of *Eucalyptus globulus* Labill.: A review. *Horticulturae* 7(11): 450.

Sharma, A., Kumar, V., Mittal, C., Rana, V., Dabral, K., and Parveen, G. 2023. Role of essential oil used pharmaceutical cosmetic product. *Journal for Research in Applied Sciences and Biotechnology* 2(3): 147–157.

Shinyuy, L. M., Loe, G. E., Jansen, O., Mamede, L., Ledoux, A., Noukimi, S. F., and Frederich, M. 2023. Secondary metabolites isolated from *Artemisia afra* and *Artemisia annua* and their anti-malarial, anti-inflammatory and immunomodulating properties— pharmacokinetics and pharmacodynamics: A review. *Metabolites* 13(5): 613.

Simic, A., Rančic, A., Sokovic, M. D., Ristic, M., Grujic-Jovanovic, S., Vukojevic, J., and Marin, P. D. 2008. Essential oil composition of *Cymbopogon winterianus* and *Carum carvi* and their antimicrobial activities. *Pharmaceutical Biology* 46(6): 437–441.

Sindhu, R. K., Upma, K. A., and Arora, S. 2010. *Santalum album* linn: A review on morphology, phytochemistry and pharmacological aspects. *International Journal of PharmTech Research* 2(1): 914–919.

Singh, S., Pathak, N., Fatima, E., and Negi, A. S. 2021. Plant isoquinoline alkaloids: Advances in the chemistry and biology of berberine. *European Journal of Medicinal Chemistry* 226: 113–839.

Sreenivasulu, N., Radchuk, V., Alawady, A., Borisjuk, L., Weier, D., Staroske, N., and Weschke, W. 2010. De-regulation of abscisic acid contents causes abnormal endosperm development in the barley mutant seg8. *The Plant Journal* 64(4): 589–603.

Sun, X., Lin, L., and Sui, N. 2019. Regulation mechanism of microRNA in plant response to abiotic stress and breeding. *Molecular Biology Reports* 46: 1447–1457.

Teoh, E. S. 2016. *Medicinal Orchids of Asia*. Springer.

Todorova, V., Ivanov, K., and Ivanova, S. 2021. Comparison between the biological active compounds in plants with adaptogenic properties (*Rhaponticum carthamoides, Lepidium meyenii, Eleutherococcus senticosus and Panax ginseng*). *Plants* 11(1): 64.

Tomilova, S. V., Globa, E. B., Demidova, E. V., and Nosov, A. M. 2023. Secondary metabolism in *Taxus* spp. plant cell culture in vitro. *Russian Journal of Plant Physiology* 70(3): 23.

Trevisan, S. C. C., Menezes, A. P. P., Barbalho, S. M., and Guiguer, É. L. 2017. Properties of Mentha piperita: A brief review. *World Journal of Pharmaceutical and Medical Research* 3(1): 309–313.

Truong, V. L., and Jeong, W. S. 2022. Red ginseng (Panax ginseng Meyer) oil: A comprehensive review of extraction technologies, chemical composition, health benefits, molecular mechanisms, and safety. *Journal of Ginseng Research* 46(2): 214–224.

Tuli, H. S., Tuorkey, M. J., Thakral, F., Sak, K., Kumar, M., Sharma, A. K., and Bishayee, A. 2019. Molecular mechanisms of action of genistein in cancer: Recent advances. *Frontiers in Pharmacology* 10: 1336.

Tuteja, N. 2007. Abscisic acid and abiotic stress signalling. *Plant Signalling and Behavior* 2(3): 135–138.

Verdeguer, M., Sánchez-Moreiras, A. M., and Araniti, F. 2020. Phytotoxic effects and mechanism of action of essential oils and terpenoids. *Plants* 9(11): 1571.

Verpoorte, R., van der Heijden, R., and Memelink, J. 2000. Engineering the plant cell factory for secondary metabolite production. *Transgenic Research* 9: 323–343.

Wahid, A., Gelani, S., Ashraf, M., and Foolad, M. R. 2007. Heat tolerance in plants: An overview. *Environmental and Experimental Botany* 61(3):199–223.

Wei, X., Xu, H., Rong, W., Ye, X., and Zhang, Z. 2019. Constitutive expression of a stabilized transcription factor group VII ethylene response factor enhances waterlogging tolerance in wheat without penalizing grain yield. *Plant, Cell and Environment* 42(5): 1471–1485.

Xia, X. J., Wang, Y. J., Zhou, Y. H., Tao, Y., Mao, W. H., Shi, K., and Yu, J. Q. 2009. Reactive oxygen species are involved in brassinosteroid-induced stress tolerance in cucumber. *Plant Physiology* 150(2): 801–814.

Yalpani, N., Enyedi, A. J., León, J., and Raskin, I. 1994. Ultraviolet light and ozone stimulate accumulation of salicylic acid, pathogenesis-related proteins and virus resistance in tobacco. *Planta* 193(3): 372–376.

Yu, F., Liang, K., Fang, T., Zhao, H., Han, X., Cai, M., and Qiu, F. 2019. A group VII ethylene response factor gene, ZmEREB180, coordinates waterlogging tolerance in maize seedlings. *Plant Biotechnology Journal* 17(12): 2286–2298.

Zehra, A., Sadaf, C., Kaiser, I. W., Naeem, M., Masroor, M., Khan, A., Tariq, A. 2020. Exogenous abscisic acid mediates ROS homeostasis and maintains glandular trichome to enhance artemisinin biosynthesis in *Artemisia annua* under copper toxicity. *Plant Physiology and Biochemistry* 156: 125–134.

Zhang, L., and Piao, X. 2023. Use of aromatic plant-derived essential oils in meat and derived products: Phytochemical compositions, functional properties, and encapsulation. *Food Bioscience* 53: 102–520.

Zhang, Z., Wang, Y., Chang, L., Zhang, T., An, J., Liu, Y., and Yang, P. 2016. MsZEP, a novel zeaxanthin epoxidase gene from alfalfa (*Medicago sativa*), confers drought and salt tolerance in transgenic tobacco. *Plant Cell Reports* 35: 439–453.

Zhu, Q., Zheng, Z. P., Cheng, K. W., Wu, J. J., Zhang, S., Tang, Y. S., and Wang, M. 2009. Natural polyphenols as direct trapping agents of lipid peroxidation-derived acrolein and 4-hydroxy-trans-2-nonenal. *Chemical Research in Toxicology* 22(10): 1721–1727.

14 Role of Rhizospheric Microflora in Medicinal Plants Metabolite Production

*Shivansh Negi, Divyansh Pathak, Daksh Kushwaha, Aman Sharma, Sonam Khan, Urja Agarwal, Pramod Rawat, Abhishek Kumar and Yashaswi Singh**

*Corresponding author

14.1 INTRODUCTION

To interpret plant-root confluence, a German Scientist, Lorenz Hiltner conceived the word, 'Rhizosphere'. He used this term for the interaction between legume roots and bacteria. This word originated from the Greek word, '*Rhiza*' meaning Root and '*Sphere*' meaning area of influence (McNear, 2013). The rhizosphere is that part of the soil ecosystem, where the soil, its biota, and the plant's roots interact, which is beneficial to the plant in most scenarios (Brimecombe et al., 2001).'Microflora' is a misleading term, which indicates the more appropriate word, 'Microbiota' (Tannock, 1999). Microbiota refers to the living microscopic community, including all the microscopic living beings (Valdes et al., 2018). Hence, Rhizospheric microflora is the collection of microorganisms in the rhizosphere, mainly bacteria, actinomycetes, fungi, and algae. Rhizospheric bacteria that benefit plant growth are called PGPR, or plant growth promoting rhizobacteria, while the bacteria that negatively affect plant growth are called DRB, that is, deleterious rhizobacteria (Barazani and Friedman, 2001). The PGPR are usually free-living microorganisms, which live in the rhizosphere of plants and give advantage to them through various ways. Though many PGPR are identified in different genera, most of the PGPR identified come from the genera *Pseudomonas* and *Bacillus* (Podile and Kishore, 2006). The effectiveness of PGPR on plant growth has also been proven in extreme conditions like high altitude and low temperature (Adhikari et al., 2021). Additionally, PGPR are effective in managing abiotic and biotic stresses cultivated plants face. PGPR also reduce reliance on chemical fertilizers while promoting the growth of plants (Khoshru et al., 2020). Selected PGPR improve plants' resistance against phytopathogens, parasites, and pests (Rabari et al., 2023).

The use of plants by humans as medicine in prehistoric times is considered instinctive, similar to other animals. For example, a male Sumatran orangutan was observed using the medicinal plant Akar Kuning (*Fibraurea tinctoria*) to treat his facial wound (Laumer et al., 2024). Their use for curing ailments was based on experience, as information about diseases and their causes was unavailable. The effectiveness of medicinal plants was later acknowledged by modern humans after which scientific explanations for their properties were discovered (Petrovska, 2012). As per written records, the first-ever evidence of medicinal plants was a clay tablet describing the use of hundreds of plants as medicine like opium and poppy, found in Iraq during the time of the Sumerian civilization, around 3000 BC (Shaheen et al., 2019). Various plants are mentioned as herbal medicine in ancient texts of India, China, and Egypt (Pterovska, 2012). In tribal communities, medicinal plants are still an important part of traditional medicine (Arya et al., 2020). For example, Ayurveda,

 DOI: 10.1201/9781003482659-14

a traditional medicine system of India, originated about 3000 years ago (Payyappallimanaand Venkatasubramanian, 2016), mentioned various plants and their use to treat many ailments, like, Neem (*Azadirachta indica*) treats chicken pox, small-pox, warts, etc; Turmeric (*Curcuma longa*) treats liver disorders, skin infections, urinary tract infections, etc; Black pepper (*Piper nigrum*) helps in the treatment of asthma, heart ailments, etc (Kaushik et al., 2023); Foxglove plant (*Digitalis purpurea*) is used for its anti-oxidant, anti-diabetic, cardio-protective properties (Tina et al., 2023); *Quercus leucotrichophora* or Banj Oak is a medicinal plant whose gum and seeds are used to treat digestive ailments (Semwal et al., 2018); Stinging nettle (*Urtica dioica*) is used as a diuretic. It treats diabetes, and hepatic and gastric disorders (Semwal et al., 2023). According to the Cambridge English Dictionary, metabolites are the substances involved in metabolic reactions. Metabolites are either primary or secondary. The primary metabolites are the substances involved directly in metabolic reactions such as simple sugars, chlorophyll, amino acids, etc, and the secondary metabolites are usually referred to as end products of metabolic reactions like alkaloids, phenolics, saponins, etc. Plant metabolites have demonstrated various therapeutic effects; hence the medicinal properties of plants are attributed to their metabolites (Hussein and El-Anssary, 2019). For example, metabolite curcumin present in *C. longa* has antioxidant, anti-cancerous, and anti-inflammatory activities. It also helps in treating cancer and neurological disorders (Kocaadam and Şanlier, 2017); Menthol present in *Mentha piperita* has anti-bacterial, anti-inflammatory, and analgesic properties (Almatroodi et al., 2021). Flavonoids, such as myricetin, that are found in various medicinal plants show antidiabetic, anticancer, and neuroprotective activities (Taheri et al., 2020). Secondary metabolites have a wide variety of therapeutic applications as they interact with various receptors and hinder the pathogenesis of many ailments. Analysis of these metabolites leads to the discovery of lead compounds for the formulation of effective drugs as popular drugs are becoming ineffective due to overuse (Velu et al., 2018). In the present scenario, with an increasing population, ailments, and dependency on allopathic drugs, the use of medicinal plants because of their pharmacological metabolites for the treatment of various diseases is also increasing because of the side effects and cost of allopathic medications. Hence, mass production of medicinal plants without or with the least usage of pesticides and fertilizers is required. Plant tissue culture has been utilized to increase the mass production of many medicinal plants, but it is expensive (Singh et al., 2024). The use of rhizospheric microflora for this objective has shown promising results in the past(Solaiman and Anawar, 2015). Moreover, the use of rhizospheric microorganisms, native to the host plant is preferred over the use of non-native microorganisms, because foreign PGPR are non-specific in function which may alter the soil microbiome and other ecological factors and could make unplanned changes in the plant environment. On the other hand, inoculation of native rhizospheric microbiota somewhat keeps the environment less changed and adapts better to the conditions of the host plant. Native microorganisms have better survival rates and they fit better in the soil environment (Raj et al., 2023; Hasan et al., 2024).

This chapter discusses some of the important medicinal plants and the effect of their native rhizospheric microflora on metabolite production and plant growth promotion.

14.2 MECHANISMS OF ACTION OF PGPR

PGPRs are categorized based on the underlying mechanism of the mode by which they act to promote plant growth, which are direct and indirect mechanisms of growth promotion of plants (Kaymak, 2011). The promotion of plant growth by PGPR without or before any pathogen outbreak is called the direct mechanism of growth promotion of plants, while during the pathogen outbreak, PGPR-mediated plant growth promotion is called the indirect mechanism of growth promotion of plants (Nazir et al., 2018; Figure 14.1).

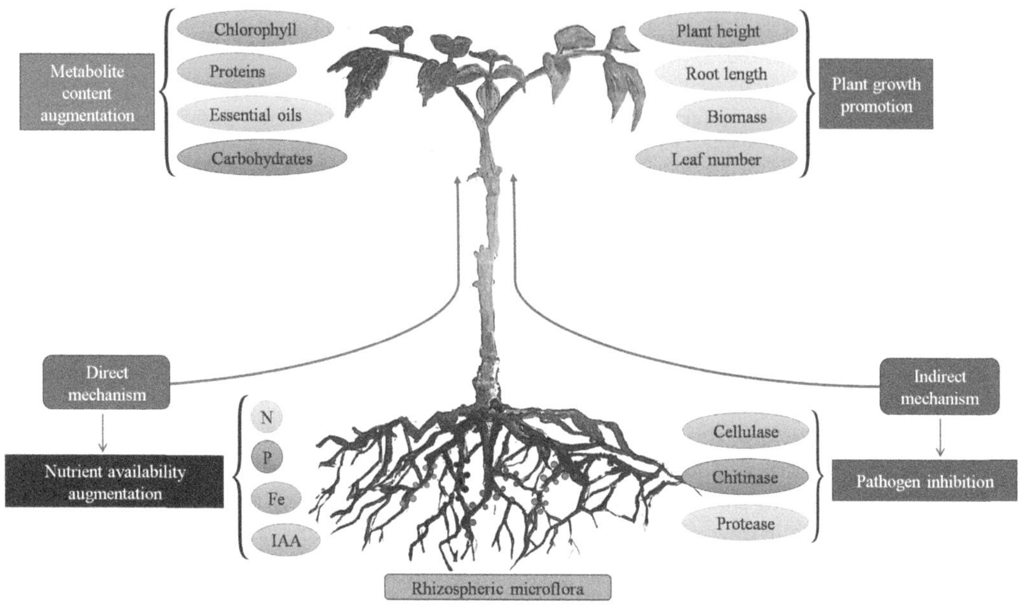

FIGURE 14.1 Effects of rhizospheric microflora on medicinal plants, a diagrammatic representation.

14.2.1 Direct Mechanism of Action of PGPR

14.2.1.1 Phosphate Solubilization

The second biggest limiting factor (after nitrogen) for plant growth is said to be phosphorus, because it cannot be absorbed directly by the plants (as in the case of nitrogen), even after the presence of large phosphorus reserves in the soil (Stevenson and Cole, 1999). After all, most plants can absorb this nutrient in 2 soluble forms, $H_2PO_4^-$ and HPO_4^{2-} (Jha and Saraf, 2015). Some rhizospheric microorganisms convert insoluble compounds to soluble compounds by producing organic acids and phosphatases (Kim et al., 1997). For example, *Azotobacter chroococcum*, *Cladosporium herbarum*, and *Bacillus circulans* are reported to solubilize phosphate from wheat rhizosphere (Kumar and Narula, 1999; Singh and Kapoor, 1999).

14.2.1.2 Nitrogen Fixation

The reduced forms of nitrogen (like Ammonia) produce many biomolecules essential for life to sustain, for example, nucleic acids and proteins. Hence, it is necessary for the survival of all living organisms, and plants are no exception. A group of nitrogen-fixing bacteria, called diazotrophs, reduce the dinitrogen biologically into soluble (to the plants) nitrogenous compounds. This process is called Biological Nitrogen Fixation (BNF) (Kim and Rees, 1994), which involves the conversion of gaseous nitrogen (N_2) in the atmosphere into nitrogenous compounds, i.e. ammonia (NH_3) or other related compounds (which can be used by the plants) in the soil and water (Hardarson and Atkins, 2003). It requires mutualism between nitrogen-fixing microbes and plants. For example, *Rhizobium* bacteria in the roots of leguminous plants multiply rapidly and form nodules in the root, where they fix nitrogen from the air in the soil pores into a soluble form. In return, they get some part of the total carbohydrates produced by the plant (Mellor and Werner, 2018).

14.2.1.3 IAA Production

Indole-3 Acetic Acid (IAA) is found in the plant itself mainly on the growth regions like shoot tips, root tips, young leaves, and developing seed embryos. It is a heterocyclic compound belonging to the carboxymethyl group (Acetic acid). IAA is responsible for the development, cell division, elongation, and differentiation of tissues in plants (Mike-Anosike et al, 2018). IAA helps in building the basic cells for the formation of organs and tissues, it may also help in fighting stress conditions. Production of IAA has been observed in Actinomycetes, in which *Streptomyces* is the most prominent genus identified to produce IAA (Khan et al. 2023).

14.2.1.4 Siderophore Production

Iron is an essential mineral for the growth of plants. But it is present in an insoluble form (which is Ferrous Fe^{3+}) in the soil like nitrogen and plants usually uptake iron in its reduced form (Ferric Fe^{2+}) (Salisbury and Ross, 1992). The word 'Siderophores' is an amalgam of two Greek words ('Sidero'- Iron, and 'Phores'- carriers) (Sah and Singh, 2015). Siderophores are small molecules produced by certain bacteria and fungi that extract iron from their surroundings (Khamna et al., 2009). For proper iron uptake from the soil, plants produce soluble organic substances called phytosiderophores (Strategy II plants) and chelators (Strategy I plants) to facilitate the reduction of iron in the more soluble forms for plants (from ferric to ferrous) (von Wirén et al., 2000). But, direct utilization of bacterial siderophores by plants for iron uptake has been reported, for example, in cucumber (Wang et al., 1993), brinjal, okra, tomato (Pahari and Mishra, 2017), and many others. Species of *Pseudomonas*, *Enterobacter*, and *Bacillus* are considered some of the important siderophore-producing microorganisms (Sayyed et al.,2012). Hence, it can be said that siderophore-producing rhizospheric microflora influences plant growth (Gangwar et al., 2014).

14.2.2 Indirect Mechanisms of PGPR

To manage soil-borne diseases of plants, biocontrol agents often use cell wall disintegrating enzymes (Chet et al., 1990). Cell wall-degrading enzymes such as -1,3-glucanase, cellulase, chitinase, and protease, which are secreted by PGPR biocontrol strains, directly suppress bacterial cells and fungal pathogen hyphal development (Compant et al., 2005). For example, the biocontrol agent, *Serratia marcescens* shows antagonistic activity against the soil-borne pathogen *Fusarium oxysporum*, by deforming its hyphae structure and induction of other abnormalities (Banerjee et al., 2020). HCN-producing rhizobacteria are also involved in pathogen resistance. In these bacteria, the HCN synthase enzyme produces Hydrogen Cyanide which inhibits cytochrome C oxidase in the respiratory chain of the pathogen, which eventually kills them (Devi et al., 2013). For example, an HCN-producing strain of *Pseudomonas brassicacearum* showed in-vitro antagonistic activity against *Fusarium oxyporium* (Singh and Varma, 2015).

14.3 EFFECT OF PGPR ON MEDICINAL PLANTS

Here, the effect of native rhizospheric microflora on the production of important metabolites and the growth of some important medicinal plants have been discussed (also refer to Table 14.1 and Figure 14.2):

1. *Agave americana* (Table 14.1; Sl. no. 1) belongs to the family of Agavaceae. This plant has been traditionally used as a food supplement to treat obesity. Its effect on obesity is attributed to its high soluble fiber content. The fiber in the plant, when interacts with digestive enzymes, expands and makes the stomach feel full, which ultimately reduces over-eating (Alonso-Castro et al., 2015). Inulin is a dietary fiber of this plant (Bouaziz et al., 2014). It acts as a prebiotic for beneficial gut bacteria, helps in weight loss, inhibits inflammatory factors, and reduces blood sugar levels (Qin et al., 2023). Its leaves are also used for the treatment of fish toxicity, to get relief from malarial fever and other fevers, and topically applied to wounds for enhanced healing (Khade et al., 2011). This plant is commonly called the Century plant and it is native to Mexico (Zwane

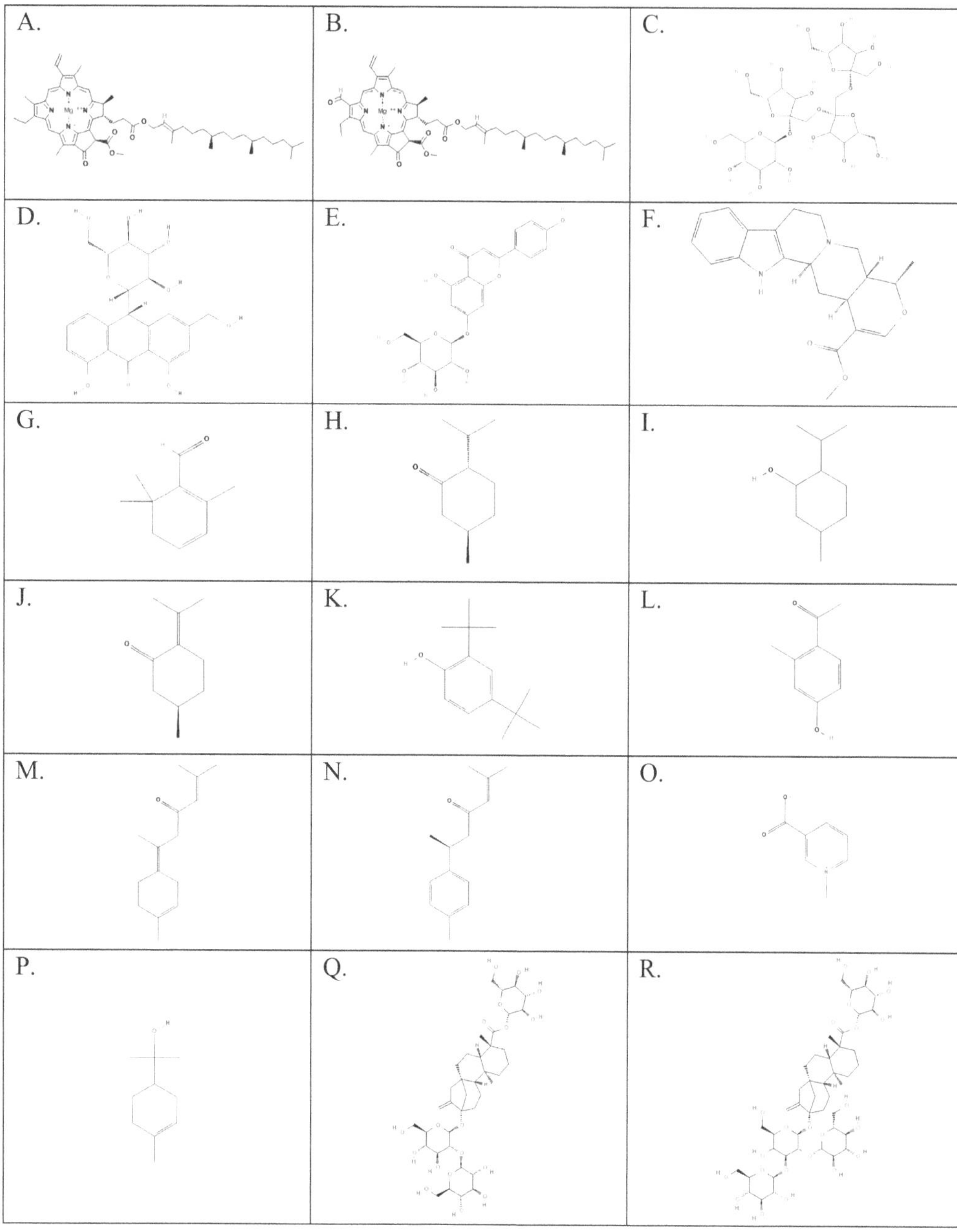

FIGURE 14.2 2D structures of metabolites augmented by the effect of rhizospheric microflora of listed medicinal plants: (A). Chlorophyll A; (B). Chlorophyll B; (C). Inulin; (D). Aloin A; (E). Apigenin-7-O-glucoside; (F). Ajmalicine; (G). Safranal; (H). Menthone; (I). Menthol; (J). Pulegone; (K). 2,4-Di-tertbutyl phenol; (L). 2, 4-Hydroxy-2-methylacetophenone; (M). Z -gamma-Atlantone; (N). aR-Turmerone; (O). Trigonelline; (P). Terpineol; (Q). Stevioside; (R). Rebaudioside-A; (S). Salvianolic acid B; (T). Rosmarinic acid; (U). Proline; (V). Picrocrocin; (W). Crocin; (X). Curcumin; (Y). Withaferin-A.

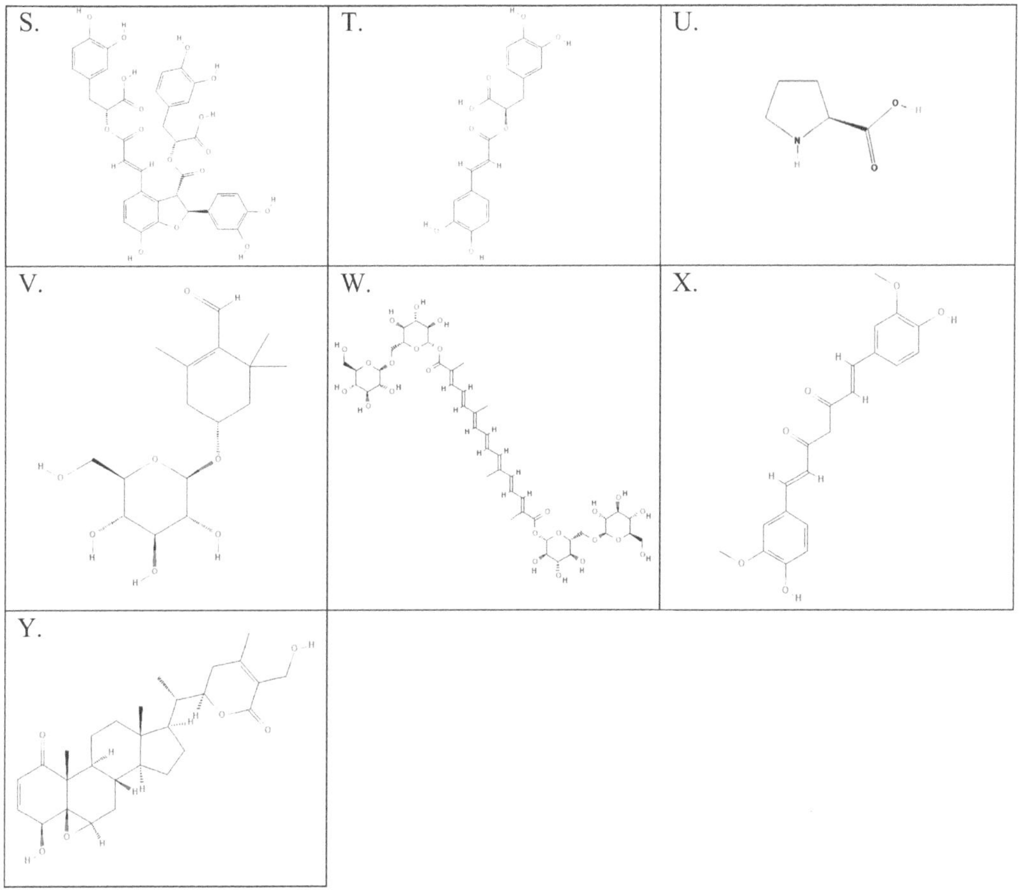

FIGURE 14.2 (Continued)

et al., 2011). In a study, Rhizospheric bacteria were isolated from this plant and were tested for their plant growth-promoting activities. All of the isolates showed plant promotion activities, in vitro. These isolates were identified as strains of *Acinetobacter calcoaceticus*, *Pseudomonas mosselii*, and *Rhizobium daejeonense*. On inoculation of these isolates on the rhizosphere of the host plant, positive effects on the plant growth were observed. These isolates increased the content of plant metabolites like inulin (Figure 14.2C) and other carbohydrates like fructose and sucrose. Moreover, significant increases in plant growth parameters like plant height, stem diameter, root length, and leaf were observed (La Torre-Ruiz et al., 2016).

2. *Aloe barbadensis* (Table 14.1;Sl. no. 2) commonly known as Aloe vera has been used as a medicinal plant for over 2000 years. It belongs to the family Asphodelaceae (Lee et al.,2016). It is suggested that this plant originated from the Arabian Peninsula (Grace et al., 2015). It has been a significant part of traditional medicine in countries like India, China, Japan, etc (Grindlayand Reynolds, 1986). Its latex collected from the leaves is used orally and topically for medicinal purposes, like healing burns and superficial skin wounds (Lee et al., 2000). One of the active components present in latex is Aloin, attributed to many medicinal properties of this plant such as, antibacterial, anticancer, and anti-inflammatory activities (Xiao et al., 2022). Aloin is also used as a laxative agent. It is composed of two diastereoisomers, Aloin-A and Aloin-B (Patel et al., 2013). Aloin-A has demonstrated antagonistic activity against many bacterial and fungal pathogens (Donkor et al., 2020). In a 2012 study, plant growth-promoting bacteria, *Burkholderia*

gladioli, *Enterobacter hormaechei*, *Pseudomonas synxantha*, and *Serratia marcescens* isolated from the rhizosphere of this plant when used as inoculants with Tricalcium phosphate during pot experiments of roots of tissue cultured *A. barbadensis* plantlets, showed a dramatic increase in aloin-A (important pharmacological metabolite) content (Figure 14.2D), and showed positive influence in biomass, leaf and root length, leaves number, total gel volume, dry gel weight, dry rind weight, and phosphorus absorption. Using these micro-organisms as a consortium gave far better results than using them individually as a bio-inoculant (Gupta et al., 2012). Another study by the same author included inoculation of these bacterial strains along with Mussoorie Rock Phosphate in the rhizosphere of the same plant for greenhouse studies. In this case, the bacterial isolates namely, *P. synxantha* and *S. marcescens* gave better results than *B. gladioli* and *E. hormaechei* in the pot experiments (Mamta et al., 2012).

3. *Arachis hypogaea* (Table 14.1; Sl. no. 3) also known as Peanut, is a food crop mainly known for its use as food, fodder, and an excellent source of vegetable oil. Apart from this, this plant is also used for its medicinal properties. Its nuts extract has been consumed by diabetics in African countries. In China, peanut peels are used in the treatment of bronchitis. Its leaves contain many medicinal phytochemicals (Cherif, 2013). *A. hypogaea* belongs to the family Fabaceae and is indigenous to South America (Hammons et al., 2016). A plant growth-promoting bacterial strain of *Alcaligenes faecalis* of the plant rhizosphere, when re-inoculated on the seeds, significantly increased the chlorophyll content (Figure 14.2A and 14.2B) of the host plant (Sayyed et al., 2010). It is important to note that chlorophyll is an important metabolite, which largely influences the energy production in the plant by photosynthesis, ultimately influencing plant growth (Li et al., 2018). A particular nitrogen-fixing bacterial strain was isolated from the rhizosphere of this plant, identified as *Bacillus amyloliquefaciens*. This isolate also tested positive for other plant growth-promoting activities as well. It demonstrated antagonism against different fungal plant pathogens like *Alternaria alternata*, *Bipolaris sorokiniana*, *Botryosphaeria dothidea*, *Cryphonectria parasitica*, *Fusarium graminearum*, *Fusarium oxysporum*, *Fusarium pseudograminearum*, *Rhizoctonia solani*, and *Sclerotium rolfsii*. It also performed well in root-colonization assay. In the field experiment, seedlings inoculated with this strain showed improved dry weight, fresh weight, and branch length. The micro-biodiversity of the rhizosphere was also increased after the treatment (Li et al., 2023).
4. *Astragalus mongolicus* (Table 14.1; Sl. no. 4) is a flowering plant used in Chinese traditional medicine. It is commonly called Mongolian milkvetch (*huángqí / běiqí* in Chinese). It belongs to the Fabaceae family (Li et al., 2014). Its roots have been used to prepare a tonic to overcome energy deficiency. It is used as a diuretic, antiperspirant, etc. (Yu et al., 2005). Flavonoids, polysaccharides, and saponins are some of the main phytoconstituents of this plant. These metabolites are attributed to many pharmacological activities, like, hepato-protection, antioxidant, and antiviral activities (Dong et al., 2023). Some plant growth-promoting bacterial strains of *Agrobacterium fabacearum*, *Pseudomonas silesiensis*, *Paenarthrobacter nitroguajacolicus*, and *Arthrobacter pascens* were isolated from rhizospheric soil. To evaluate the effect of these isolates on the plant itself, field trials were conducted in the same study, in which these isolates were inoculated with the seeds of this plant. These isolates positively affected the plant growth. The biomass, plant height, root length, dry weight of root and shoot, content of nitrogen, phosphorus, potassium, and various metabolites (Flavonoids, Polysaccharides, and Saponins) were significantly augmented (Shi et al., 2023).
5. *Catharanthus roseus* (Table 14.1; Sl. no.5) commonly known as Madagascar periwinkle. Sadabahar and Nayantara are some of its other names. It was used as a diuretic in traditional European medicine (Kaushik et al., 2017). Its diuretic ability is attributed to the presence of an anti-hypersensitive alkaloid named Ajmalicine in this plant. This alkaloid is extracted from the roots of *C. roseus*. It also has antioxidant and antimicrobial properties (Ambrin et al., 2020). It is used as a mouthwash to relieve toothache, dental decay, and scurvy. It also treats diabetic ulcers

(Balaji et al., 2014). It belongs to the Apocynaceae family (Ajaib et al., 2010). This plant is native to Madagascar (Moudi et al., 2013). A strain of *Azotobacter* (nitrogen-fixing bacteria) was isolated from the rhizosphere of *C. roseus*. In the seed germination assay, this isolate effectively increased the seed germination efficiency of this plant. In the field studies, inoculation of this bacterial strain significantly improved root and shoot length, dry matter, vigor index, and plant antioxidant activities (Karthikeyan et al., 2007). Similarly, native plant growth-promoting strains of *Azotobacter chroococcum*, *Bacillus subtilis*, and *Pseudomonas fluorescens* as a consortium significantly increased the Ajmalicine (Figure 14.2F) content in the roots of this medicinal plant (Karthikeyan et al., 2010).

6. *Chamomilla recutita* (or *Matricaria chamomilla*; Table 14.1; Sl. no. 6) is a traditional medicinal plant with many health benefits. It belongs to the Asteraceae family (Repčák et al., 2001). As a component of herbal teas, it helps in relieving flatulence. Applying its oil to affected areas heals smooth muscle spasms and skin irritation. Its flower is mainly used for its medicinal properties (Zadeh et al., 2014). This plant is native to Europe (Franke et al., 2006). Plant growth-promoting bacterial strains of *Bacillus subtilis*, *Streptomyces subrutilus*, and *Paenibacillus polymyxa* isolated from the *C. recutita* on reinoculation in the seedlings of the host plant, significantly increased the content of important metabolite apigenin-7-O-glucoside (Figure 14.2E; Schmidt et al., 2014). It is interesting to note that, apigenin-7-O-glucoside is a pharmacologically important metabolite. Its anti-inflammatory and anti-oxidant effects are at par with Trolox, astandard anti-oxidant (Wang et al., 2020).
7. *Chenopodium quinoa* (Table 14.1; Sl. no. 7) or Quinoa, a popular pseudo-cereal, belongs to the family Amaranthaceae (Bravi et al., 2024). South America is considered as its place of origin (Bruno, 2017). Traditionally, aside from being used as a food plant, the seed and leaves are consumed by Andean people to improve bone, mental, and women's health (Repo-Carrasco et al., 2003). Strains of phosphate solubilizing bacteria namely, *Bacillus licheniformis* and *Enterobacter asburiae* were isolated from the rhizosphere of this plant, applied to the seeds of the host plant, increased the chlorophyll content, phosphorus and potassium uptake of the plant. Moreover, this physical growth parameters were also influenced positively (Mahdi et al., 2020). A study was conducted for the isolation and evaluation of rhizospheric bacteria from Quinoa under salt stress and heavy-metal toxicity. They isolated plant growth-promoting strains of *Pseudomonas frederiksbergensis*, *Pseudomonas stutzeri*, *Bacillus atrophaeus*, *Enterobacter asburiae*, and *Enterobacter hormaechei*. These isolates were also tested for their salt and heavy metal tolerance. Pot experiments were performed on seeds by the seed coating method. Germination rate, plant length and biomass was measured. All the isolates performed better than those of control. The PSBs, *B. atrophaeus*, and *E. asburiae* were the best-performing isolates in the pot experiments (Mahdi et al., 2021).
8. *Codonopsis pilosula* (Table 14.1; Sl. no. 8) or Dangshen is a well-known Chinese traditional medicinal plant. This plant belongs to the Campanulaceae family. Its edible roots are used in the treatment of the insufficient activity of the spleen which causes digestive issues, coughs, eating disorders, etc. It also improves overall physique (Luan et al., 2021). According to the Kew Royal Botanic Garden database, this plant is native to Asian countries like China, Korea, and Mongolia. About 24 bacterial isolates were derived from the rhizosphere of *C. pilosula*. Out of these isolates, one isolate was a strain of *Klebsiella michiganensis*. This strain's plant growth-promoting activity was demonstrated in-vitro, along with other plant growth-promoting activities. On inoculation of this strain in the field experiments, on the seedlings of this plant, improvement in chlorophyll content was observed. Also, the height, biomass, and seedlings' ground diameter were significantly augmented. Positive effects on soil micro-biodiversity and enzyme activity were also observed. This strain also showed antagonistic activities against many fungal pathogens: *Colletotrichum melliae*, *Cytospora chrysosperma*, *Phomopsis macrospore*, and *Rhizoctonia solani* (Jin et al., 2024).

9. *Crocus sativus* (Table 14.1; Sl. no. 9) or Saffron is a traditional medicinal plant. Its origin is traced back to Greece. It belongs to the Iridaceae family. Traditionally, their stigma is used as an aphrodisiac in various cultures. This plant also has antidepressant, antioxidant, and analgesic agents. It also treats digestive issues (Mzabri et al., 2019). The quality of saffron is based on the absorbance of the sample at specific wavelengths which corresponds to the content of particular metabolites; absorbance of the sample at 440nm wavelength gives crocin (Figure 14.2W) content (attributed to saffron color), in which the absorbance of at least 200 is placed at category 1 (best) according to ISO 3632. Similarly, the absorbance of at least 70 at 257nm indicates the picrocrocin (Figure 14.2V) content (flavor of saffron), while an absorbance range of 20–50 at 330 nm indicates the safranal (Figure 14.2G) content (aroma of saffron) of category 1 saffron (Giorgi et al., 2017). In a recent study, strains of *Rahnella aquatilis* and *Variovorax paradoxus* were isolated from the Saffron rhizosphere, and these strains were inoculated in the host plant rhizosphere individually and in a consortium. Inoculation of these strains increased the chlorophyll content of the plant and produced category 1 saffron (Chamkhi et al., 2023). In another study, bacterial strains of *Actinobacteria calcoaceticus*, *Bacillus aryabhattai*, *Chryseobacterium elymi*, *Pseudomonas kilonensis*, *Pseudomonas koreensis*, and *Pseudomonas tremae*; all produced IAA. They were found non-toxic to each other in culture, hence used as a consortium for pot trials for the host plant as an inoculant. The isolates were applied to the corms of the plant. Many positive effects of the inoculant were recorded in the experiment. A significant increase in the average number of roots and shoots and many cornlets was observed. Moreover, the incidence of corn-rot disease was also reduced as compared to control (Ambardar et al., 2013).
10. *Curcuma longa* (Table 14.1; Sl. no. 10) or Turmeric, belongs to the Zingiberaceae family (Larsen et al., 2005). Its rhizome is popular for its culinary and traditional medicinal uses, throughout the Indian sub-continent (Reiter et al., 2006). It is believed to have originated from South Asia (Velayudhan et al., 2012). Traditionally its rhizome is used to treat Jaundice, diminish pain and swelling, toothache, diarrhea, skin infections, etc (Rezvanirad et al., 2016). Many of the medicinal properties of this plant are attributed to the presence of Curcumin (metabolite) in the plant (Jyotirmayee et al., 2022). For example, curcumin helps control Cystic fibrosis (Quispe et al., 2022). In a study published in the year 2014, a strain of nitrogen-fixing bacteria named, *Azotobacter chroococcum* was isolated from the Turmeric rhizosphere and demonstrated various PGPR activities such as Phosphate solubilization, HCN production, nitrogen fixation, and IAA production. This bacterial strain was re-inoculated in the rhizome of this plant. It significantly increased the fresh biomass and height of the shoot and rhizome. Moreover, the concentration of Curcumin (Figure 14.2A) in the plant also increased dramatically (Kumar et al., 2014). Likewise, inoculation of the nitrogen-fixing native strain of *Rhizobium pusense* in the rhizotomous buds of turmeric positively influenced the chlorophyll content and various physical parameters of the plant such as, height, harvest index, root length and biomass, rhizome size, weight and yield, leaf area, number, length, breadth and biomass. In this field experiment, this strain significantly increased the plant growth. This is why, the authors proposed this strain as a potential biofertilizer for this plant (Friesen et al., 2019). Another study claimed that inoculation of native strains of *Pseudomonas plecoglossicida* and *Serratia nematodiphila* significantly increased the content of many metabolites like aR-Turmerone,4-Di-tertbutyl phenol, 2, 4-Hydroxy-2-methylacetophenone, and (Z) -gamma-Atlantone (Figure 14.2K-N) when applied to the plant in consortium (Jagtap et al., 2023).
11. *Foeniculum vulgare* (Table 14.1; Sl. no. 11) or Fennel is a native Mediterranean and Southern European plant. It is a medicinal plant native to China (Tanira et al., 1996). It is a plant of the Apiaceae family (Akbar and Akbar, 2020). In China, its common name is 'Xiaohuixiang'. Its fruit is used in the treatment of indigestion in infants. It also helps in treating coughs,

bronchitis, vomiting, and diarrhea (He et al., 2011). PSB was isolated from the rhizosphere of *F. vulgare* from different districts of Rajasthan, India, and their phosphate solubilizing abilities were tested in vitro. All these isolates belonged to 2 genera, *Pseudomonas* and *Bacillus* (Mishra et al., 2015). In the follow-up study, nine in vitro best-performing isolates in PGPR tests were subjected to field experiments of the native plant using the seed coating technique. These bacteria positively influenced the production of essential oil and various growth traits, like the branch number, flowers, and seeds. According to this study, a strain of *Bacillus subtilis* performed best among other isolates (Mishra et al., 2016).

12. *Hypericum perforatum* (Table 14.1; Sl. no. 12) with a common name, St. John's wort, belongs to the Hypericaceae family (Sharma et al., 2020). It is a traditional medicine in Chinese, Islamic, and Greek cultures. It is used to treat dermal ailments, wounds, stomach ulcers, bile-related disorders, and respiratory and genital infections. Nowadays, it is also considered effective in the treatment of symptoms of depression. *H. perforatum* is indigenous to Europe, Northern Africa, and Western Asia (Nobakht et al., 2022). Mainly the plant part above ground is used to prepare herbal preparations for medicinal purposes (Oliveira et al., 2016). Its most important biochemical constituents are carotenoids, chlorophylls, flavonoids, hypericin, and phenols (Rahnavard et al., 2012). Plant growth-promoting rhizospheric bacterial strains were isolated from the host plant, namely, *Curtobacterium albidum*, *Enterobacter huaxiensis*, *Flavobacterium acidificum*, *Kocuria palustris*, *Kosakonia cowanii*, *Rahnella woolbedingensis*, *Rahnella variigena*, and *Staphylococcus edaphicus*. Their effect as rhizospheric inoculants on the host plant seedlings was studied in the field experiments. These inoculants effectively increased the chlorophyll-a (Figure 14.2A), carotenoids, content in the leaves of plants, especially on the treatment of *K. cowanii*. The increase in the chlorophyll-b (Figure 14.2B) content was highest in the treatment of *R. variigena*. These treatments as expected, improved the process of photosynthesis as well. The shoot and root length, above-ground dry biomass (AGB), below-ground dry biomass (BGB), and root volume were also increased. According to the authors of this study, native bacterial strains of *K. cowanii* and *R. variigena* performed the best among other isolates and hence have the potential of an effective biofertilizer for this medicinal plant (Raj et al., 2023).
13. *Mentha piperita* (Table 14.1; Sl. no. 13) or Peppermint belongs to the family Lamiaceae (Güntert et al., 2001). This plant is native to Europe. Its leaves are used to treat symptoms of cold, irritation in the gastrointestinal tract, and nausea. It is topically applied to relieve headaches (Shah et al., 2004). Apart from contributing to the characteristic smell of this plant, Menthol is also one of the major chemical constituents of the essential oil of *M. piperita*. Moreover, this constituent is attributed to many of the medicinal properties of this plant, like analgesic properties, pain relief, anti-inflammation, anti-microbial, and anti-cancer activities. Menthol is also effective in reducing nasal congestion (Almatroodi et al., 2021). Three native strains of *Mentha piperita* were isolated and identified as *Pseudomonas putida*. All the strains tested positive for various plant growth promotion activities. On inoculation of these strains on the roots developed on the host plant nodes, all three strains influenced the increase in the constituents of essential oils, like terpineol, menthone, menthol, and pulegone (Figure 14.2 H-J, 14.2P) and overall quantity of essential oils. An increase in dry root weight, shoot ramification number, shoot weight, and trichome number was also observed after treatment of each strain (Santoro et al., 2015).
14. *Panax ginseng* (Table 14.1; Sl. no. 14) is a traditional medicinal plant that is indigenous to Korea, China, and Russia. Some of its common names are Korean Ginseng, Asian Ginseng, and Oriental Ginseng. Its fleshy roots are used as a tonic to relieve symptoms of weakness and fatigue, especially during the recovery time of any chronic illness. This plant also helps treat respiratory ailments (Mahady et al., 2000). Many of its medicinal properties are attributed to the phenolic compounds like gallic acid, coumaric acid, etc; possessing antioxidant activity

(Morshed et al., 2023). *P. ginseng* belongs to the Araliaceae family (Liu et al., 2020). In a study, seedlings of Asian Ginseng were dipped in the culture of a plant growth-promoting bacterial strain of *Rhizobium panacihumi* (isolated from host plant soil), significantly increased the proline (Figure 14.2U) content, phenolic content, and plant biomass. It is important to note that proline plays an important role in plant growth, like adaptation in stressful environments, cell wall transduction, seed development, etc (Kavi et al., 2015). This bacterial isolate was proclaimed as a potential PGPR for *P. ginseng* (Kang et al., 2021). A bacterial strain of *Arthrobacter nicotinovorans* was isolated from the rhizosphere of this plant. This strain of *A. nicotinovorans* demonstrated high IAA-producing ability. In the seed germination assay, the average length of seed buds (inoculated with *A. nicotinovorans* culture) was much better than the control (uninoculated). Likewise, in the field experiments, the average fresh weight of ginseng plants (inoculated with *A. nicotinovorans* culture) was much more than that of control (uninoculated). The soil enzyme activity after its treatment with the isolated strain was significantly increased (Jiang et al., 2022).

15. *Rehmannia glutinosa* (Table 14.1; Sl. no. 15) is a traditional Chinese medicinal plant whose roots are used to treat different ailments. It has many medicinal properties, such as liver protection, anti-inflammatory, anti-fungal, diuretic, hypoglycemic, etc (Du et al., 2020). It belongs to the family Scrophulariaceae. It is commonly called Chinese Foxglove (Zhou et al, 2007). The polysaccharides present in this plant inhibit cytokines related to inflammation and cancer (Lu et al., 2022). China and Japan are considered as its places of origin (Qi et al., 2008). Like many medicinal plants, *R. glutinosa* also has a consecutive monoculture problem, hence it is subjected to low, poor-quality yield and diseases (Yang et al., 2011). To overcome this, a nitrogen-fixing bacterial strain of *Bacillus tequilensis* was isolated from the rhizosphere of *R. glutinosa* which showed in-vitro antagonistic activity against *Fusarium oxyporum* (Root-rot causing fungal pathogen of *R. glutinosa*). Moreover, inoculating this strain in the plants' rhizosphere increased the fresh and dry weight of leaves and roots, content, and yield of sterol and polysaccharides in the treated plants (Du et al., 2020).
16. *Ricinus communis* (Table 14.1; Sl. no. 16) is a traditional medicinal plant from the Euphorbiaceae family. It is commonly called the Castor bean plant. It is Indigenous to Africa. Even though this plant has been known for its toxicity for hundreds of years, its seeds were used as a folk medicine (Kumar, 2017). It has antioxidant, anti-inflammatory, anti-ulcer, and injury-healing properties (Bhakta and Das, 2015). Phenolic compounds in leaves are responsible for their insecticidal effects (Usha Rani and Pratyusha, 2014). Its seed protein demonstrates significant antibacterial and anticancerous activity (Al-Mamun et al., 2016). A bacterial strain of *Pseudomonas aeruginosa* demonstrating various PGPR activities was isolated from the plant rhizosphere influencing the biochemical and chlorophyll content of the leaves. Inoculation of this isolate on the seeds of *R. communis* significantly increased the content of carbohydrates, lipids, phenols, and soluble proteins in the plant (Sandilya et al., 2017). In a study, many nitrogen-fixing isolates were derived from the rhizosphere of *R. communis*. These isolates also demonstrated other PGPR activities as well. These isolates effectively promoted plant height, root weight (fresh and dry), shoot weight (fresh and dry), and rhizospheric microbial biodiversity in the field experiments of *R. communis* plants. Out of these isolates, a strain of *Klebsiella variicola* was the best-performing isolate. This isolate was pitched as the potential biofertilizer for this plant by the authors (Rashmi et al., 2021).
17. *Salvia miltiorrhiza* (Table 14.1; Sl. no. 17) or Chinese sage or the red sage root is traditional Chinese medicine (Chen et al., 2007). Its dried roots have been used to treat heart and brain-related ailments throughout Aisa (Shi et al., 2019). It is native to China and Japan and it belongs to the Labiatae family (Wang, 2010). Because of its medicinal importance, demand for this plant has increased significantly over the years, but its production is hampered due to its susceptibility to the root rot disease caused by the fungal pathogen *Fusarium oxysporum*. Moreover,

the use of pesticides for the eradication of this pathogen has resulted in adverse effects on the environment, hence the use of biofertilizers is required (Wang et al., 2018). In a study conducted in 2017, strains of *Pseudomonas chlororaphis*, *Pantoea rodasii*, and *Pseudomonas vancouverensis* were isolated. The culture filtrate with bacterial elicitors of these isolates was inoculated in the hairy root culture of the host plant. It significantly the content of important metabolites like Rosmarinic acid (Figure 14.2T; Precursor of Salvianolic acid B or SAB) and SAB (Figure 14.2S; Quality indicator of the plant), and root dry weight, and biomass (You et al., 2018). To account for the problem of root-rot disease, actinomycetes were isolated from the rhizospheric soil of *S. miltiorrhiza* and the antagonistic activity of these isolates was tested in vitro. The best-performing isolate was a strain of *Streptomyces abidoflavus*. It also tested positive for nitrogen fixation along with other growth-promoting activities. In the field experiments, this particular isolate effectively reduced disease incidence and also promoted the growth of the plants. It increased the length, weight, and dry weight of the root and shoot. A significant increase in root diameter was also observed (Du et al., 2022).

18. *Stevia rebaudiana* (Table 14.1; Sl. no. 18) belongs to the family Asteraceae. It is an annual, herbaceous plant Indigenous to Paraguay, South America (Kinghorn and Soejarto, 1985). The leaves of this plant produce Stevioside, a sweetener that does not get digested in the human body and keeps blood sugar levels unchanged (Strauss, 1995). Hence it is called, Sweet-leaf, Honey-leaf and Sugar-leaf (Singh and Rao, 2005). This medicinal plant is used as a sugar alternative by diabetic and obese individuals to control blood sugar levels. It has also shown antibacterial, anti-inflammatory, and diuretic properties (Hossain et al., 2017). In a study, researchers isolated some phosphate-solubilizing bacteria from the rhizosphere of this plant, namely, *Burkholderia gladioli*, *Enterobacter aerogenes*, and *Serratia marcescens*. For studying their effect on plant growth, these isolates were re-inoculated individually, and also as a consortium with phosphorus fertilizer Tricalcium phosphate in the rhizosphere of their native plant. The metabolite Rebaudioside-A (Figure 14.2R; steviol glycoside of *S. rebaudiana*) content in the leaves was significantly increased by the inoculation of *B. gladioli* and *S. marcescens*. An increase in the plant growth was also observed, which was checked through various parameters like, shoot-root length, dry weight of stem and roots, and total shoot biomass. The bacterial strain of *B. gladioli* performed better than other bacterial strains. However, consortium enhanced the plant growth even more. However, no significant increase in the amount of stevioside (Figure 14.2Q; glycoside of *S. rebaudiana*) was observed. In their follow-up study, using the same bacterial isolates, they conducted a similar pot experiment with Mussoorie Rock Phosphate as phosphorus fertilizer, instead of TCP. They found results similar to the previous experiment but this time, a significant increase in the amount of stevioside was also observed (Gupta et al., 2011).
19. *Trigonella foenum-graecum* (Table 14.1; Sl. no. 19) also known as, Fenugreek, belongs to the family of Leguminosae. This traditional medicinal plant is native to Iran. Its leaves and seeds improve the gastrointestinal system, cardiovascular system, muscles, bones, sore throat condition, acid reflux, etc. (Sun et al., 2021). The medicinal properties of this plant are attributed to its chemical constituents, such as trigonelline, flavonoids, alkaloids, and many others. Metabolites like trigonelline regulate glucose metabolism, ß cell regeneration, and neural activity (Ghosal et al., 2016). In a study, a strain of *Priestia endophytica* was isolated from the rhizosphere of *T. foenum-graecum*. When applied as inoculum to the fenugreek plant, this isolate promoted nodulation and increased the content of nitrogen, phosphorus, metabolites like trigonelline (Figure 14.2O), and overall biomass in high concentrations of salts (Sharma et al., 2022).
20. *Withania somnifera* (Table 14.1; Sl. no. 20) or Indian ginseng, popularly known as Ashwagandha in the Indian subcontinent, is found worldwide, from the Mediterranean to south-east Asia. It is a part of traditional medicine in India and its roots and flowers are used as an aphrodisiac (Ehigiator et al., 2024). It has also been used to treat asthma, ulcers, and senile dementia

(Verma et al., 2011). It belongs to the Solanaceae family (Pandey et al., 2017). This plant is indigenous to India (Mikulska et al., 2023). This plant has many alkaloids, like somniferin, witanin, anaferin, withaferin-A etc, which are attributed to its medicinal properties (John, 2014). For example, withaferin-A has demonstrated neuroprotective activities, and inhibition of inflammatory cytokines, and hence helps in improving the conditions of patients suffering from Alzheimer's disease, and Parkinson's disease (Mikulska et al., 2023). In a study, strains of *Azotobacter chrococcum*, *Pseudomonas fluorescens*, *Azospirillum*, *Bacillus megaterium*, were isolated from the rhizosphere of *Withania somnifera* or Ashwagandha, used as a biofertilizer on the host plant, individually as well as in consortium, and this improved root and shoot height. These isolates also affected the withaferin-A (Figure 14.2Y) content of the roots positively. The best results were obtained by the consortium of *Pseudomonas fluorescens* and *Bacillus megaterium* (Rajasekar and Elango, 2011). In a research article published in 2011, the authors isolated plant growth-promoting rhizospheric bacteria from this plant, which were 3 bacterial strains of *Pseudomonas* sp. and 3 bacterial strains of *Bacillus* sp. which gave prominent results in the in-vivo experiments on the rooted cuttings of this plant. They significantly increased plant growth. The performance of *Bacillus* strains was better than that of *Pseudomonas* strains in terms of plant height (Kaur et al., 2011).

14.4 CHALLENGES AND FUTURE PROSPECTS

The positive effects of PGPR on plant growth have been reported numerous times, hence their impact on plants is quite evident (Bergottini et al., 2015). At the same time, commercial use of rhizospheric microflora as PGPR is still challenging. Some of the challenges are discussed below:

1. Effectiveness of Biofertilizers: In general, plants are susceptible to numerous diseases caused by numerous pathogens. Hence, the biofertilizers should possess antagonistic activity against a broad spectrum of pathogens. But, in reality, only a few commercial biocontrol agents are available in the market. Also, an ideal biofertilizer should have compatibility with the native rhizosphere, and it should withstand all ecological stresses of the environment (Nakkeeran et al., 2006). However, native rhizospheric isolates are quite compatible with the environment and rhizospheric microflora (Raj et al., 2023).
2. Tendency of usage of chemical fertilizers: Despite the visible ill-effects of usage of chemical fertilizers in the environment, conventional farmers still prefer using them over biofertilizers because of their relatively faster results. On the other hand, the result of using biofertilizer is still uncertain (Moser et al., 2008). Also, farmers are devoid of the right knowledge about biofertilizers. Moreover, the cost of these bioinoculants is not very economical in comparison to chemical fertilizers (Tabassum et al., 2017).
3. Handling of biofertilizers: Gram-positive bacteria are much easier to handle than Gram-negative bacteria as they do not produce spores and hence, they get killed easily and have a short shelf life. Biofertilizers underperform in fields in comparison to the controlled environment mainly because of mishandling by the farmers as they require specific conditions for their long-term survival. Moreover, re-inoculation of biofertilizer is required as these microbes do not live forever and their activity gets diminished over time (Tabassum et al., 2017).
4. Rules and regulations: Formulation of a bioinoculant requires many lab and field tests, after which the efficacy of a biofertilizer is proved, which is a tedious process. Adding to this, complex rules and regulations for registering a microorganism or a group of microorganisms as a commercial biofertilizer make the commercialization of these biofertilizers even more difficult and hinder their market expansion (Ravensberg, 2011).

14.5 CONCLUSION

We have discussed the role of rhizospheric microflora in promoting the growth and metabolite production of medicinal plants in this chapter. Examples of various medicinal plants and the positive effects of specific rhizospheric bacteria on their growth and metabolite content were also discussed. The influence of various metabolites in the medicinal properties of plants is also pointed out. It is implied that rhizospheric bacteria as biofertilizers have potential benefits, especially in terms of sustainable agriculture. Rhizospheric bacteria can effectively increase metabolite production, promote plant growth, increase soil nutrients, and reduce the use of chemical fertilizers, which have adverse effects on the environment. Native rhizospheric bacteria are well-adapted to local conditions and can withstand ecological stresses than that of non-native rhizospheric bacteria. Though, it is important to note that, studies related to the effect of native rhizospheric microflora in metabolites production are relatively less than that of non-native microorganisms. Challenges such as the scarcity of bioinoculants with broad-spectrum antagonistic activity against pathogens, compatibility of inoculants with rhizospheres, and handling and storage issues must be addressed. With advancements in biotechnology and a better understanding of microbial interactions, the commercial use of rhizospheric bacteria could become more understood and reachable to the growers, which could help in sustainable agriculture.

ACKNOWLEDGEMENTS

The authors would like to acknowledge Mr. Rahul Negi, Department of Fine Arts, Graphic Era Hill University, Dehradun, Uttarakhand, for drawing the coloured portrait of the plant for this chapter.

REFERENCES

Adhikari, P., Jain, R., Sharma, A., and Pandey, A., 2021. Plant growth promotion at low temperature by phosphate-solubilizing Pseudomonas spp. isolated from high-altitude Himalayan soil. *Microbial Ecology*, *82*(3): 677–687.

Ajaib, M., Khan, Z., Khan, N., and Wahab, M., 2010. Ethnobotanical studies on useful shrubs of district Kotli, Azad Jammu & Kashmir, Pakistan. *Pakistan Journal of Botany*, *42*(3): 1407–1415.

Akbar, S., and Akbar, S., 2020. Foeniculum vulgare Mill.(Apiaceae/Umbelliferae) (Syns.: F. officinale All.; F. commune Bubani; F. foeniculum (L.) H. Karst.; Anethum foeniculum L.; Ligusticum foeniculum (L.) Crantz.; Meum foeniculum (L.) Spreng.). In Akbar, S. (Ed.), *Handbook of 200 Medicinal Plants: A Comprehensive Review of Their Traditional Medical Uses and Scientific Justifications*: 931–945. Springer.

Al-Mamun, M. A., Akter, Z., Uddin, M. J., Ferdaus, K. M., Hoque, K. M. F., Ferdousi, Z., and Reza, M. A., 2016. Characterization and evaluation of antibacterial and antiproliferative activities of crude protein extracts isolated from the seed of Ricinus communis in Bangladesh. *BMC Complementary and Alternative Medicine*, *16*: 1–10.

Almatroodi, S. A., Alsahli, M. A., Almatroudi, A., Khan, A. A., and Rahmani, A. H., 2021. Peppermint,(Mentha× piperita): Role in management of diseases through modulating various biological activities. *Pharmacognosy Journal*, *13*(3): 822–827.

Alonso-Castro, A. J., Domínguez, F., Zapata-Morales, J. R., and Carranza-Álvarez, C., 2015. Plants used in the traditional medicine of Mesoamerica (Mexico and Central America) and the Caribbean for the treatment of obesity. *Journal of ethnopharmacology*, *175*: 335–345.

Ambardar, S., and Vakhlu, J., 2013. Plant growth promoting bacteria from Crocus sativus rhizosphere. *World Journal of Microbiology and Biotechnology*, *29*: 2271–2279.

Ambrin, G., Ali, H. M., and Ahmad, A., 2020. Metabolic regulation analysis of ajmalicine biosynthesis pathway in Catharanthus roseus (L.) G. Don suspension culture using nanosensor. *Processes*, *8*(5): 589.

Arya, O. P., Mylliemngap, W., and Pandey, A., 2020. Ethnomedicinal plants used by Adi community of upper Siang district of Arunachal Pradesh in north-east India. *Pleione*, *14*(2): 265–276.

Balaji, D. H., 2014. Versatile. Therapeutic effects of Vinca rosea Linn. *International Journal of Pharmaceutical Science and Health Care, 1*(4): 59–76.

Banerjee, S., Singh, S., Pandey, S., Bhandari, M. S., Pandey, A., and Giri, K., 2020. Biocontrol potential of Pseudomonas azotoformans, Serratia marcescens and Trichoderma virens against Fusarium wilt of Dalbergia sissoo. *Forest Pathology, 50*(2): e12581.

Barazani, O., and Friedman, J., 2001. Allelopathic bacteria and their impact on higher plants. *Critical Reviews in Microbiology, 27*(1): 41–55.

Bergottini, V. M., Otegui, M. B., Sosa, D. A., Zapata, P. D., Mulot, M., Rebord, M., Zopfi, J., Wiss, F., Berney, B., and Junier, P., 2015. Bio-inoculation of yerba mate seedlings (Ilex paraguariensis St. Hill.) with native plant growth-promoting rhizobacteria: A sustainable alternative to improve crop yield. *Biology and Fertility of Soils, 51*: 749–755.

Bhakta, S., and Das, S. K., 2015. In praise of the medicinal plant Ricinus communis L.: A review. *Global Journal of Research on Medicinal Plants & Indigenous Medicine, 4*(5): 95.

Bouaziz, M. A., Rassaoui, R., and Besbes, S., 2014. Chemical composition, functional properties, and effect of inulin from Tunisian Agave americana L. leaves on textural qualities of pectin gel. *Journal of Chemistry, 2014*(1): 758697.

Bravi, E., Sileoni, V., and Marconi, O., 2024. Quinoa (Chenopodium Quinoa Willd.) as functional ingredient for the formulation of gluten-free shortbreads. *Foods, 13*(3): 377.

Brimecombe, M. J., Leij, F. A., and Lynch, J. M., 2001. Nematode community structure as a sensitive indicator of microbial perturbations induced by a genetically modified Pseudomonas fluorescens strain. *Biology and Fertility of Soils*, 34: 270–275.

Bruno, M. C., 2017. Quinoa: Origins and development. *Encyclopedia of Global Archeology, 2017*(2): 9039–9044.

Chamkhi, I., Sbabou, L., and Aurag, J., 2023. Improved growth and quality of saffron (Crocus sativus L.) in the field conditions through inoculation with selected native plant growth-promoting rhizobacteria (PGPR). *Industrial Crops and Products, 197*: 116606.

Chandran, S. J. G., 2019. Isolation and characterization of N2 fixing bacteria from curcuma rhizosphere soil of western ghats. *International Journal of Analytical Reviews 6*: 487–491.

Chen, J., Lee, F. C., Li, L., Yang, B., and Wang, X., 2007. Standardized extracts of Chinese medicinal herbs: Case study of Danshen (Salvia miltiorrhiza Bunge). *Journal of Food and Drug Analysis, 15*(4): 1.

Cherif, A., 2013. *The Peanut Used As a Source of Phytochemicals,* 77–177. Nova Science Publishers, Inc.: Hauppauge, NY, USA.

Chet, I., Ordentlich, A., Shapira, R., and Oppenheim, A., 1990. Mechanisms of biocontrol of soil-borne plant pathogens by rhizobacteria. *Plant and soil, 129*: 85–92.

Compant, S., Duffy, B., Nowak, J., Clément, C., and Barka, E. A., 2005. Use of plant growth-promoting bacteria for biocontrol of plant diseases: Principles, mechanisms of action, and future prospects. *Applied and Environmental Microbiology, 71*(9): 4951–4959.

Devi, K. K., Deepika, S., Bhaduri, A., and Kothamasi, D., 2013. Polymorphism in hcnAB gene in Pseudomonas species reveals ecologically distinct hydrogen cyanide-producing populations. *Geomicrobiology Journal, 30*(2): 131–139.

Dong, M., Li, J., Yang, D., Li, M., and Wei, J., 2023. Biosynthesis and pharmacological activities of flavonoids, Triterpene Saponins and polysaccharides derived from Astragalus membranaceus. *Molecules, 28*(13): 5018.

Donkor, A. M., Donkor, M. N., and Kuubabongnaa, N., 2020. Evaluation of anti-infective potencies of formulated aloin A ointment and aloin A isolated from Aloe barbadensis Miller. *BMC Chemistry, 14*: 1–8.

Du, J. F., Li, X. Z., Liu, M. Y., Zhang, Z. Y., and Wang, M., 2020. Screening for antagonistic bacteria against Fusarium oxysporum and their influence on the consecutive monoculture problem of Rehmanniaglutinosa. *Journal of Plant Pathology, 102*: 489–497.

Du, Y., Wang, T., Jiang, J., Wang, Y., Lv, C., Sun, K., Sun, J., Yan, B., Kang, C., Guo, L., and Huang, L., 2022. Biological control and plant growth promotion properties of Streptomyces albidoflavus St-220 isolated from Salvia miltiorrhiza rhizosphere. *Frontiers in Plant Science, 13*: 976813.

Ehigiator, B. E., and Ozolua, R. I., 2024. Reported Aphrodisiac agents of plant origin and the mechanistic basis of their actions in erectile dysfunction. *Plant Specialized Metabolites: Phytochemistry, Ecology and Biotechnology*: 1–26.

Franke, R., and Schilcher, H., 2006. Relevance and use of chamomile (Matricaria recutita L.). I International Symposium on Chamomile Research. *Development and Production, 749*: 29–43.

Friesen, J. B., Liu, Y., Chen, S. N., McAlpine, J. B., and Pauli, G. F., 2019. Selective depletion and enrichment of constituents in "curcumin" and other Curcuma longa preparations. *Journal of Natural Products, 82*(3): 621–630.

Gangwar, M., Dogra, S., Gupta, U. P., and Kharwar, R. N., 2014. Diversity and biopotential of endophytic actinomycetes from three medicinal plants in India. *African Journal of Microbiology Research, 8*(2): 184–191.

Ghosal, N., Firdaus, S. B., Naaz, S., Paul, S., Ghosh, A. K., Chattopadhyay, A., Mohan, V., Thakurdesai, P., Bhaskaran, S., Pattari, S., and Bandyopadhyay, D., 2016. Gastroprotective effect of Fenugreek 4-hydroxyisoleucine and trigonelline enriched fraction (TF4H (28%)) Sugaheal® against indomethacin induced ulcer in male wistar rats. *Journal of Pharmacy Research, 10*(6): 351–364.

Giorgi, A., Pentimalli, D., Giupponi, L., and Panseri, S., 2017. Quality traits of saffron (Crocus sativus L.) produced in the Italian Alps. *Open Agriculture, 2*(1): 52–57.

Grace, O. M., Buerki, S., Symonds, M. R., Forest, F., van Wyk, A. E., Smith, G. F., Klopper, R. R., Bjorå, C. S., Neale, S., Demissew, S., and Simmonds, M. S., 2015. Evolutionary history and leaf succulence as explanations for medicinal use in aloes and the global popularity of Aloe vera. *BMC Evolutionary Biology, 15*: 1–12.

Grindlay, D., and Reynolds, T., 1986. The Aloe vera phenomenon: A review of the properties and modern uses of the leaf parenchyma gel. *Journal of Ethnopharmacology, 16*(2–3): 117–151.

Güntert, M., Krammer, G., Lambrecht, S., Sommer, H., Surburg, H., and Werkhoff, P., 2001. Flavor chemistry of peppermint oil (Mentha piperita L.). *Aroma Active Compounds in Foods*: 119–137.

Gupta, M., Bisht, S., Singh, B., Gulati, A., and Tewari, R., 2011. Enhanced biomass and steviol glycosides in Stevia rebaudiana treated with phosphate-solubilizing bacteria and rock phosphate. *Plant Growth Regulation, 65*: 449–457.

Gupta, M., Kiran, S., Gulati, A., Singh, B., and Tewari, R., 2012. Isolation and identification of phosphate solubilizing bacteria able to enhance the growth and aloin-A biosynthesis of Aloe barbadensis Miller. *Microbiological Research, 167*(6): 358–363.

Hammons, R. O., Herman, D., and Stalker, H. T., 2016. Origin and early history of the peanut. In: Stalker, H. Thomas, Wilson, Richard F (eds.) *Peanuts*, Aocs Press, 1–26.

Hardarson, G., and Atkins, C., 2003. Optimising biological N 2 fixation by legumes in farming systems. *Plant and Soil, 252*: 41–54.

Hasan, A., Tabassum, B., Hashim, M., and Khan, N., 2024. Role of plant growth promoting rhizobacteria (PGPR) as a plant growth enhancer for sustainable agriculture: A review. *Bacteria, 3*(2): 59–75.

He, W., and Huang, B., 2011. A review of chemistry and bioactivities of a medicinal spice: Foeniculum vulgare. *Journal of Medicinal Plants Research, 5*(16): 3595–3600.

Hossain, M. F., Islam, M. T., Islam, M. A., and Akhtar, S. J. A. J. F., 2017. Cultivation and uses of stevia (Stevia rebaudiana Bertoni): A review. *African Journal of Food, Agriculture, Nutrition and Development, 17*(4): 12745–12757.

Hussein, R. A., and El-Anssary, A. A., 2019. Plants secondary metabolites: The key drivers of the pharmacological actions of medicinal plants. *Herbal Medicine, 1*(3): 11–30.

Jagtap, R. R., Mali, G. V., Waghmare, S. R., Nadaf, N. H., Nimbalkar, M. S., and Sonawane, K. D., 2023. Impact of plant growth promoting rhizobacteria Serratia nematodiphila RGK and Pseudomonas plecoglossicida RGK on secondary metabolites of turmeric rhizome. *Biocatalysis and Agricultural Biotechnology, 47*: 102622.

Jha, C. K., and Saraf, M., 2015. Plant growth promoting rhizobacteria (PGPR): A review. *Journal of Agricultural Research and Development, 5*(2): 108–119.

Jiang, Y., Song, Y., Jiang, C., Li, X., Liu, T., Wang, J., Chen, C., and Gao, J., 2022. Identification and characterization of Arthrobacter nicotinovorans JI39, a novel plant growth-promoting rhizobacteria strain from panax ginseng. *Frontiers in Plant Science, 13*: 873621.

Jin, T., Ren, J., Bai, B., Wu, W., Cao, Y., Meng, J., and Zhang, L., 2024. Effects of Klebsiella michiganensis LDS17 on Codonopsispilosula growth, rhizosphere soil enzyme activities, and microflora, and genome-wide analysis of plant growth-promoting genes. *Microbiology Spectrum, 12*(5): e04056–23.

John, J., 2014. Therapeutic potential of Withaniasomnifera: A report on phyto-pharmacological properties. *International Journal of Pharmaceutical Sciences and Research, 5*(6): 2131–2148.

Jyotirmayee, B., and Mahalik, G., 2022. A review on selected pharmacological activities of Curcuma longa L. *International Journal of Food Properties*, *25*(1): 1377–1398.

Kang, J. P., Huo, Y., Yang, D. U., and Yang, D. C., 2021. Influence of the plant growth promoting Rhizobium panacihumi on aluminum resistance in Panax ginseng. *Journal of Ginseng Research*, *45*(3): 442–449.

Karthikeyan, B., Jaleel, C. A., Gopi, R., and Deiveekasundaram, M., 2007. Alterations in seedling vigour and antioxidant enzyme activities in Catharanthus roseus under seed priming with native diazotrophs. *Journal of Zhejiang University Science B*, *8*: 453–457.

Karthikeyan, B., Joe, M. M., Jaleel, C. A., and Deiveekasundaram, M., 2010. Effect of root inoculation with plant growth promoting rhizobacteria (PGPR) on plant growth, alkaloid content and nutrient control of Catharanthus roseus (L.) G. Don. *Natura Croatica: Periodicum Musei Historiae Naturalis Croatici*, *19*(1): 205–212.

Kaur, M., Sharma, S., and Mishra, A., 2011. Influence of phosphate solubilizing pseudomonas and bacillus strains on the growth of Ashvagandha (Withaniasomnifera). *Indian Journal of Agricultural Research*, *45*(2): 128–133.

Kaushik, P., Ahlawat, P., Singh, K., and Singh, R., 2023. Chemical constituents, pharmacological activities, and uses of common ayurvedic medicinal plants: A future source of new drugs. *Advances in Traditional Medicine*, *23*(3): 673–714.

Kaushik, S., Tomar, R. S., Gupta, M., and Mishra, R. K., 2017. An overview of Catharanthus roseus and medicinal properties of their metabolites against important diseases. *European Academic Research*, *5*(2): 1237–1247.

Kavi Kishor, P. B., Hima Kumari, P., Sunita, M. S. L., and Sreenivasulu, N., 2015. Role of proline in cell wall synthesis and plant development and its implications in plant ontogeny. *Frontiers in Plant Science*, *6*: 544.

Kaymak, H. C., 2011. Potential of PGPR in agricultural innovations. *Plant Growth and Health Promoting Bacteria*, 18: 45–79.

Khade, K. V., Dubey, H., Tenpe, C. R., Yeole, P. G., Patole, A. M., Anajwala, C. C., and Patel, R. M., 2011. Anticancer activity of the ethanolic extracts of Agave americana leaves. *Pharmacologyonline*, *2*: 53–68.

Khamna, S., Yokota, A., and Lumyong, S., 2009. Actinomycetes isolated from medicinal plant rhizosphere soils: Diversity and screening of antifungal compounds, indole-3-acetic acid and siderophore production. *World Journal of Microbiology and Biotechnology*, *25*: 649–655.

Khan, S., Ambika, R. K., Sharma, S., Kumar, A., Singh, S., Thapliyal, M., Rawat, P., Thakur A., Pandey S., Thapliyal A., Pal, M., and Singh, Y., 2023. Rhizobacterial mediated interactions in Curcuma longa for plant growth and enhanced crop productivity: A systematic review. *Frontiers in Plant Science*, *14*: 1231676.

Khoshru, B., Mitra, D., Khoshmanzar, E., Myo, E. M., Uniyal, N., Mahakur, B., Mohapatra, P. K. D., Panneerselvam, P., Boutaj, H., Alizadeh, M., and Cely, M. V. T., 2020. Current scenario and future prospects of plant growth-promoting rhizobacteria: An economic valuable resource for the agriculture revival under stressful conditions. *Journal of Plant Nutrition*, *43*(20): 3062–3092.

Kim, J., and Rees, D. C., 1994. Nitrogenase and biological nitrogen fixation. *Biochemistry*, *33*(2): 389–397.

Kim, K. Y., Jordan, D., and McDonald, G. A., 1997. Effect of phosphate-solubilizing bacteria and vesicular-arbuscular mycorrhizae on tomato growth and soil microbial activity. *Biology and Fertility of Soils*, *26*: 79–87.

Kinghorn, A. D., and Soejarto, D. D., 1985. Current status of stevioside as a sweetening agent for human use. *Economic and Medicinal Plant Research*, 1:1–52.

Kocaadam, B., and Şanlier, N., 2017. Curcumin, an active component of turmeric (Curcuma longa), and its effects on health. *Critical Reviews in Food Science and Nutrition*, *57*(13): 2889–2895.

Kumar, A., Singh, R., Giri, D. D., Singh, P. K., and Pandey, K. D., 2014. Effect of Azotobacter chroococcum CL13 inoculation on growth and curcumin content of turmeric (Curcuma longa L.). *International Journal of Current Microbiology and Applied Sciences*, *3*(9): 275–283.

Kumar, M., 2017. A review on phytochemical constituents and pharmacological activities of Ricinus communis L. Plant. *International Journal of Pharmacognosy and Phytochemical Research*, *9*(4): 466–472.

Kumar, V., and Narula, N., 1999. Solubilization of inorganic phosphates and growth emergence of wheat as affected by Azotobacter chroococcum mutants. *Biology and Fertility of Soils*, *28*: 301–305.

La Torre-Ruiz, D., Ruiz-Valdiviezo, V. M., Rincón-Molina, C. I., Rodríguez-Mendiola, M., Arias-Castro, C., Gutiérrez-Miceli, F. A., Palomeque-Dominguez, H., and Rincón-Rosales, R., 2016. Effect of plant growth-promoting bacteria on the growth and fructan production of Agave americana L. *Brazilian Journal of Microbiology*, *47*: 587–596.

Larsen, K., Friis, I., and Balslev, H., 2005. Distribution patterns and diversity centres of Zingiberaceae in SE Asia. *Biology Skr*, *55*: 219–228.

Laumer, I. B., Rahman, A., Rahmaeti, T., Azhari, U., Hermansyah, Atmoko, S. S. U., and Schuppli, C., 2024. Active self-treatment of a facial wound with a biologically active plant by a male Sumatran orangutan. *Scientific Reports*, *14*(1): 8932.

Lee, K. Y., Weintraub, S. T., and Yu, B. P., 2000. Isolation and identification of a phenolic antioxidant from Aloe barbadensis. *Free Radical Biology and Medicine*, *28*(2): 261–265.

Lee, Y. S., Park, H. M., Kim, N. H., Waminal, N. E., Kim, Y. J., Lim, K. B., Baek, J. H., Kim, H. H., and Yang, T. J., 2016. Phylogenetic relationship of 40 species of genus Aloe L. and the origin of an allodiploid species revealed by nucleotide sequence variation in chloroplast intergenic space and cytogenetic in situ hybridization. *Genetic Resources and Crop Evolution*, *63*: 235–242.

Li, H., Li, C., Song, X., Li, J., Zhang, P., Sun, F., Geng, Z., and Liu, X., 2023. Isolation and identification of antagonistic Bacillus amyloliquefaciens HSE-12 and its effects on peanut growth and rhizosphere microbial community. *Frontiers in Microbiology*, *14*: 1274346.

Li, X., Qu, L., Dong, Y., Han, L., Liu, E., Fang, S., Zhang, Y., and Wang, T., 2014. A review of recent research progress on the astragalus genus. *Molecules*, *19*(11): 18850–18880.

Li, Y., He, N., Hou, J., Xu, L., Liu, C., Zhang, J., Wang, Q., Zhang, X., and Wu, X., 2018. Factors influencing leaf chlorophyll content in natural forests at the biome scale. *Frontiers in Ecology and Evolution*, 64(6): 1–10.

Liu, H., Lu, X., Hu, Y., and Fan, X., 2020. Chemical constituents of Panax ginseng and Panax notoginseng explain why they differ in therapeutic efficacy. *Pharmacological Research*, *161*: 105263.

Lu, M. K., Chang, C. C., Chao, C. H., and Hsu, Y. C., 2022. Structural changes, and anti-inflammatory, anticancer potential of polysaccharides from multiple processing of Rehmanniaglutinosa. *International Journal of Biological Macromolecules*, *206*: 621–632.

Luan, F., Ji, Y., Peng, L., Liu, Q., Cao, H., Yang, Y., He, X., and Zeng, N., 2021. Extraction, purification, structural characteristics and biological properties of the polysaccharides from Codonopsispilosula: A review. *Carbohydrate Polymers*, *261*: 117863.

Mahady, G. B., Gyllenhaal, C., Fong, H. H., and Farnsworth, N. R., 2000. Ginsengs: A review of safety and efficacy. *Nutrition in Clinical Care,* 3: 90–101.

Mahdi, I., Fahsi, N., Hafidi, M., Allaoui, A., and Biskri, L., 2020. Plant growth enhancement using rhizospheric halotolerant phosphate solubilizing bacterium Bacillus licheniformis QA1 and Enterobacter asburiae QF11 isolated from Chenopodium quinoa willd. *Microorganisms*, *8*(6): 948.

Mahdi, I., Fahsi, N., Hafidi, M., Benjelloun, S., Allaoui, A., and Biskri, L., 2021. Rhizospheric phosphate solubilizing Bacillus atrophaeus GQJK17 S8 increases quinoa seedling, withstands heavy metals, and mitigates salt stress. *Sustainability*, *13*(6): 3307.

Mamta, G., Rahi, P., Pathania, V., Gulati, A., Singh, B., Bhanwra, R. K., and Tewari, R., 2012. Comparative efficiency of phosphate-solubilizing bacteria under greenhouse conditions for promoting growth and aloin-A content of Aloe barbadensis. *Archives of Agronomy and Soil Science*, *58*(4): 437–449.

McNear Jr, D. H., 2013. The rhizosphere-roots, soil and everything in between. *Nature Education Knowledge*, *4*(3): 1.

Mellor, R. B., and Werner, D., 2018. Legume nodule biochemistry and function. In: Gresshoff, P.M. (Ed.), *Molecular Biology of Symbiotic Nitrogen Fixation*: 111–129. CRC Press/Taylor & Francis.

Mike-Anosike, E. E., Braide, W., and Adeleye, S. A., 2018. Studies on indole acetic acid (IAA) production by rhizobacteria and growth promoting potentials. *International Journal of Advanced Research in Biological Sciences*, *5*(2): 133–140.

Mikulska, P., Malinowska, M., Ignacyk, M., Szustowski, P., Nowak, J., Pesta, K., Szeląg, M., Szklanny, D., Judasz, E., Kaczmarek, G., and Ejiohuo, O. P., 2023. Ashwagandha (Withaniasomnifera)—Current research on the health-promoting activities: A narrative review. *Pharmaceutics*, *15*(4): 1057.

Mishra, B. K., Meena, K. K., Dubey, P. N., Aishwath, O. P., Kant, K., Sorty, A. M., and Bitla, U., 2016. Influence on yield and quality of fennel (Foeniculum vulgare Mill.) grown under semi-arid saline soil, due to application of native phosphate solubilizing rhizobacterial isolates. *Ecological Engineering*, *97*: 327–333.

Mishra, B. K., Singh, B., Singh, P., Rathore, S. S., Aishwath, O. P., Kant, K., and Dubey, P. N., 2015. Isolation and evaluation of phosphate solubilizing microorganisms from fennel (Foeniculum vulgare Mill.) rhizospheric soils of Rajasthan. *International Journal of Seed Spices*, *5*(1): 71–75.

Morshed, M. N., Ahn, J. C., Mathiyalagan, R., Rupa, E. J., Akter, R., Karim, M. R., Jung, D. H., Yang, D. U., Yang, D. C., and Jung, S. K., 2023. Antioxidant activity of Panax ginseng to regulate ROS in various chronic diseases. *Applied Sciences*, *13*(5): 2893.

Moser, R., Pertot, I., Elad, Y., and Raffaelli, R., 2008. Farmers' attitudes toward the use of biocontrol agents in IPM strawberry production in three countries. *Biological Control*, *47*(2): 125–132.

Moudi, M., Go, R., Yien, C. Y. S., and Nazre, M., 2013. Vinca alkaloids. *International Journal of Preventive Medicine*, *4*(11): 1231.

Mzabri, I., Addi, M., and Berrichi, A., 2019. Traditional and modern uses of saffron (Crocus sativus). *Cosmetics*, *6*(4): 63.

Nakkeeran, S., Fernando, W. D., and Siddiqui, Z. A., 2006. Plant growth promoting rhizobacteria formulations and its scope in commercialization for the management of pests and diseases. In: Siddiqui, Z.A. (eds.) *PGPR: Biocontrol and Biofertilization*, Springer, Dordrecht, 257–296.

Nazir, N., Kamili, A. N., and Shah, D., 2018. Mechanism of plant growth promoting rhizobacteria (PGPR) in enhancing plant growth-A review. *International Journal of the Management Technology Engineering*, *8*: 709–721.

Nobakht, S. Z., Akaberi, M., Mohammadpour, A. H., Moghadam, A. T., and Emami, S. A., 2022. Hypericum perforatum: Traditional uses, clinical trials, and drug interactions. *Iranian Journal of Basic Medical Sciences*, *25*(9): 1045.

Oliveira, A. I., Pinho, C., Sarmento, B., and Dias, A. C., 2016. Neuroprotective activity of Hypericum perforatum and its major components. *Frontiers in Plant Science*, *7*: 1004.

Pahari, A., and Mishra, B. B., 2017. Characterization of siderophore producing rhizobacteria and its effect on growth performance of different vegetables. *International Journal Current Microbiology Applied Science*, *6*(5): 1398–1405.

Pandey, V., Ansari, W. A., Misra, P., and Atri, N., 2017. Withaniasomnifera: Advances and implementation of molecular and tissue culture techniques to enhance its application. *Frontiers in Plant Science*, *8*: 1390.

Patel, K., and Patel, D. K., 2013. Medicinal importance, pharmacological activities, and analytical aspects of aloin: A concise report. *Journal of Acute Disease*, *2*(4): 262–269.

Payyappallimana, U., and Venkatasubramanian, P., 2016. Exploring ayurvedic knowledge on food and health for providing innovative solutions to contemporary healthcare. *Frontiers in Public Health*, *4*: 57.

Petrovska, B. B., 2012. Historical review of medicinal plants' usage. *Pharmacognosy Reviews*, *6*(11): 1.

Podile, A., and Kishore, G., 2006. Plant growth-promoting rhizobacteria. In: Gnanamanickam, S.S. (eds.) *Plant-Associated Bacteria*. Springer, Dordrecht, 195–230.

Qi, J., Li, X., Song, J., Eneji, A. E., and Ma, X., 2008. Genetic relationships among Rehmanniaglutinosa cultivars and varieties. *Planta Medica*, *74*(15): 1846–1852.

Qin, Y. Q., Wang, L. Y., Yang, X. Y., Xu, Y. J., Fan, G., Fan, Y. G., Ren, J. N., An, Q., and Li, X., 2023. Inulin: Properties and health benefits. *Food & Function*, *14*(7): 2948–2968.

Quispe, C., Herrera-Bravo, J., Khan, K., Javed, Z., Semwal, P., Painuli, S., Kamiloglu, S., Martorell, M., Calina, D., and Sharifi-Rad, J., 2022. Therapeutic applications of curcumin nanomedicine formulations in cystic fibrosis. *Progress in Biomaterials*, *11*(4): 321–329.

Rabari, A., Ruparelia, J., Jha, C. K., Sayyed, R. Z., Mitra, D., Priyadarshini, A., Senapati, A., Panneerselvam, P., and Mohapatra, P. K. D., 2023. Articulating beneficial rhizobacteria-mediated plant defenses through induced systemic resistance: A review. *Pedosphere*, *33*(4): 556–566.

Rahnavard, A., Daneshian, J., Heravan, E., Valadabadi, S., and Golein, B., 2012. Investigation of the most important secondary metabolites of St. John's wort (Hypericum perforatum L.) in Caspian climate. *Journal of Food Agricultural Environment*, *10*: 375–381.

Raj, Y., Kumar, A., Kumari, S., Kumar, R., and Kumar, R., 2023. Comparative genomics and physiological investigations supported multifaceted plant growth-promoting activities in two Hypericum perforatum L.-Associated plant growth-promoting Rhizobacteria for microbe-assisted cultivation. *Microbiology Spectrum*, *11*(3): e00607–e00623.

Rajasekar, S., and Elango, R., 2011. Effect of microbial consortium on plant growth and improvement of alkaloid content in Withaniasomnifera (Ashwagandha). *Current Botany*, *2*(8): 27–30.

Rashmi, Y., Veer, P. D., and Rakesh, K., 2021. Effect of nitrogen fixing bioinoculants resistant to inhibitors in castor bean (Ricinus communis L.) on the crop. *International Journal of Agricultural Sciences and Veterinary Medicine, 9*(4): 7–14.

Ravensberg, W. J., 2011. A roadmap to the successful development and commercialization of microbial pest control products for control of arthropods. *10*: 151–154.

Reiter, B., and Sessitsch, A., 2006. Bacterial endophytes of the wildflower Crocus albiflorusanalyzed by characterization of isolates and by a cultivation-independent approach. *Canadian Journal of Microbiology, 52*(2): 140–149.

Repčák, M., Imrich, J., and Franeková, M., 2001. Umbelliferone, a stress metabolite of Chamomilla recutita (L.) Rauschert. *Journal of Plant Physiology, 158*(8): 1085–1087.

Repo-Carrasco, R., Espinoza, C., and Jacobsen, S. E., 2003. Nutritional value and use of the Andean crops quinoa (Chenopodium quinoa) and kañiwa (Chenopodium pallidicaule). *Food Reviews International, 19*(1–2): 179–189.

Rezvanirad, A., Mardani, M., Shirzad, H., Ahmadzadeh, S. M., Asgary, S., Naimi, A., and Mahmoudi, G., 2016. Curcuma longa: A review of therapeutic effects in traditional and modern medical references. *Journal of Chemical and Pharmaceutical Sciences, 9*(4): 3438–3448.

Sah, S., and Singh, R., 2015. Siderophore: Structural and functional characterisation–A comprehensive review. *Agriculture (Pol'nohospodárstvo), 61*(3): 97–114.

Salisbury, F. B., and Ross, C. W., 1992. *Plant Physiology*, Wadsworth Pub. Com., Inc.

Sandilya, S. P., Bhuyan, P. M., Nageshappa, V., Gogoi, D. K., and Kardong, D., 2017. Impact of Pseudomonas aeruginosa MAJ PIA03 affecting the growth and phytonutrient production of castor, a primary host-plant of Samia ricini. *Journal of Soil Science and Plant Nutrition, 17*(2): 499–514.

Santoro, M. V., Cappellari, L. D. R., Giordano, W., and Banchio, E., 2015. Plant growth-promoting effects of native Pseudomonas strains on Mentha piperita (peppermint): An in vitro study. *Plant Biology, 17*(6): 1218–1226.

Sayyed, R. Z., Chincholkar, S. B., Reddy, M. S., Gangurde, N. S., and Patel, P. R., 2012. Siderophore producing PGPR for crop nutrition and phytopathogen suppression. In: Gnanamanickam, S.S. (eds.). *Bacteria in Agrobiology: Disease Management*. Springer, Dordrecht, 449–471.

Sayyed, R. Z., Gangurde, N. S., Patel, P. R., Josh, S. A., and Chincholkar, S. B., 2010. Siderophore production by Alcaligenes faecalis and its application for growth promotion in Arachis hypogaea. *Indian Journal of Biotechnology, 9*: 302–307.

Schmidt, R., Köberl, M., Mostafa, A., Ramadan, E. M., Monschein, M., Jensen, K. B., Bauer, R., and Berg, G., 2014. Effects of bacterial inoculants on the indigenous microbiome and secondary metabolites of chamomile plants. *Frontiers in Microbiology, 5*: 64.

Semwal, P., Painuli, S., Badoni, H., and Bacheti, R. K., 2018. Screening of phytoconstituents and antibacterial activity of leaves and bark of Quercus leucotrichophora A. Camus from Uttarakhand Himalaya. *Clinical Phytoscience, 4*: 1–6.

Semwal, P., Rauf, A., Olatunde, A., Singh, P., Zaky, M. Y., Islam, M. M., Khalil, A. A., Aljohani, A. S., Al Abdulmonem, W., and Ribaudo, G., 2023. The medicinal chemistry of Urtica dioica L.: From preliminary evidence to clinical studies supporting its neuroprotective activity. *Natural Products and Bioprospecting, 13*(1): 16.

Shah, P. P., and Mello, P. M. D., 2004. A review of medicinal uses and pharmacological effects of Mentha piperita. *Indian Journal of Natural Products and Resources, 3*(4): 214–221.

Shaheen, S., Ramzan, S., Khan, F., and Ahmad, M., 2019. *Adulteration in Herbal Drugs: A Burning Issue*: 1–8. Springer.

Sharma, K., Sharma, S., Vaishnav, A., Jain, R., Singh, D., Singh, H. B., Goel, A., and Singh, S., 2022. Salt-tolerant PGPR strain Priestiaendophytica SK1 promotes fenugreek growth under salt stress by inducing nitrogen assimilation and secondary metabolites. *Journal of Applied Microbiology, 133*(5): 2802–2813.

Sharma, S., Walia, S., Rathore, S., Kumar, P., and Kumar, R., 2020. Combined effect of elevated CO2 and temperature on growth, biomass and secondary metabolite of Hypericum perforatum L. in a western Himalayan region. *Journal of Applied Research on Medicinal and Aromatic Plants, 16*: 100239.

Shi, M., Huang, F., Deng, C., Wang, Y., and Kai, G., 2019. Bioactivities, biosynthesis and biotechnological production of phenolic acids in Salvia miltiorrhiza. *Critical Reviews in Food Science and Nutrition, 59*(6): 953–964.

Shi, Z., Guo, X., Lei, Z., Wang, Y., Yang, Z., Niu, J., and Liang, J., 2023. Screening of high-efficiency nitrogen-fixing bacteria from the traditional Chinese medicine plant Astragalus mongolicus and its effect on plant growth promotion and bacterial communities in the rhizosphere. *BMC Microbiology*, *23*(1): 292.

Singh, A., Pal, V., Chauhan, K., Pant, M., and Husen, A., 2024. Tissue culture of medicinal spice and condiment crops for mass propagation. In: Husen, A. (ed.) *Medicinal Spice and Condiment Crops*, CRC Press, 42–61.

Singh, N., and Varma, A., 2015. Antagonistic activity of siderophore producing rhizobacteria isolated from the semi-arid regions of Southern India *International Journal of Current Microbiology Applied Sciences*, *4*(9): 501–510.

Singh, S., and Kapoor, K. K., 1999. Inoculation with phosphate-solubilizing microorganisms and a vesicular-arbuscular mycorrhizal fungus improves dry matter yield and nutrient uptake by wheat grown in a sandy soil. *Biology and Fertility of Soils*, *28*: 139–144.

Singh, S. D., and Rao, G. P., 2005. Stevia: The herbal sugar of 21st century. *Sugar Tech*, *7*(1): 17–24.

Solaiman, Z. M., and Anawar, H. M., 2015. Rhizosphere microbes interactions in medicinal plants. In: Egamberdieva, D., Shrivastava, S., Varma, A. (eds.) *Plant-Growth-Promoting Rhizobacteria (PGPR) and Medicinal Plants*. Vol 42. Springer, Cham, 19–41.

Stevenson, F. J., and Cole, M. A., 1999. *Cycles of Soils: Carbon, Nitrogen, Phosphorus, Sulfur, Micronutrients*. Wiley.

Strauss, S. J. T. R., 1995. The perfect sweetener. *Technology Review*, *98*(8): 18–20.

Sun, W., Shahrajabian, M. H., and Cheng, Q., 2021. Fenugreek cultivation with emphasis on historical aspects and its uses in traditional medicine and modern pharmaceutical science. *Mini Reviews in Medicinal Chemistry*, *21*(6): 724–730.

Tabassum, B., Khan, A., Tariq, M., Ramzan, M., Khan, M. S. I., Shahid, N., and Aaliya, K., 2017. Bottlenecks in commercialisation and future prospects of PGPR. *Applied Soil Ecology*, *121*: 102–117.

Taheri, Y., Suleria, H. A. R., Martins, N., Sytar, O., Beyatli, A., Yeskaliyeva, B., Seitimova, G., Salehi, B., Semwal, P., Painuli, S., and Kumar, A., 2020. Myricetin bioactive effects: Moving from preclinical evidence to potential clinical applications. *BMC Complementary Medicine and Therapies*, *20*: 1–14.

Tanira, M. O. M., Shah, A. H., Mohsin, A., Ageel, A. M., and Qureshi, S., 1996. Pharmacological and toxicological investigations on Foeniculum vulgare dried fruit extract in experimental animals. *Phytotherapy Research*, *10*(1): 33–36.

Tannock, G. W., 1999. The normal microflora: An introduction. In: Tannock, G.W. (eds.) *Medical Importance of the Normal Microflora.* Springer, Boston, MA, 1–23.

Tina, Pal, V., and Pant, M., 2023. Digitalis purpurea (Foxglove Plant). In Husen, A. (Ed.), *Exploring Poisonous Plants*: 241–254. CRC Press.

Usha Rani, P., and Pratyusha, S., 2014. Role of castor plant phenolics on performance of its two herbivores and their impact on egg parasitoid behaviour. *BioControl*, *59*: 513–524.

Valdes, A. M., Walter, J., Segal, E., and Spector, T. D., 2018. Role of the gut microbiota in nutrition and health. *BMJ*, *361*: 36–44.

Velayudhan, K. C., Dikshit, N., and Nizar, M. A., 2012. Ethnobotany of turmeric (Curcuma longa L.). *Indian Journal of Traditional Knowledge*, *11*(4): 607–614.

Velu, G., Palanichamy, V., and Rajan, A. P., 2018. Phytochemical and pharmacological importance of plant secondary metabolites in modern medicine. In: Roopan, S., Madhumitha, G. (eds.) *Bioorganic Phase in Natural Food: An Overview* Springer, Cham, 135–156.

Verma, S. K., and Kumar, A., 2011. Therapeutic uses of Withaniasomnifera (Ashwagandha) with a note on withanolides and its pharmacological actions. *Asian Journal of Pharmaceutical and Clinical Research*, *4*(1): 1–4.

von Wirén, N., Khodr, H., and Hider, R. C., 2000. Hydroxylated phytosiderophore species possess an enhanced chelate stability and affinity for iron (III). *Plant Physiology*, *124*(3): 1149–1158.

Wang, B. Q., 2010. Salvia miltiorrhiza: Chemical and pharmacological review of a medicinal plant. *Journal of Medicinal Plants Research*, *4*(25): 2813–2820.

Wang, T. L., Guan, W., Sun, K., Wang, S., Chi, X. L., and Guo, L. P., 2018. Progress in researches on pathogens, epidemiology and integrated control of diseases on Salvia miltiorrhiza in China. *Zhongguo Zhong yao za zhi= ZhongguoZhongyaoZazhi= China Journal of Chinese Materia Medica*, *43*(11): 2402–2406.

Wang, W., Yue, R. F., Jin, Z., He, L. M., Shen, R., Du, D., and Tang, Y. Z., 2020. Efficiency comparison of apigenin-7-O-glucoside and trolox in antioxidative stress and anti-inflammatory properties. *Journal of Pharmacy and Pharmacology*, *72*(11): 1645–1656.

Wang, Y., Brown, H. N., Crowley, D. E., and Szaniszlo, P. J., 1993. Evidence for direct utilization of a siderophore, ferrioxamine B, in axenically grown cucumber. *Plant, Cell & Environment*, *16*(5): 579–585.

Xiao, J., Chen, S., Chen, Y., and Su, J., 2022. The potential health benefits of aloin from genus Aloe. *Phytotherapy Research*, *36*(2): 873–890.

Yang, Y., Chen, X., Chen, J., Xu, H., Li, J., and Zhang, Z., 2011. Differential miRNA expression in Rehmanniaglutinosa plants subjected to continuous cropping. *BMC Plant Biology*, 11: 1–11.

You, H., Yang, S., Zhang, L., Hu, X., and Li, O., 2018. Promotion of phenolic compounds production in Salvia miltiorrhiza hairy roots by six strains of rhizosphere bacteria. *Engineering in Life Sciences*, *18*(3): 160–168.

Yu, D. H., Bao, Y. M., Wei, C. L., and An, L. J., 2005. Studies of chemical constituents and their antioxidant activities from Astragalus mongholicus Bunge. *Biomedical and Environmental Sciences*, *18*(5): 297.

Zadeh, J. B., Kor, N. M., and Kor, Z. M., 2014. Chamomile (Matricaria recutita) as a valuable medicinal plant. *International Journal of Advanced Biological and Biomedical Research*, *2*(3): 823–829

Zhou, Y. Q., Li, J. J., Wang, F., Gu, F. P., Zhou, C. E., Zhang, Z. Y., and Gao, Z. M., 2007. Genetic diversity of different cultivars in RehmanniaglutinosaLibosch. f. hueichingensis (Chao et Schih) Hsiao. *Life Science Journal*, *4*(2): 69–75.

Zwane, P. E., Masarirambi, M. T., Magagula, N. T., Dlamini, A. M., and Bhebhe, E., 2011. Exploitation of Agave americana L. plant for food security in Swaziland. *American Journal of Food and Nutrition*, *1*(2): 82–88.

TABLE 14.1
List of medicinal plants, their use, and effects of their rhizospheric microflora on metabolites production and plant growth

Sl. no.	Scientific Name	Common Name	Place of origin	Medicinal uses	Plant part used	Bacterial strain inoculated	Stage at which inoculum applied	Effect of microorganisms on plant metabolites	Effect of microorganism on plant growth	Reference
01.	*Agave americana*	Century plant	Mexico	Used for treatment of fish toxicity, malarial, and other fevers. Topically applied on wounds, etc.	Leaves	*Acinetobacter calcoaceticus*, *Pseudomonas mosselii*, and *Rhizobium daejeonense*	Plantlets	Increase the inulin, fructose, sucrose content in the leaves.	Increase in plant height, stem diameter, root length, leaves number, sugar content in leaves	La Torre-Ruiz et al., 2016
02.	*Aloe barbadensis*	Aloe Vera	Arabian Peninsula	Helps in healing burns and superficial skin wounds	Leaves	*Burkholderia gladioli*, *Enterobacter hormaechei*, *Pseudomonas synxantha*, and *Serratia marcescens*	Plantlets	Increase in aloin-A production	Increase in Phosphorus absorption, and biomass.	Gupta et al., 2012; Mamta et al., 2012
03.	*Arachis hypogaea*	Peanut	South America	Improves condition in diabetes, used in the treatment of bronchitis	Leaves, skin, nut	*Bacillus amyloliquefaciens*	Seedlings	Increase in chlorophyl content.	Antagonistic activity against *A. alternata*, *B. sorokiniana*, *B. dothidea*, *C. parasitica*, *F. graminearum*, *F. oxysporum*, *F. pseudograminearum*, *R. solani*, and *S. rolfsii*. Increase in dry weight, fresh weight and branch length, rhizospheric micro-biodiversity	Sayyed et al., 2010; Li et al., 2023
04.	*Astragalus mongolicus*	Mongolian milkvetch, Huáng qí, Běiqí	China	It helps in maintaining energy level, used as a diuretic, and antiperspirant	Roots	*Agrobacterium fabacearum*, *Pseudomonas silesiensis*, *Paenarthro bacternitroguaj acolicus, and Arthrobacter pascens*	Seeds	Increase in content of Flavonoids, Polysaccharides, and Saponins	Increase in biomass, plant height, root length, dry weight of root and shoot, nitrogen, phosphorus, and potassium content.	Shi et al., 2023

05.	*Catharanthus roseus*	Madagascar periwinkle, Sadabahar, Nayantara	Madagascar	Used as a diuretic, and mouthwash to relieve toothache, dental decay, and scurvy. It is also used for treating diabetic ulcers.	Leaves	*Azotobacter chroococcum, Bacillus subtilis, and Pseudomonas fluorescens*	Seeds, seedlings	Increase in the content of Ajmalicine	Increase in germination efficiency, root and shoot length, dry matter, and vigour index, antioxidant activities.	Karthikeya et al., 2007; Karthikeyan et al., 2010
06.	*Chamomilla recutita*	Chamomile	Europe	It helps in treating skin irritation, relieves from flatulence, smooth muscle spasms.	Flowers	*Bacillus subtilis, Streptomyces subrutilus, and Paenibacillus polymyxa*	Seedlings	Increase in content of apigenin-7-O-glucoside	NA	Wang et al., 2020
07.	*Chenopodium quinoa*	Quinoa	South America	Improves bone health, mental health, and treats women's health related problems	Seeds and leaves	*Pseudomonas frederiksbergensis, Pseudomonas stutzeri, Bacillus atrophaeus, Enterobacter asburiae, Enterobacter hormaechei, Bacillus licheniformis and Enterobacter asburiae*	Seeds	Increase in chlorophyll content.	Increase in germination rate and plant length, biomass	Mahdi et al., 2021 Mahdi et al., 2020
08..	*Codonopsispilosula*	Dangshen	China, Korea and Mongolia	Improvement of overall physique, treatment of spleen inactivity	Roots	*Klebsiella michiganensis*	Seedlings	Increase in chlorophyll content.	Antagonistic activity against *C. camelliae*, *C. chrysosperma*, *P macrospore*, and *R. solani*, increase in plant height, biomass, and ground diameter.	Jin et al., 2024
09.	*Crocus sativus*	Saffron	Greece	Used as aphrodisiac, antidepressant, antioxidant, and analgesic. It also treats digestive issues.	Stigma	*Acinetobacter calcoaceticus, Bacillus aryabhattai, Chryseobacterium elymi, Pseudomonas kilonensis, Pseudomonas koreensis, Pseudomonas tremae, Rahnella aquatilis and Variovorax paradoxus*	Corn	Increase in chlorophyll, crocin, picrocrocin, and safranal content.	Increase in average number of roots and shoots, number of cornlets. Decrease in incidence of corm rot.	Ambardar et al., 2013; Chamkhi et al., 2023

(continued)

TABLE 14.1 (Continued)
List of medicinal plants, their use, and effects of their rhizospheric microflora on metabolites production and plant growth

Sl. no.	Scientific Name	Common Name	Place of origin	Medicinal uses	Plant part used	Bacterial strain inoculated	Stage at which inoculum applied	Effect of microorganisms on plant metabolites	Effect of microorganism on plant growth	Reference
10.	*Curcuma longa*	Turmeric	South Asia	Used to treat Jaundice, diminish pain and swelling, tooth-ache, diarrhea, skin infections	Rhizome	*Azotobacter chroococcum* and *Rhizobiumpusense*	Rhizomatous buds, rhizome	Increase in curcumin,aR-Turmerone, 2,4-Di-tertbutyl phenol2,Hydroxy-2-methylacetophenone, and (Z) -gamma-Atlantoneand chlorophyll content	Increase in plant height, harvest index, root length and biomass, rhizome size, weight and yield, leaf area, number, length, breadth and biomass.	Kumar et al., 2014; Chandran et al., 2019
11.	*Foeniculum vulgare*	Fennel	Mediterranean and Southern Europe	Used to treat symptoms of coughs, bronchitis, vomiting, indigestion and diarrhea	Fruits	*Bacillus subtilis*	Seeds	Increase in essential oil production	Increase in branch number, flowers, and seeds.	Mishra et al., 2016
12.	*Hypericum perforatum*	St. John's wort	Europe, Northern Africa, and Western Asia	Used to treat dermal ailments, wounds, stomach ulcers, bile-related disorders, respiratory and genital infections	Plant parts above ground	*Curtobacterium albidum, Enterobacter huaxiensis, Flavobacterium acidificum, Kocuria palustris, Kosakonia cowanii, Rahnella woolbedingensis, Rahnella varıgena, and Staphylococcus edaphicus*	Seedlings	Increase in the content of chlorophyll-a, chlorophyll-b, and cartenoids	Increase in shoot and root length, above-ground dry biomass (AGB), below-ground dry biomass (BGB), and root volume	Raj et al., 2023
13.	*Mentha piperita*	Peppermint	Europe	Used to treat Cold, irritation in the gastro-intestinal tract, nausea, headache	Leaves	*Pseudomonas putida*	Nodes	Increase in total Essential oil content, terpineol, menthone, menthol, and pulegone	Increase in dry root weight, shoot ramification number, shoot weight, trichome number	Santoro et al., 2015

14.	*Panax ginseng*	Korean Ginseng, Asian Ginseng and Oriental Ginseng	Korea, China and Russia	Used in treatment of symptoms of fatigue during recovery from diseases, respiratory ailments	Roots	*Rhizobium panacihumi, Arthrobacter' nicotinovorans*	Seeds, seedlings and plants	Increase in proline content and phenolic content	Increase in average length of seed buds, average fresh weight, soil enzyme activity	Kang et al., 2021; Jiang et al., 2022
15.	*Rehmannia glutinosa*	Chinese Foxglove	China and Japan	Helps in liver protection. Shows anti-inflammatory, anti-fungal, diuretic, hypoglycemic activity.	Roots	*Bacillus tequilensis*	Plants	Increase in content, and yield of sterol and polysaccharides	Increase in fresh and dry weight of leaves and roots.	Du et al., 2020
16.	*Ricinus communis*	Castor bean plant	Africa	Shows antioxidant, anti-inflammatory, antiulcer, and injury-healing properties	Seeds	*Klebsiella variicola, Pseudomonas aeruginosa*	Plants, seeds	Increase in content of phenol, soluble protein, lipids, carbohydrates, and chlorophyll	Increase in plant height, root weight (fresh and dry), and shoot weight (fresh and dry), rhizospheric microbial-biodiversity	Sandilya et al., 2017; Rashmi et al., 2021
17.	*Salvia miltiorrhiza*	Chinese sage or Red sage root	China and Japan	Helps in treatment of heart and brain-related ailments	Roots	*Streptomyces abidoflavus, Pseudomonas chlororaphis, Pantoea rodasii, and Pseudomonas vancouverensis*	Hairy roots, seeds	Increase in content of Rosmarinic acid, and Salvianolic acid B	Increase in length, weight, and dry weight of the root and shoot, root diameter and antagonistic activity against *Fusarium oxyporum*	Du et al., 2022
18.	*Stevia rebaudiana*	Sweet-leaf, Honey-leaf and Sugar-leaf	Paraguay, South America	Shows antibacterial, anti-inflammatory activities. Used as diuretic and sugar alternative.	Leaves	*Burkholderia gladioli, Enterobacter aerogenes, and Serratia marcescens*	Plantlets	Rebaudioside-A and Stevioside content increased	Increase in Shoot-root length, dry weight of stem and roots, and total shoot biomass	Gupta et al., 2011

(continued)

V

W

Z

For Product Safety Concerns and Information please contact our EU representative GPSR@taylorandfrancis.com Taylor & Francis Verlag GmbH, Kaufingerstraße 24, 80331 München, Germany